Vogel's Qualitative Inorganic Analysis

Vogel's
Qualitative Inorganic Analysis

Seventh Edition

Revised by
G. Svehla PhD, DSC, FRSC
formerly Professor of Analytical Chemistry,
University College, Cork

 **LONGMAN**

Addison Wesley Longman Ltd
Edinburgh Gate, Harlow
Essex CM20 2JE, England
and Associated Companies throughout the world

Trademarks
Throughout this book trademarked names are used. Rather than put a trademark symbol in every occurrence of a trademarked name, we state that we are using the names only in editorial fashion and to the benefit of the trademark owner with no intention of infringement of the trademark.

First published under the title *A Text-book of Qualitative Chemical Analysis* 1937
Second edition 1941
Reissue with Appendix 1943
Third edition under the title *A Text-book of Qualitative Chemical Analysis including Semimicro Qualitative Analysis* 1945
Fourth edition under the title *A Text-book of Macro and Semimicro Qualitative Inorganic Analysis* 1954
Fifth edition 1979
Sixth edition under the title *Vogel's Qualitative Inorganic Analysis* 1987
Seventh edition 1996
Reprinted 1997

British Library Cataloguing in Publication Data
A catalogue entry for this title is available from the British Library.

ISBN 0-582-21866-7

Library of Congress Cataloging-in-Publication Data
A catalog entry for this title is available from the Library of Congress.

Set by 22 in 10/12pt Times
Printed & bound by Antony Rowe Ltd, Eastbourne
Transferred to digital print on demand 2003

Contents

Preface to the seventh edition

The seventh edition of Vogel's *Qualitative Inorganic Analysis* contains some new material, which has been added in response to readers' comments.

First of all, Chapter 5 was extended to include more on preliminary tests, dissolution and fusion of samples. This will help users who wish to tackle real samples – industrial, agricultural, environmental, etc.

Furthermore, atomic spectrometric tests have been added to the reactions of metal ions, as the necessary instrumentation is nowadays commonly found in laboratories; testing for metals in this way can be carried out quickly and reliably. Only the essential details are given – fuel, wavelength, light source as well as preparation of standards. It was felt necessary to add a short section on the theory and instrumentation of atomic spectrometry.

A number of health and hazard warnings have also been included. Nomenclature (especially that of organic reagents) has been modified so that readers should be able to find such chemicals in modern catalogues. Figures have been updated, though I was reluctant to change them to show only full-glass apparatus (as some have suggested), because this can be quite expensive and beyond the budgets of certain users. Instead, I have given information on commercially available all-glass devices in the text. To save space, preparations of reagents have been removed from the main text; the reader will find them in the Appendix.

I have also added comments in the margins, highlighting the most important facts and points described in the adjoining text. Hopefully, all these additions will enhance the practical value of the book, which first appeared about sixty years ago.

I would like to acknowledge many friends and students for their helpful suggestions. I am especially indebted to Dr Colin Graham (Birmingham) and to Mr Alan Robinson (Belfast) for reading and correcting the manuscript. Two postgraduate students in Cork, Miss Drewan McCaul and Mr Ronan Ennis, have carried out most of the atomic spectrometric tests experimentally; I wish to thank them for their efforts and I hope they will find the experience useful in their professional life.

As in the past, any suggestions for improvement will be welcomed by the revisor.

1996.

G. Svehla
Department of Chemistry
University College
Cork
Ireland

From preface to the first edition (1937)

Experience of teaching qualitative analysis over a number of years to large numbers of students has provided the nucleus around which this book has been written. The ultimate object was to provide a text-book at moderate cost which can be employed by the student continuously throughout his study of the subject.

It is the author's opinion that the theoretical basis of qualitative analysis, often neglected or very sparsely dealt with in the smaller texts, merits equally detailed treatment with the purely practical side; only in this way can the true spirit of qualitative analysis be acquired. The book accordingly opens with a long Chapter entitled 'The Theoretical Basis of Qualitative Analysis', in which most of the theoretical principles which find application in the science are discussed.

The writer would be glad to hear from teachers and others of any errors which may have escaped his notice: any suggestions whereby the book can be improved will be welcomed.

A. I. Vogel
Woolwich Polytechnic   London S.E.18

From preface to the fifth edition (1979)

Vogel's *Qualitative Inorganic Analysis* has been in continuous use in universities and polytechnics since its first publication in 1937. The fourth edition has now been in use for more than twenty years, and it was therefore considered appropriate to prepare a new one.

Since Dr Vogel died in 1966, the publishers asked me to prepare this new edition. When undertaking the enormous task, I wanted to preserve all that made this popular book so good: the detailed theoretical introduction, the well illustrated laboratory instructions and the rich selection of reactions. Nevertheless, because of the age of the text and changes in emphasis and style of the teaching of qualitative inorganic analysis, some rather drastic changes have had to be made.

When launching this new edition I would like to repeat what the author said in the first edition: any suggestions whereby the book can be improved will be welcomed.

G. Svehla
Department of Chemistry
Queen's University
Belfast, N. Ireland

From preface to the sixth edition (1987)

Previous editions of Vogel's *Qualitative Inorganic Analysis* were used mainly as university textbooks.... With the advent of more complex, mainly instrumental techniques, teaching of the reactions of ions and traditional qualitative analysis has moved from the classroom almost entirely into the laboratory. When preparing the sixth edition, this fact had to be taken into consideration.

The new edition is aimed more to be a laboratory manual than a textbook. With this aim in mind, the text has been rearranged and shortened. Furthermore, dangerous experiments and those involving carcinogenic materials have been removed.

G. Svehla
Department of Chemistry
University College
Cork
Ireland

Chapter 1

Introduction

The purpose of chemical analysis is to establish the composition of naturally occurring or artificially manufactured substances. This is usually done in two distinct steps. First, *qualitative analysis* is used to identify the sample components. This is followed with *quantitative analysis*, by which the relative amounts of these components are determined.

The present book describes the traditional methods of qualitative inorganic analysis. These can be divided broadly into two categories: *dry tests* which are carried out on solid samples, usually at higher temperatures; and, more commonly, *wet reactions* involving dissolved samples and reagent solutions. The chemical change (or its absence) is observed and used for the elucidation of sample composition.

In order to be able to do such analyses it is essential to study these reactions in a systematic way. Chapter 2 describes the laboratory equipment and skills necessary for such tests. Chapter 3 contains the reactions of the most common cations, while in Chapter 4 those of the most common anions are introduced with some explanations. Once the reader is familiar with these, he/she can attempt analyses of unknown samples, as described in Chapter 5. The book is concluded with the reactions of the less common ions (Chapter 6).

It must be emphasized that the study of 'classical' qualitative inorganic analysis is invaluable for any intending chemist, as this is where he/she first comes across and handles materials which are discussed in the traditional courses of inorganic chemistry. After a few weeks spent in the qualitative analytical laboratory the young chemists become familiar with solids, liquids, gases, acids, bases, salts – the bread and butter of every chemist's knowledge and skills.

Theoretical background

The intelligent study of qualitative inorganic analysis requires a certain level of theoretical background in general and inorganic chemistry. These, being taught in the introductory courses of chemistry at colleges and universities, will not be described here. Such a background involves chemical symbols, formulae, equations; theory of electrolytes, equilibria in electrolyte solutions; acid–base theory, strength of acids, pH, buffer systems, hydrolysis; morphology and structure of precipitates and colloids, precipitation equilibria; formation, structure and stability of complexes; balancing redox equations, redox potentials; the principles of solvent extraction and the distribution law. These are described to the necessary extent in most introductory chemistry texts and also in the first chapter of the 5th Edition of this book.[1]

[1] *Vogel's Textbook of Macro and Semimicro Qualitative Inorganic Analysis* (5th edn) revised by G. Svehla. Longman 1979.

Users of this book should be aware of the fact that almost all the reagents involved in qualitative inorganic analysis are **poisonous,** and some of the operations can be **dangerous**. It should be remembered that samples themselves often contain substances which are **environmentally hazardous**. It is important to consult appropriate safety legislation that is extant in your country. This normally relates not only to the use of certain toxic substances but also to their labelling and storage. The disposal of waste solutions must also be given serious consideration. In many countries there are special regulations governing the acquisition and disposal of radioactive materials (among which uranium and thorium are mentioned in this book). When special care is needed (e.g. handling cyanides, compounds of mercury, cadmium and arsenic and working with hydrogen sulphide) a warning is included in the text. Students and experienced chemists alike should always follow good laboratory practice (GLP), work in tidy, clean, well-lit laboratories. If necessary or if in doubt one must use fume cupboards with good ventilation. Safety glasses must be worn at all times and in many cases suitable gloves should be used. A person should never work alone in a laboratory, especially beyond normal working hours.

Students must be supervised during practicals by adequately trained personnel. It is worthwhile to spend some time at the beginning of a practical course on explaining safety measures in the laboratory, including practice with fire extinguishers. It is prudent to keep a written record of such safety courses, with participants acknowledging by their signature that they have attended.

Chapter 2

Experimental techniques

2.1 Introduction

Although it is assumed that the reader is familiar with basic laboratory operations such as dissolution, evaporation, crystallization, distillation, precipitation, filtration, decantation, bending of glass tubes, preparation of ignition tubes, boring of corks, etc., a brief discussion of those laboratory operations which are of special importance in qualitative inorganic analysis, will be given here.

Qualitative analysis may be carried out on various scales. In *macro analysis* the quantity of the substance employed is 0.1 to 0.5 g and the volume taken for analysis is about 20 ml. In what is usually termed *semimicro analysis* these quantities are reduced by a factor of 10–20, i.e. about 0.05 g material and 1 ml solution are employed. For *micro analysis* the factor is of the order of 100–200. The special operations needed for semimicro and micro work will be discussed in somewhat more detail, together with the apparatus needed to perform these.

Advantages of semimicro scale

It must be said that *the semimicro scale is most appropriate for the study of qualitative inorganic analysis*; some of the special advantages being as follows:

(a) Reduced consumption of chemicals with a considerable saving in the laboratory budget.

(b) The greater speed of analysis, due to working with smaller quantities of materials and the saving of time in carrying out the various standard operations of filtration, washing, evaporation, saturation with hydrogen sulphide, etc.

(c) Increased sharpness of separation, e.g. washing of precipitates, can be carried out rapidly and efficiently when a centrifuge replaces a filter.

(d) The amount of hydrogen sulphide used is considerably reduced.

(e) Much space is saved both on the reagent shelves and more especially in the lockers provided immediately below the bench for the housing of the individual person's apparatus; this latter merit may be turned to good use by reducing the size of the bench lockers considerably and thus effectively increasing the accommodation of the laboratory.

(f) The desirability of securing a training in the manipulation of small amounts of material.

Macro, semimicro and micro procedures will be discussed separately in this book to cater for all requirements. The readers should familiarize themselves first with macro operations, even if they adopt the semimicro scale throughout. It may be said that when the general technique of semi-micro analysis has been mastered and appreciated, no serious difficulty should be encountered in adapting a macro procedure to the semimicro scale. Among micro scale operations, spot tests are of special importance; for this reason these will be described in a separate section.

Qualitative analysis utilizes two kinds of tests, *dry tests* and *wet reactions*. The former are applicable to solid substances, while the latter apply to solutions. We begin with the discussion of the most important dry tests.

2.2   Dry tests

A number of useful tests can be carried out in the dry, that is before dissolving the sample for wet analysis. The information obtained in a comparatively short time often provides a clue to the presence or absence of certain substances. With this knowledge the course of wet analysis may be modified and shortened. Instructions for some important dry tests are given below.

Heating

1. *Heating*   the substance is placed in a small ignition tube (bulb tube), prepared from soft glass tubing, and heated in a Bunsen flame, gently at first and then more strongly. Small test-tubes, 60–70 mm × 7–8 mm, which are readily obtainable and are cheap, may also be employed. Sublimation may take place, or the material may melt or may decompose with an attendant change in colour, or a gas may be evolved which can be recognized by certain characteristic properties.

Sublimation

Sublimation yields a deposit of solid substance in the upper, colder parts of the test-tube. A **white** deposit occurs if ammonium salts, mercury(I) or mercury(II) chloride, or arsenic(III), antimony(III) or selenium(IV) oxides are present. A **yellow** deposit originates from sulphur, arsenic(III) sulphide and mercury(I) or (II) iodide. The deposit can be **black** if mercury(I) or (II) sulphides are present, and a **metallic** deposit may occur as the result of decomposition of other mercury salts, amalgams, arsenic(III) and cadmium compounds.

Colour change

Colour change can occur, usually indicating decomposition. The most common changes are associated with the removal of water of crystallization.

Gases

Gases produced may be colourless or coloured, occasionally with a characteristic smell. They can be identified with appropriate reagents, as described later under the reactions of cations and anions. **Carbon dioxide** can form if certain carbonates or organic materials are present. **Sulphur dioxide** can result when heating sulphites, thiosulphates or sulphides. **Chlorine** occurs if certain chlorides (e.g. magnesium chloride) are heated, while **bromine** or **iodine** can be formed when heating bromides or iodides in the presence of oxidizing agents. Nitrates yield **nitrogen dioxide**; formates and oxalates, **carbon monoxide**. Heating cyanides yields **cyanogen**. When heating ammonium salts mixed with alkalis, **ammonia** can occur. **Oxygen** is obtained if chlorates, perchlorates, bromates or iodates are heated.

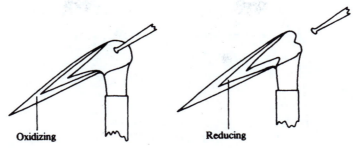

Oxidizing   Reducing

Fig. 2.1

Blowpipe tests

2. *Blowpipe tests*   a luminous Bunsen flame (air holes completely closed), about 5 cm long, is employed for these tests. A **reducing flame** is produced by placing the nozzle of a mouth blowpipe just outside the flame, and blowing gently so as to cause the inner cone to play on the substance under examination. An **oxidizing flame** is obtained by holding the nozzle of the blowpipe about one-third within the flame and blowing somewhat more vigorously in a direction parallel with the burner top; the extreme tip of the flame is allowed to play upon the substance. Figure 2.1 illustrates the oxidizing and reducing flames.

The tests are carried out upon a clean charcoal block in which a small cavity has been made with a penknife or with a small coin. A little of the substance is placed in the cavity and heated in the oxidizing flame. Crystalline salts break into smaller pieces; burning indicates the presence of an oxidizing agent (nitrate, nitrite, chlorate, etc.). More frequently the powdered substance is mixed with twice its bulk of anhydrous sodium carbonate or, preferably, with 'fusion mixture' (an equimolecular mixture of sodium and potassium carbonates; this has a lower melting point than sodium carbonate alone) in a reducing flame. The initial reaction consists of the formation of the carbonates of the cations present and the alkali salts of the anions. The alkali salts are largely adsorbed by the porous charcoal, and the carbonates are, for the most part, decomposed into the oxides and carbon dioxide. The oxides of the metals may further decompose, or be reduced to the metals, or they may remain unchanged. The final products of the reaction are therefore either the metals alone, metals and their oxides, or oxides. The oxides of the noble metals (silver and gold) are decomposed, without the aid of the charcoal, to the metal, which is often obtained as a globule, and oxygen. The oxides of lead, copper, bismuth, antimony, tin, iron, nickel, and cobalt are reduced either to a fused metallic globule (lead, bismuth, tin, and antimony) or to a sintered mass (copper) or to glistening metallic fragments (iron, nickel, and cobalt). The oxides of cadmium, arsenic, and zinc are readily reduced to the metal, but these are so volatile that they vaporize and are carried from the reducing to the oxidizing zone of the flame, where they are converted into sparingly volatile oxides. The oxides thus formed are deposited as an incrustation round the cavity of the charcoal block. Zinc yields an incrustation which is yellow while hot and white when cold; that of cadmium is brown and is moderately volatile; that of arsenic is

5

Bunsen flame

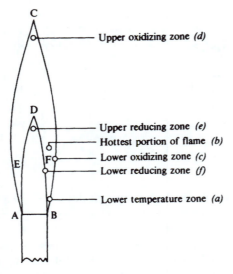

Upper oxidizing zone *(d)*

Upper reducing zone *(e)*
Hottest portion of flame *(b)*
Lower oxidizing zone *(c)*
Lower reducing zone *(f)*

Lower temperature zone *(a)*

Fig. 2.2

white and is accompanied by a garlic odour due to the volatilization of the arsenic. A characteristic incrustation accompanies the globules of lead, bismuth, and antimony.

The oxides of aluminium, calcium, strontium, barium, and magnesium are not reduced by charcoal; they are infusible and glow brightly when strongly heated. If the white residue or white incrustation left on a charcoal block is treated with a drop of cobalt nitrate solution and again heated, a bright-blue colour, which probably consists of either a compound or a solid solution of cobalt(II) and aluminium oxides (Thenard's blue), indicates the presence of aluminium;[1] a pale-green colour, probably of similar composition (Rinmann's green), is indicative of zinc oxide; and a pale pink mass is formed when magnesium oxide is present.

Flame tests on dry samples

3. *Flame tests* in order to understand the operations involved in the flame colour tests and the various bead tests to be described subsequently, it is necessary to have some knowledge of the structure of the non-luminous Bunsen flame (Fig. 2.2).

The non-luminous Bunsen flame consist of thee parts: (1) an inner blue cone ADB consisting largely of unburnt gas; (2) a luminous tip at D (this is only visible when the air holes are slightly closed); and (3) an outer mantle ACBD in which complete combustion of the gas occurs. The principal parts of the flame, according to Bunsen, are clearly indicated in Fig. 2.2. The lowest temperature is at the base of the flame *(a)*; this is employed for testing volatile substances to determine whether they impart any colour to the flame. The hottest part of the flame is the fusion zone at *b* and lies at about one-third of the height of the flame and approximately equidistant

---

[1] A blue colour is also given by phosphates, arsenates, silicates, or borates.

from the outside and inside of the mantle; it is employed for testing the fusibility of substances, and also, in conjunction with *a*, in testing the relative volatilities of substances or of a mixture of substances. The **lower oxidizing zone** (*c*) is situated on the outer border of *b* and may be used for the oxidation of substances dissolved in beads of borax, sodium carbonate, or microcosmic salt. The **upper oxidizing zone** (*d*) consists of the non-luminous tip of the flame; here a large excess of oxygen is present and the flame is not so hot as at *c*. It may be used for all oxidation processes in which the highest temperature is not required. The **upper reducing zone** (*e*) is at the tip of the inner blue cone and is rich in incandescent carbon; it is especially useful for reducing oxide incrustations to the metal. The **lower reducing zone** (*f*) is situated in the inner edge of the mantle next to the blue cone and it is here that the reducing gases mix with the oxygen of the air; it is a less powerful reducing zone than *e*, and may be employed for the reduction of fused borax and similar beads.

We can now return to the flame tests. Compounds of certain metals are volatilized in a non-luminous Bunsen flame and impart characteristic colours to the flame. The chlorides are among the most volatile compounds, and these are prepared *in situ* by mixing the compound with a little concentrated hydrochloric acid before carrying out the tests. The procedure is as follows. A thin platinum wire about 5 cm long and 0.03–0.05 mm diameter, fused into the end of a short piece of glass tubing or glass rod which serves as a handle, is employed. This is first thoroughly cleaned by dipping it into concentrated hydrochloric acid contained in a watch glass and then heating it in the fusion zone (*b*) of the Bunsen flame; the wire is clean when it imparts no colour to the flame. The wire is dipped into concentrated hydrochloric acid on a watch glass, then into a little of the substance being investigated so that a little adheres to the wire. It is then introduced into the lower oxidizing zone (*c*), and the colour imparted to the flame observed. Less volatile substances are heated in the fusion zone (*b*); in this way it is possible to make use of the difference in volatilities for the separation of the constituents of a mixture.

The colours imparted to the flame by salts of different metals are shown in Table 2.1.

Carry out flame tests with the chlorides of sodium, potassium, calcium, strontium, and barium and record the colours you observe. Repeat the test

Table 2.1	**Flame tests**
Important applications of flame tests	

Observation	Caused by
Persistent golden-yellow flame	Sodium
Violet (lilac) flame	Potassium
Carmine-red flame	Lithium
Brick-red (yellowish) flame	Calcium
Crimson flame	Strontium
Yellowish-green flame	Barium, molybdenum
Green flame	Borates, copper, thallium
Blue flame (wire slowly corroded)	Lead, arsenic, antimony, bismuth, copper

Table 2.2   **Flame tests with cobalt glass**

Flame colouration	Flame colouration through cobalt glass	Caused by
Golden-yellow	Nil	Sodium
Violet	Crimson	Potassium
Brick-red	Light-green	Calcium
Crimson	Purple	Strontium
Yellowish-green	Bluish-green	Barium

with a mixture of sodium and potassium chlorides. The yellow colouration due to the sodium masks that of the potassium. View the flame through two thicknesses of cobalt glass; the yellow sodium colour is absorbed and the potassium flame appears crimson.

Potassium chloride is much more volatile than the chlorides of the alkaline earth metals. It is therefore possible to detect potassium in the lower oxidizing flame and the calcium, strontium, and barium in the fusion zone.

Repeat all the tests with a cobalt glass and note your observations. Table 2.2 indicates the results to be expected.

After all the tests, the platinum wire should be cleaned with concentrated hydrochloric acid. It is a good plan to store the wire permanently in the acid. A cork is selected that just fits into a test-tube, and a hole is bored through the cork through which the glass holder of the platinum wire is passed. The test-tube is about half filled with concentrated hydrochloric acid so that when the cork is placed in position, the platinum wire is immersed in the acid.

A platinum wire sometimes acquires a deposit which is removed with difficulty by hydrochloric acid and heat. It is then best to employ fused potassium hydrogen sulphate. A coating of potassium hydrogen sulphate is made to adhere to the wire by drawing the hot wire across a piece of the solid salt. Upon passing the wire slowly through a flame, the bead of potassium pyrosulphate which forms travels along the wire, dissolving the contaminating deposits. When cool, the bead is readily dislodged. Any small residue of pyrosulphate dissolves at once in water, whilst the last traces are usually removed by a single moistening with concentrated hydrochloric acid, followed by heating. The resulting bright clean platinum wire imparts no colour to the flame.

4. *Spectroscopic tests (flame spectra)*   the best way to employ flame tests in analysis is to resolve the light into its component tints and to identify the cations present by their characteristic sets of tints. The instrument employed

to resolve light into its component colours is called a **spectroscope**. A simple form is shown in Fig. 2.3. It consists of a collimator A which throws a beam of parallel rays on the prism B, mounted on a turntable; the telescope C through which the spectrum is observed; and a tube D, which contains a scale of reference lines which may be superimposed upon the spectrum. The spectroscope is calibrated by observing the spectra of known substances, such as sodium chloride, potassium chloride, thallium chloride, and lithium chloride. The conspicuous lines are located on a graph drawn

Components of a
spectroscope:
collimator (A)
prism (B)
telescope (C)
scale (D)

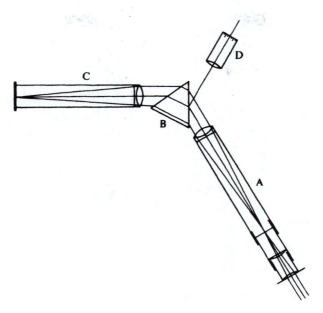

Fig. 2.3

with wavelengths as ordinates and scale divisions as abscissae. The wavelength curve may then be employed in obtaining the wavelength of all
intermediate positions and also in establishing the identity of the component
elements of a mixture.

To adjust the simple table spectroscope described above (which is always
mounted on a rigid stand), a lighted Bunsen burner is placed in front of the
collimator A at a distance of about 10 cm from the slit. Some sodium chloride
is introduced by means of a clean platinum wire into the lower part of the
flame, and the tube containing the adjustable slit rotated until the sodium
line, as seen through the telescope C, is in a vertical position. (If available,
it is more convenient to employ an electric discharge 'sodium lamp': this
constitutes a high-intensity sodium light source.) The sodium line is then
sharply focused by suitably adjusting the sliding tubes of the collimator
and the telescope. Finally, the scale D is illuminated by placing a small
electric lamp in front of it, and the scale sharply focused. The slit should
also be made narrow in order that the position of the lines on the scale can
be noted accurately.

A smaller, relatively inexpensive, and more compact instrument, which is
more useful for routine tests in qualitative analysis, is the **direct vision
spectroscope** with comparison prism, shown in Fig. 2.4.

The light from the flame passes along the central axis of the instrument
through the slit, which is adjustable by a milled knob at the side. When the
comparison prism is interposed, half the length of the slit is covered and
thus light from a source in a position at right angles to the axis of the
instrument will fall on one-half of the slit adjacent to the direct light which
enters the other half. This light passes through an achromatic objective lens

A 'pocket'
spectroscope

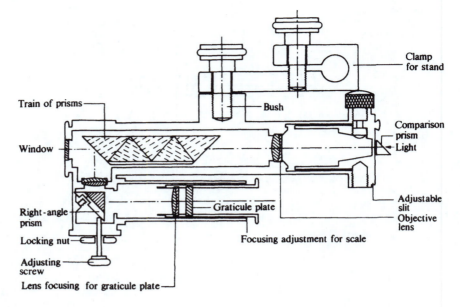

Fig. 2.4

and enters a train of five prisms of the 60° type, three being of crown glass and the alternate two of flint glass. The train of prisms gives an angular dispersion of about 11° between the red and the blue ends of the spectrum. The resulting spectrum, which can be focused by means of a sliding tube adjustment, is observed through the window. There is a subsidiary tube adjacent to the main tube: the former contains a graticule on a glass disc, which is illuminated either from the same source of light as that being observed or from a small subsidiary source (e.g. a flash lamp bulb). It is focused by means of a lens system comprising two achromatic combinations between which is a right angle prism. This prism turns the beam of light so that it falls on the face of the end prism (of the train of five prisms) and is reflected into the observer's eye, where it is seen superimposed upon the spectrum. An adjusting screw is provided to alter the position of the right angle prism in order to adjust the scale relative to the spectrum. The scale is calibrated directly into divisions of 10 nanometers (or 100 Å in older instruments) and has also an indication mark at the D-line: it is 'calibrated' by means of a sodium source, and the adjusting screw is locked into position by means of a locking nut. The instrument can be mounted on a special stand.

If a sodium compound is introduced into the colourless Bunsen flame, it colours it yellow; if the light is examined by means of a spectroscope, a bright yellow line is visible. By narrowing the slit, two fine yellow lines may be seen. The mean wavelength of these two lines is 5.893×10^{-7} m. [Wavelengths are generally expressed in nanometers (nm). $1\,nm = 10^{-9}$ m. The old units of ångstrom (Å, 10^{-10} m) and millimicron (m$\mu = 10^{-9}$ m, identical to nm) are now obsolete.] The mean wavelength of the two sodium lines is therefore 589.3 nm. The elements which are usually identified by the

Table 2.3   **Commonly occurring spectrum lines**

Element	Description of line(s)	Wavelength in nm
Sodium	Double yellow	589.0, 589.6
Potassium	Double red	766.5, 769.9
	Double violet	404.4, 404.7
Lithium	Red	670.8
	Orange (faint)	610.3
Thallium	Green	535.0
Calcium	Orange band	618.2–620.3
	Yellowish-green	555.4
	Violet (faint)	422.7
Strontium	Red band	674.4, 662.8
	Orange	606.0
	Blue	460.7
Barium	Green band	553.6, 534.7, 524.3, 513.7
	Blue (faint)	487.4

spectroscope in qualitative analysis are: sodium, potassium, lithium, thallium and, less frequently, because of the comparative complexity of their spectra, calcium, strontium, and barium. The wavelengths of the brightest lines, visible through a good-quality direct vision spectroscope, are collected in Table 2.3. As already stated, the spectra of the alkaline earth metals are relatively complex and consist of a number of fine lines; the wavelengths of the brightest of these are given. If the resolution of the spectroscope is small, they will appear as bands.

The spectra of the various elements are shown diagrammatically in Fig. 2.5; the positions of the lines have been drawn to scale.

Tests for certain metals can be carried out conveniently on solutions using **flame atomic emission spectrometry (FAES)** or **flame atomic absorption spectrometry (FAAS)**, if the appropriate apparatus is available. These tests will be described under Section 2.7.

Borax bead test   5. *Borax bead tests*   a platinum wire, similar to that referred to under flame tests, is used for the borax bead tests. The free end of the platinum wire is coiled into a small loop through which an ordinary match will barely pass. The loop is heated in the Bunsen flame until it is red hot and then quickly dipped into powdered borax, $Na_2B_4O_7 \cdot 10H_2O$. The adhering solid is held in the hottest part of the flame; the salt swells up as it loses its water of crystallization and shrinks upon the loop forming a colourless, transparent, glass-like bead consisting of a mixture of sodium metaborate and boric anhydride.[2]

$$Na_2B_4O_7 = 2NaBO_2 + B_2O_3$$

---

[2] Some authors do not recommend the use of a loop on the platinum wire as it is considered that too large a surface of the platinum is thereby exposed. According to their procedure, the alternate dipping into borax and heating is repeated until a bead 1.5–2 mm diameter is obtained. The danger of the bead falling off is reduced by holding the wire horizontally. It is the author's experience that the loop method is far more satisfactory, especially in the hands of beginners, and is less time consuming.

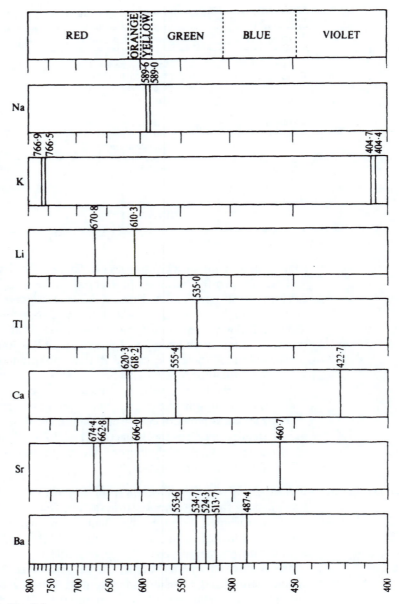

Fig. 2.5

The bead is moistened and dipped into the finely powdered substance so that a minute amount of it adheres to the bead. It is important to employ a minute amount of substance as otherwise the bead will become dark and opaque in the subsequent heating. The bead and adhering substance are first heated in the lower reducing flame, allowed to cool and the colour observed. They are then heated in the lower oxidizing flame, allowed to cool and the colour observed again.

Characteristic coloured beads are produced with salts of copper, iron, chromium, manganese, cobalt, and nickel. Carry out borax bead tests with salts of these metals and compare results with those given in Chapter 3.

After each test, the bead is removed from the wire by heating it again to fusion, and then jerking it off the wire into a vessel of water. The borax bead also provides an excellent method for cleaning a platinum wire; a bead is run backwards and forwards along the wire by heating, and is then shaken off by a sudden jerk.

The coloured borax beads are due to the formation of coloured borates; in those cases where different coloured beads are obtained in the oxidizing and the reducing flames, borates corresponding to varying stages of oxidation of the metal are produced. Thus with copper salts in the oxidizing flame, one has:

Chemical reactions in borax bead tests

$$Na_2B_4O_7 = 2NaBO_2 + B_2O_3$$

$$CuO + B_2O_3 = Cu(BO_2)_2 \text{ (copper(II) metaborate)}$$

The reaction:

$$CuO + NaBO_2 = NaCuBO_3 \text{ (orthoborate)}$$

probably also occurs. In the reducing flame (i.e. in the presence of carbon), two reactions may take place:

(1) the coloured copper(II) is reduced to colourless copper(I) metaborate:

$$2Cu(BO_2)_2 + 2NaBO_2 + C = 2CuBO_2 + Na_2B_4O_7 + CO\uparrow$$

(2) the copper(II) borate is reduced to metallic copper, so that the bead appears red and opaque:

$$2Cu(BO_2)_2 + 4NaBO_2 + 2C = 2Cu + 2Na_2B_4O_7 + 2CO\uparrow$$

With iron salts, $Fe(BO_2)_2$ and $Fe(BO_2)_3$ are formed in the reducing and oxidizing flames respectively.

Some authors assume that the metal metaborate combines with sodium metaborate to give complex borates of the type $Na_2[Cu(BO_2)_4]$, $Na_2[Ni(BO_2)_4]$ and $Na_2[Co(BO_2)_4]$:

$$Cu(BO_2)_2 + 2NaBO_2 = Na_2[Cu(BO_2)_4]$$

Results obtainable with the borax-bead test are summarized in Table 2.4. In certain circumstances the borax-bead test is inconclusive, but it can be supplemented with the phosphate (or microcosmic salt) bead test (see below).

Microcosmic salt bead test

6. *Phosphate (or microcosmic salt) bead tests*   the bead is produced similarly to the borax bead except that microcosmic salt, sodium ammonium hydrogen phosphate tetrahydrate $Na(NH_4)HPO_4 \cdot 4H_2O$, is used. The colourless, transparent bead contains sodium metaphosphate:

$$Na(NH_4)HPO_4 = NaPO_3 + H_2O\uparrow + NH_3\uparrow$$

This combines with metallic oxides forming orthophosphates, which are often coloured. Thus a blue phosphate bead is obtained with cobalt salts:

$$NaPO_3 + CoO = NaCoPO_4$$

Table 2.4 **Colours of borax beads**

Oxidizing flame		Reducing flame		Metal
Hot	*Cold*	*Hot*	*Cold*	
Green	Blue	Colourless	Opaque red-brown	Copper
Yellowish-brown	Yellow	Green	Green	Iron
Yellow	Green	Green	Green	Chromium
Violet (amethyst)	Amethyst	Colourless	Colourless	Manganese
Blue	Blue	Blue	Blue	Cobalt
Violet	Reddish-brown	Grey	Grey	Nickel
Yellow	Colourless	Brown	Brown	Molybdenum
Rose-violet	Rose-violet	Red	Violet	Gold
Yellow	Colourless	Yellow	Yellowish-brown	Tungsten
Yellow	Pale yellow	Green	Bottle-green	Uranium
Yellow	Greenish-yellow	Brownish	Emerald-green	Vanadium
Yellow	Colourless	Grey	Pale violet	Titanium
Orange-red	Colourless	Colourless	Colourless	Cerium

The sodium metaphosphate glass exhibits little tendency to combine with acidic oxides. Silica, in particular, is not dissolved by the phosphate bead. When a silicate is strongly heated in the bead, silica is liberated and this remains suspended in the bead in the form of a semi-translucent mass; the so-called silica 'skeleton' is seen in the bead during and after fusion. This reaction is employed for the detection of silicates:

$$CaSiO_3 + NaPO_3 = NaCaPO_4 + SiO_2$$

It must, however, be pointed out that many silicates dissolve completely in the bead so that the absence of a silica 'skeleton' does not conclusively prove that a silicate is not present.

In general, it may be stated that the borax beads are more viscous than the phosphate beads. They accordingly adhere better to the platinum wire loop.

Table 2.5 **Microcosmic salt bead tests**

Oxidizing flame	Reducing flame	Metal
Green when hot, blue when cold.	Colourless when hot, red when cold.	Copper
Yellowish- or reddish-brown when hot, yellow when cold.	Yellow when hot, colourless to green when cold.	Iron
Green, hot and cold.	Green, hot and cold.	Chromium
Violet, hot and cold.	Colourless, hot and cold.	Manganese
Blue, hot and cold.	Blue, hot and cold.	Cobalt
Brown, hot and cold.	Grey when cold.	Nickel
Yellow, hot and cold.	Green when cold.	Vanadium
Yellow when hot, yellow-green when cold.	Green, hot and cold	Uranium
Pale yellow when hot, colourless when cold.	Green when hot, blue[a] when cold.	Tungsten
Colourless, hot and cold.	Yellow when hot, violet[a] when cold.	Titanium

[a] Blood red when fused with a trace of iron(II) sulphate.

The colours of the phosphates, which are generally similar to those of the borax beads, are usually more pronounced. The various colours of the phosphate beads are collected in Table 2.5.

Na$_2$CO$_3$ bead

7. *Sodium carbonate bead tests*   the sodium carbonate bead is prepared by fusing a small quantity of sodium carbonate on a platinum wire loop in the Bunsen flame; a white, opaque bead is produced. If this is moistened, dipped into a little potassium nitrate and then into a small quantity of a manganese compound, and the whole heated in the oxidizing flame, a green bead of sodium manganate is formed:

$$MnO + Na_2CO_3 + O_2 = Na_2MnO_4 + CO_2\uparrow$$

A yellow bead is obtained with chromium compounds, due to the production of sodium chromate:

$$2Cr_2O_3 + 4Na_2CO_3 + 3O_2 = 4Na_2CrO_4 + 4CO_2\uparrow$$

2.3   Wet reactions; macro apparatus and analytical operations on a macro scale

Wet reactions are carried out with dissolved substances. Such reactions lead to: (a) the formation of a precipitate; (b) the evolution of a gas; or (c) a change of colour. The majority of reactions in qualitative analysis are carried out in solution, and details of these are given in later chapters. The following general notes on basic laboratory operations should be carefully studied.

Equipment

1. *Test-tubes*   the best size for general use is 15×2 cm with 25 ml total capacity. It is useful to remember that 10 ml of liquid fills a test-tube of this size to a depth of about 5.5 cm. Smaller test-tubes are sometimes used for special tests. For heating moderate volumes of liquids a somewhat larger tube, about 18×2.5 cm, the so-called 'boiling tube' is recommended. A test-tube brush should be available for cleaning the tubes.

2. *Beakers*   those of 50, 100, and 250 ml capacity and of the Griffin form are the most useful in qualitative analysis. Clock glasses of the appropriate size should be provided. For evaporations and chemical reactions which are likely to become vigorous, the clock glass should be supported on the rim of the beaker by means of V-shaped glass rods.

3. *Conical or Erlenmeyer flasks*   these should be of 50, 100, and 250 ml capacity, and are useful for decompositions and evaporations. The introduction of a funnel, whose stem has been cut off, prevents loss of liquid through the neck of the flask and permits the escape of steam.

4. *Stirring rods*   a length of glass rod, of about 4 mm diameter, is cut into suitable lengths and the ends rounded in the Bunsen flame. The rods should be about 20 cm long for use with test-tubes and 8–10 cm long for work with basins and small beakers. Open glass tubes must not be used as stirring rods. A rod pointed at one end, prepared by heating a glass rod in

the flame, drawing it out when soft as in the preparation of a glass jet and then cutting it into two, is employed for piercing the apex of a filter paper to enable one to transfer the contents of a filter paper by means of a stream of water from a wash bottle into another vessel.

A rubber-tipped glass rod or 'policeman' is employed for removing any solid from the sides of glass vessels. A stirring rod of polythene (polyethylene) with a thin fan-shaped paddle on each end is available commercially and functions as a satisfactory 'policeman' at laboratory temperature: it can be bent into any form.

5. *Wash bottle*   this may consist of a 500 ml flat-bottomed flask, and the stopper carrying the two tubes should be preferably of glass or rubber (cf. Fig. 2.19*a* and *b*). It is recommended that the wash bottle be kept ready for use filled with hot water as it is usual to wash precipitates with hot water; this runs through the filter paper rapidly and has a greater solvent power than cold water so that less is required for efficient washing. Insulating cloth should be wound round the neck of the flask in order to protect the hand.

A rubber bulb may be attached to the short tube; this form of wash bottle is highly convenient and is more hygienic than the type requiring blowing by the mouth.

Polyethylene wash bottles (Fig. 2.19*c* and *d*) can be operated by pressing the elastic walls of the bottle. The pressure pushes the liquid out through the tube. The flow rate can easily be regulated by changing the pressure exerted by the hand. When set aside the walls flatten out again and air is sucked into the bottle through the tube. As the air bubbles through the liquid, the latter will always be saturated with dissolved oxygen and this might interfere with some of the tests. These flasks cannot be heated, and although they will take hot liquids, it is very inconvenient to use them because one has to grip the hot walls.

Precipitation

6. *Precipitation*   when excess of a reagent is to be used in the formation of a precipitate, this does not mean that an excessive amount should be employed. In most cases, unless specifically stated, only a moderate amount over that required to bring about the reaction is necessary. This is usually best detected by filtering a little of the mixture and testing the filtrate with the reagent; if no further precipitation occurs, a sufficient excess of the reagent has been added. It should always be borne in mind that a large excess of the precipitating agent may lead to the formation of complex ions and consequent partial solution of the precipitate; furthermore, an unnecessary excess of the reagent is wasteful and may lead to complications at a subsequent stage of the analysis. When studying the reactions of ions the concentrations of the reagents are known and it is possible to judge the required volume of the reagent by quick mental calculation.

How to use H_2S safely and sparingly

7. *Precipitation with hydrogen sulphide*   operations with hydrogen sulphide are of such importance in qualitative inorganic analysis that they merit detailed discussion. **Hydrogen sulphide gas is highly poisonous**, its lethal dose

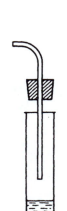

Fig. 2.6

is the same as that of hydrogen cyanide. Fortunately its unpleasant smell repels people from breathing in dangerous amounts. Hydrogen sulphide must be handled **in a fume cupboard with good ventilation**.

One method, which is sometimes employed, consists of passing a stream of gas in the form of bubbles through the solution contained in an open beaker, test-tube, or conical flask; this procedure is sometimes termed the 'bubbling' method. The efficiency of the method is low, particularly in acid solution; absorption of the gas takes place at the surface of the bubbles and, since the gas is absorbed slowly, most of it escapes into the air of the fume chamber and is wasted. The 'bubbling' method is not recommended and should not be used for macro analysis. The most satisfactory procedure (the 'pressure' method) is best described with the aid of Fig. 2.6. The solution is contained in a small conical flask A, which is provided with a stopper and lead-in tube; B is a wash bottle containing water and serves to remove any hydrochloric acid spray that might be carried over from the Kipp's apparatus in the gas stream; C is a stopcock which controls the flow of gas from the generator, whilst stopcock D provides an additional control[3] for the gas flow. The conical flask is connected to the wash bottle by a short length of rubber tubing. The stopper is first loosened in the neck of the flask and the gas stream turned on (C first, followed by D) so as to displace most of the air in the flask: this will take not more than about 30 seconds. With the gas flowing, the stopper is inserted tightly and the flask is gently shaken with a rotary motion; splashing of the liquid on to the hydrogen sulphide delivery tube should be avoided. In order to ensure that all the air has been expelled, it is advisable to loosen the stopper in A again to sweep out the gas and then to stopper the flask tightly. Passage of the gas is continued

---

[3] Stopcock D is optional; it prevents diffusion of air into the wash bottle and thus ensures an almost immediate supply of hydrogen sulphide.

with gentle rotation of the flask until the bubbling of the gas in B has almost ceased.[4] At this point the solution in A should be saturated with hydrogen sulphide and precipitation of the sulphides should be complete: this will normally require only a few minutes. Complete precipitation should be tested for by separating the precipitate by filtration, and repeating the procedure with the filtrate until the hydrogen sulphide produces no further precipitate. Occasionally a test-tube replaces the conical flask; it is shown (with the stopper) in (*b*). The delivery tube must be thoroughly cleaned after each precipitation. The same device can be put together with glass joints. The appropriate Quickfit apparatus consists of the MF 28/3/125 wash bottle, FE 100/1 conical flask, MF 15/1B/SC inlet tube and MF 24/1 test tube. The advantages of the 'pressure' method are: (1) a large surface of the liquid is presented to the gas and (2) it prevents the escape of large amounts of unused gas.

Alternatively, a saturated aqueous solution of hydrogen sulphide can be used as a reagent. This can most easily be prepared in the bottle B of the apparatus shown in Fig. 2.6. Such a reagent can most conveniently be used in teaching laboratories or classroom demonstrations when studying the reactions of ions. For a quantitative precipitation of sulphides (e.g. for separation of metals) the use of hydrogen sulphide gas is however recommended.

Filtration

8. *Filtration and washing*   the purpose of **filtration** is, of course, to separate the mother liquor and excess of reagent from the precipitate. A moderately fine-textured filter paper is generally employed. The size of the filter paper is controlled by the quantity of precipitate and not by the volume of the solution. The upper edge of the filter paper should be about 1 cm from the upper rim of the glass funnel. It should never be more than about two-thirds full of the solution. Liquids containing precipitates should be heated before filtration except in special cases, for example that of lead chloride which is markedly more soluble in hot than cold water. Gelatinous precipitates, which usually clog the pores of the filter paper and thus considerably reduce the rate of filtration, may be filtered through fluted filter paper or through a pad of filter papers resting on the plate of a Buchner funnel (Fig. 2.9*b* or *c*); this procedure may be used when the quantity of the precipitate is large and it is to be discarded. The best method is to add a little filter paper pulp (macerated filter paper) to the solution and then to filter in the normal manner. Filter paper pulp increases the speed of filtration by retaining part of the precipitate and thus preventing the clogging of the pores of the filter paper.

When a precipitate tends to pass through the filter paper, it is often a good plan to add an ammonium salt, such as ammonium chloride or nitrate, to the solution; this will help to prevent the formation of colloidal solutions.

Washing

A precipitate may be **washed** by decantation, as much as possible being retained in the vessel during the first two or three washings, and the precipitate then transferred to the filter paper. This procedure is unnecessary for

---

[4] It must be borne in mind that the gas may contain a small proportion of hydrogen due to the iron usually present in the commercial iron(II) sulphide. Displacement of the gas in the flask (by loosening the stopper) when the bubbling has diminished considerably will ensure complete precipitation.

coarse, crystalline, easy filterable precipitates as the washing can be carried out directly on the filter paper. This is best done by directing a stream of water from a wash bottle first around the upper rim of the filter and following this down in a spiral towards the precipitate in the apex; the filter is filled about one-half to two-thirds full at each washing. The completion of washing, i.e. the removal of the precipitating agent, is tested for by chemical means; thus if a chloride is to be removed, silver nitrate solution is used. If the solution is to be tested for acidity or alkalinity, a drop of the thoroughly stirred solution, removed upon the end of a glass rod, is placed in contact with a small strip of 'neutral' litmus paper or of 'wide range' or 'universal' test paper on a watch glass. Other test papers are employed similarly.

Removal of precipitate from the filter

9. *Removal of the precipitate from the filter*   if the precipitate is bulky, sufficient amounts for examination can be removed with the aid of a small nickel or stainless steel spatula. If the amount of precipitate is small, one of two methods may be employed. In the first, a small hole is pierced in the base of the filter paper with a pointed glass rod and the precipitate washed into a test-tube or a small beaker with a stream of water from the wash bottle. In the second, the filter paper is removed from the funnel, opened out on a clock glass, and scraped with a spatula.

It is frequently necessary to dissolve a precipitate completely. This is most readily done by pouring the solvent, preferably whilst hot, onto the filter and repeating the process, if necessary, until all the precipitate has passed into solution. If it is desired to maintain a small volume of the liquid, the filtrate may be poured repeatedly through the filter until all the precipitate has passed into solution. When only a small quantity of the precipitate is available, the filter paper and precipitate may be heated with the solvent and filtered.

Filtration aids and techniques

10. *Aids to filtration*   the simplest device is to use a funnel with a long stem, or better to attach a narrow-bored glass tube, about 45 cm long and bent as shown in Fig. 2.7, to the funnel by means of rubber tubing. The lower end of the tube or of the funnel should touch the side of the vessel in which the filtrate is being collected in order to avoid splashing. The speed of filtration depends *inter alia* upon the length of the water column.

Where large quantities of liquids and/or precipitates are to be handled, or if rapid filtration is desired, filtration under diminished pressure is employed: a metal or glass water pump may be used to provide the reduced pressure. A filter flask of 250–500 ml capacity is fitted with a two-holed rubber bung; a long glass tube is passed through one hole and a short glass tube, carrying a glass stopcock at its upper end, through the other hole. The side arm of the flask is connected by means of thick-walled rubber tubing ('pressure' tubing) to another flask, into the mouth of which a glass funnel is fitted by

Using suction

means of a rubber bung (Fig. 2.9*a*). Upon applying suction to the filter paper fitted into the funnel in the usual way, it will be punctured or sucked through, particularly when the volume of the liquid in the funnel is small. To surmount the difficulty, the filter paper must be supported in the funnel. For this purpose either a Whatman filter cone (No. 51) made of a specially hardened filter paper, or a Schleicher and Schuell (U.S.A.) filter

Fig. 2.7

Evaporation

paper support (No. 123), made from a muslin-type material which will not retard filtration, may be used. Both types of support are folded with the filter paper to form the normal type of cone (Fig. 2.8). After the filter paper has been supported in the funnel, filtration may be carried out in the usual manner under the partial vacuum created by the pump, the stopcock T being closed. When filtration is complete, the stopcock T is opened, air thereby entering the apparatus which thus attains atmospheric pressure; the filter funnel may now be removed from the filter flask.

For a large quantity of precipitate, a small **Buchner funnel** (*b* in Fig. 2.9, shown enlarged for the sake of clarity) is employed. This consists of a porcelain funnel in which a perforated plate is incorporated. Two thicknesses of well-fitting filter paper cover the plate. The Buchner funnel is fitted into the filter flask by means of a cork. When the volume of liquid is small, it may be collected in a test-tube placed inside the filter flask. The **Jena 'slit sieve' funnel**,[5] shown in *c*, is essentially a transparent Buchner funnel; its great advantage over the porcelain Buchner funnel is that it is easy to see whether the funnel is perfectly clean.

Strongly acidic or alkaline solutions cannot be filtered through ordinary filter paper. They may be filtered through a small pad of glass wool placed in the apex of a glass funnel. A more convenient method, applicable to strongly acidic and mildly basic solutions, is to employ a **sintered glass funnel** (Fig. 2.9*d*); the filter plate, which is available in various porosities, is fused into a resistance glass (borosilicate) funnel. Filtration is carried out under reduced pressure exactly as with a Buchner funnel.

Filtration with suction can be achieved by components with glass joints. The Quickfit FB 100/2 Buchner flask with FC 15/2 suction tubes, combined with SF 42/2 conical filter with perforated plate is adequate for the purpose. The SF 4A 32 conical filter with sintered disc or the Buchner filter SF 3A 32 are also available for filtration.

11. *Evaporation*   the analytical procedure may specify evaporation to a smaller volume or evaporation to dryness. Both operations can be conveniently carried out in a porcelain evaporating dish or casserole; the capacity of the vessel should be as small as possible for the amount of liquid being reduced in volume. The most rapid evaporation is achieved by heating the dish directly on a wire gauze. For many purposes a water bath (a beaker half-filled with water maintained at the boiling point is quite suitable) will

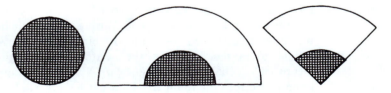

Fig. 2.8

[5] A Pyrex 'slit sieve' funnel, with 65 mm disc, is available commercially.

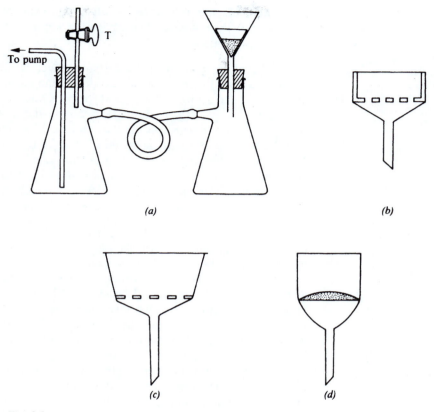

Fig. 2.9

serve as a source of heat; the rate of evaporation will of course be slower than by direct heating with a flame. Should corrosive fumes be evolved during the evaporation, the process must be carried out in the fume cupboard. When evaporating to dryness, it is frequently desirable, in order to minimize spattering and bumping, to remove the dish whilst there is still a little liquid left; the heat capacity of the evaporating dish is usually sufficient to complete the operation without further heating.

The reduction in volume of a solution may also be accomplished by direct heating in a small beaker over a wire gauze or by heating in a wide test-tube ('boiling-tube'), held in a holder, by a free flame; in the latter case care must be taken that the liquid does not bump violently. A useful **anti-bumping device**, applicable to solutions from which gases (hydrogen sulphide, sulphur dioxide, etc.) are to be removed by boiling, is shown in Fig. 2.10. It consists

1 cm

Fig. 2.10

of a length of glass tubing sealed off about 1 cm from one end, which is inserted into the solution. The device must not be used in solutions that contain a precipitate.

Drying precipitates

12. *Drying of precipitates*   partial drying, which is sufficient for many purposes, is accomplished by opening out the filter, laying it upon several dry filter papers and allowing them to absorb the water. More complete drying is obtained by placing the funnel containing the filter paper in a 'drying cone' (a hollow tinned-iron cone or cylinder), which rests either upon a sand bath or upon a wire gauze and is heated by means of a small flame. The funnel is thus exposed to a current of hot air, which rapidly dries the filter and precipitate. Great care must be taken not to char the filter paper. A safer but slower method is to place the funnel and filter paper, or the filter paper alone resting upon a clock glass, inside a drying oven.

Cleaning of apparatus

13. *Cleaning of apparatus*   the importance of using clean apparatus cannot be too strongly stressed. All glassware should be put away clean. A few minutes should be devoted at the end of the day's work to 'cleaning up'; one should remember that wet dirt is very much easier to remove than dry dirt. A test-tube brush should be used to clean test-tubes and other glass apparatus. Test-tubes may be inverted in the test-tube stand and allowed to drain. Other apparatus, after rinsing with distilled water, should be wiped dry with a 'glass cloth', that is a cloth which has been washed at least once and contains no dressing.

Glass apparatus which appears to be particularly dirty or greasy is cleaned by soaking in chromosulphuric acid (concentrated sulphuric acid containing about 100 g of potassium dichromate per litre), followed by a liberal washing with tap water, and then with distilled water. **The cleaning agent is highly corrosive!**

14. *Some working hints*

Tidiness

(a) Always work in a tidy, systematic manner. Remember a tidy bench is indicative of a methodical mind. A string duster is useful to wipe up liquids split upon the bench. All glass and porcelain apparatus must be scrupulously clean.

(b) Reagent bottles and their stoppers should not be put upon the bench. They should be returned to their correct places upon the shelves immediately after use. If a reagent bottle is empty, it should be returned to the store-room for filling.

Cleanliness

(c) When carrying out a test which depends upon the formation of a precipitate, make sure that both the solution to be tested and the reagent are absolutely free from suspended particles. If this is not the case, filter the solutions first.

Not wasting chemicals

(d) Do not waste gas or chemicals. The size of the Bunsen flame should be no larger than is absolutely necessary. It should be extinguished when no longer required. Avoid using unnecessary excess of reagents. Reagents should always be added portion-wise.

Waste disposal

(e) Pay particular attention to the disposal of waste. Neither strong acids nor strong alkalis should be thrown into the sink; they must be well diluted first, and the sink flushed with much water. Solids (corks, filter paper, etc.) should be placed in the special boxes provided for them in the laboratory. On no account may they be thrown into the sink.

Fume cupboard

(f) All operations involving (1) the passage of hydrogen sulphide into a solution, (2) the evaporation of concentrated acids, (3) the evaporation of solutions for the removal of ammonium salts, and (4) the evolution of poisonous or disagreeable vapours or gases, must be conducted in the fume cupboard.

Taking notes

(g) All results, whether positive, negative, or inconclusive, must be recorded neatly in a notebook at the time they are made. The writing up of experiments should not be postponed until after one has left the laboratory. Apart from inaccuracies which may thus creep in, the habit of performing experiments and recording them immediately is one that should be developed from the very outset.

(h) If the analysis is incomplete at the end of the laboratory period, label all solutions and precipitates clearly. It is a good plan to cover these with filter paper to prevent the entrance of dust, etc.

2.4 Semimicro apparatus and semimicro analytical operations

The essential technique of semimicro analysis does not differ very greatly from that of macro analysis. Since volumes of the order of 1 ml are dealt with, the scale of the apparatus is reduced; it may be said at once that it is just as easy to manipulate these small volumes and quantities as to work with larger volumes and quantities. The various operations occupy less time and the consumption of chemicals and glassware is reduced considerably; these two factors are of great importance when time and money are limited. Particular care must be directed to having both the apparatus and the work bench scrupulously clean.

Test-tubes

1. *Test-tubes and centrifuge tubes* small Pyrex test-tubes (usually 75×10 mm, 4 ml, sometimes 100×12 mm, 8 ml) are used for reactions which do not require boiling (Fig. 2.11*a* and *b*). When a precipitate is to be separated by centrifuging, a **centrifuge tube** (Fig. 2.11*d*) is generally employed; here, also, the contents cannot be boiled as 'bumping' will occur. Various sizes are available; the 3 ml centrifuge tube is the most widely used and will be adopted as standard throughout this book. For rapid concentration of a solution by means of a free flame, the **semimicro boiling tube** (60×25 mm, Pyrex; Fig. 2.11*c*) will be found convenient.

Stirring

2. *Stirring rods* solutions do not mix readily in semimicro test-tubes and centrifuge tubes; mixing is effected by means of stirring rods. These can readily be made by cutting 2 mm diameter glass rod into 12 cm lengths. A handle may be formed, if desired, by heating about 1 cm from the end and bending it back through 135° to an angle of 45° (see Fig. 2.12*b*). The sharp edges are fire-polished by heating momentarily in a flame. In washing a precipitate with water or other liquid, it is essential to stir the precipitate so that every particle is brought into contact with as large a volume of liquid

Test tubes

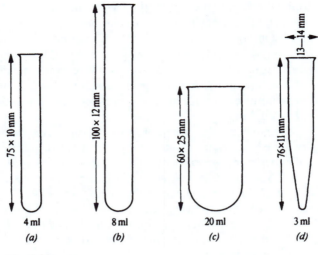

Fig. 2.11

as possible: this is best done by holding the tube almost horizontal to spread the precipitate over a larger surface, and then stirring the suspension.

Droppers

3. *Droppers* for handling liquids in semimicro analysis, a **dropper** (also termed a dropper pipette) is generally employed. Two varieties are shown in Fig. 2.13*a* and *b*. The former one finds application in 30 or 60 ml reagent bottles and may therefore be called a **reagent dropper**; the capillary of the latter (*b*) is long enough to reach to the bottom of a 3 ml centrifuge tube, and is used for removing supernatant liquids from test-tubes and centrifuge tubes and for the quantitative addition of reagents. Dropper *b* will be referred to as a **capillary dropper**.

Before use the droppers must be calibrated, i.e. the volume of the drop delivered must be known. Introduce some distilled water into the clean dropper by dipping the capillary end into some distilled water in a beaker and compressing and then releasing the rubber teat or bulb. Hold the dropper vertically over a clean dry 5 ml measuring cylinder, and gently press the

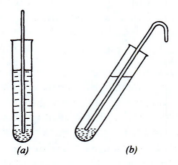

(a)   *(b)*

Fig. 2.12

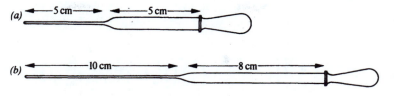

Fig. 2.13

rubber bulb. Count the number of drops until the meniscus reaches the 2 ml mark. Repeat the calibration until two results are obtained which do not differ by more than 2 drops. Calculate the volume of a single drop. The dropper should deliver between 30 and 40 drops per ml. Attach a small label to the upper part of the dropper giving the number of drops per ml.

The standard commercial form of **medicine dropper**, with a tip of 1.5 mm inside diameter and 3 mm outside diameter, delivers drops of dilute aqueous solutions about 0.05 ml in volume, i.e. about 20 drops per ml. This dropper is somewhat more robust than that shown in Fig. 2.13*a*, as it has a shorter and thicker capillary; the size of the drop (*c.* 20 per ml) may be slightly too large when working with volumes of the order of 1 ml. However, if this dropper is used, it should be calibrated as described in the previous paragraph.

It must be remembered that the volume of the drop delivered by a dropper pipette depends upon the density, surface tension, etc., of the liquid. If the dropper delivers 20 drops of distilled water, the number of drops per ml of other liquids will be very approximately as follows: dilute aqueous solutions, 20–22; concentrated hydrochloric acid, 23–24; concentrated nitric acid, 36–37; concentrated sulphuric acid, 36–37; acetic acid, 63; and concentrated ammonia solution, 24–25.

Reagent bottles

4. *Reagent bottles and reagents* a **semimicro reagent bottle** may be easily constructed by inserting a reagent dropper through a cork or rubber stopper that fits a 30 or 60 ml bottle – as in Fig. 2.14*a*. These dropping bottles (Fig. 2.14*b*) may be purchased and are inexpensive; the stoppers of these bottles are usually made of plastic, and this, as well as the rubber teat (or bulb), is attacked by concentrated inorganic acids. A dropping bottle of 30 ml capacity with an interchangeable glass cap (Fig. 2.14*c*) is also marketed. The bottles *a* and *b* cannot be used for concentrated acids and other corrosive liquids because of their action upon the stoppers. The simplest containers for these corrosive liquids are 30 or 60 ml dropping bottles with glass stopper containing a groove (the so-called 'TK bottle', Fig. 2.14*d*).

For the study of reactions it is best to divide reagents into two groups. Those reagents which are most frequently used should be kept by each worker separately; solid wooden stands are available commercially to house these. A list of such reagents is given in Table 2.6.

The other reagents, which are used less frequently, are kept in 60 or 125 ml dropping bottles (30 ml for expensive or unstable reagents) on the reagent shelf (**side shelf** or **side rack reagents**); further details of these will be found in the Appendix. When using these side shelf reagents, great care should be taken that the droppers do not come into contact with the test solutions,

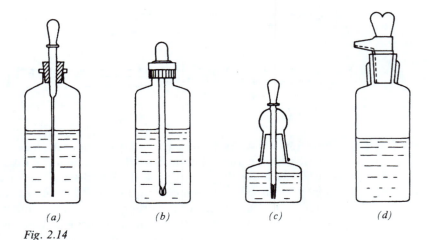

(a)   *(b)*   *(c)*   *(d)*

Fig. 2.14

thus contaminating the reagents. If accidental contact should be made, the droppers must be thoroughly rinsed with distilled water and then dried. Under no circumstances should the capillary end of the dropper be dipped into any foreign solution.

Centrifuge

5. *The centrifuge* the separation of a precipitate from a supernatant liquid is carried out with the aid of a centrifuge. This is an apparatus for the separation of two substances of different density by the application of centrifugal force which may be several times that of gravity. In practice, the liquid containing the suspended precipitate is placed in a semimicro centrifuge tube. The tube and its contents, and a similar tube containing an equal weight of water, are placed in diagonally opposite buckets of the centrifuge, and the cover is placed in position; upon rotation for a short time and after allowing the buckets to come to rest and removing the cover, it will be found that the precipitate has separated at the bottom of the tube. This operation (**centrifugation**) replaces filtration in macro analysis. The supernatant

Table 2.6
Most common reagents

Reagents for semimicro work

Reagent Bottles Fitted with Reagent Droppers (Fig. 2.14a, b or c)

Sodium hydroxide	2M	Acetic acid, dil.	2M
Ammonium sulphide	M	Ammonia solution, conc.	15M
Potassium hydroxide	2M	Ammonia solution, dil.	2M
Barium chloride	0.25M	Potassium hexacyanoferrate(II)	0.025M
Silver nitrate	0.1M	Potassium chromate	0.1M
Iron(III) chloride	0.5M	Ammonium carbonate	M

TK Dropper Bottles (Fig. 2.14d)

Hydrochloric acid, conc.	12M	Hydrochloric acid, dil.	2M
Sulphuric acid, conc.	18M	Sulphuric acid, dil.	M
Nitric acid, conc.	16M	Nitric acid, dil.	2M

Hand-driven centrifuge

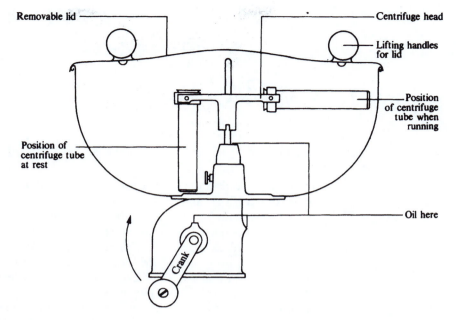

Fig. 2.15

liquid can be readily removed by means of a capillary dropper; the clear liquid may be called the centrifugate or 'centrate'.

The advantages of centrifugation are: (1) speed, (2) the precipitate is concentrated into a small volume so that small precipitates are observed readily and their relative magnitudes estimated, (3) the washing of the precipitate can be carried out rapidly and efficiently, and (4) concentrated acids, bases, and other corrosive liquids can be manipulated easily.

Several types of centrifuge are available for semimicro analysis. These are:

A. A small 2-tube hand centrifuge with protecting bowl and cover (Fig. 2.15); if properly constructed it will give speeds up to $2-3000\,\text{rev}\,\text{min}^{-1}$ with 3 ml centrifuge tubes. The central spindle should be provided with a locking screw or nut: this is an additional safeguard against the possibility of the head carrying the buckets flying off – an extremely rare occurrence. The hand-driven centrifuge is inexpensive and is satisfactory in most cases.

B. An inexpensive constant speed, electrically-driven centrifuge (see Fig. 2.16) has a working speed of $1450\,\text{rev}\,\text{min}^{-1}$, sometimes supplied with a dual purpose head. The buckets can either 'swing out' to the horizontal position or, by means of a rubber adaptor, the buckets can be held at a fixed angle. In the latter case, the instrument acts as an 'angle' centrifuge.

Operations with the centrifuge

When using a centrifuge, the following points should be borne in mind:

(a) The two tubes should have approximately the same size and weight.

(b) The tube should not be filled beyond 1 cm from the top. Spilling may corrode the buckets and produce an unbalanced head.

(c) Before centrifuging a precipitate contained in a centrifuge tube, prepare a balance tube by adding sufficient distilled water from a dropper

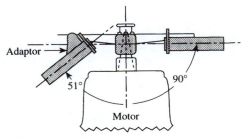

Electric centrifuge

Adaptor

51°   90°

Motor

Fig. 2.16

to an empty tube of the same capacity until the liquid levels in both tubes are approximately the same.

(d) Insert the tubes in diametrically opposite positions in the centrifuge; the head (sometimes known as a rotor) will then be balanced and vibration will be reduced to a minimum. Fix the cover in place.

(e) When using a hand-driven centrifuge, start the centrifuge slowly and smoothly, and bring it to the maximum speed with a few turns of the handle. Maintain the maximum speed for 30–45 seconds, and then allow the centrifuge to come to rest of its own accord by releasing the handle. Do not attempt to retard the speed of the centrifuge with the hand. A little practice will enable one to judge the exact time required to pack the precipitate tightly at the bottom of the tube. It is of the utmost importance to avoid strains or vibrations as these may result in stirring up the mixture and may damage the apparatus.

(f) Before commencing a centrifugation, see whether any particles are floating on the surface of the liquid or adhering to the side of the tube. Surface tension effects prevent surface particles from settling readily. Agitate the surface with a stirring rod if necessary, and wash down the side of the centrifuge tube using a capillary dropper and a small volume of water or appropriate solution.

(g) Never use centrifuge tubes with broken or cracked lips.

(h) When using a motor-driven centrifuge, never leave it while it is in motion. If a suspicious sound is heard, or you observe that the instrument is vibrating or becomes unduly hot, turn off the current at once and report the matter to the supervisor. An unusual sound may be due to the breaking of a tube; vibration suggests an unbalanced condition.

Most semimicro centrifuges will accommodate both semimicro test-tubes (75 × 10 mm) and centrifuge tubes (up to 5 ml capacity). The advantages of the latter include easier removal of the mother liquor with a dropper, and, with small quantities of solids, the precipitate is more clearly visible (and the relative quantity is therefore more easily estimated) in a centrifuge tube (Fig. 2.17).

Separation of solid and liquid

To remove the supernatant liquid, a capillary dropper is generally used. The centrifuge tube is held at an angle in the left hand, the rubber teat or nipple of the capillary dropper, held in the right hand, is compressed to expel the air and the capillary end is lowered into the tube until it is just below the liquid (Fig. 2.18). As the pressure is very slowly released the liquid rises in the dropper and the latter is lowered further into the liquid

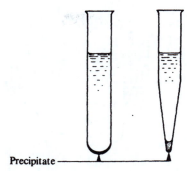

Precipitate ——————————

Fig. 2.17

until all the liquid is removed. Great care should be taken as the capillary approaches the bottom of the centrifuge tube that its tip does not touch the precipitate. The solution in the dropper should be perfectly clear; it can be transferred to another vessel by merely compressing the rubber bulb. In difficult cases, a little cotton wool may be inserted in the tip of the dropper and allowed to protrude about 2 mm below the glass tip; any excess of cotton wool should be cut off with scissors.

Washing

6. *Washing the precipitates*   it is essential to wash all precipitates in order to remove the small amount of solution present in the precipitate, otherwise it will be contaminated with the ions present in the centrifugate. It is best to wash the precipitate at least twice, and to combine the first washing with the centrifugate. The wash liquid is a solvent which does not dissolve the precipitate but dilutes the quantity of mother liquor adhering to it. The wash liquid is usually water, but may be water containing a small amount of the precipitant (common ion effect) or a dilute solution of an electrolyte (such as an ammonium salt) since water sometimes tends to produce colloidal solutions, i.e. to peptize the precipitate.

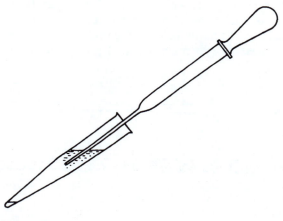

Fig. 2.18

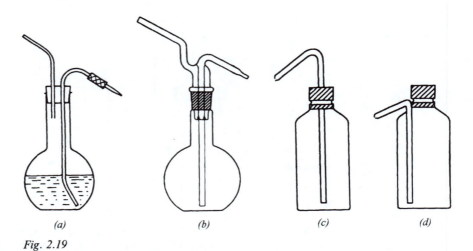

Fig. 2.19

To wash a precipitate in a centrifuge tube, 5–10 drops of water or other reagent are added and the mixture thoroughly stirred (stirring rod or platinum wire); the centrifuge tube is then counterbalanced against another similar tube containing water to the same level and centrifuged. The supernatant liquid is removed by a capillary dropper, and the washing is repeated at least once.

Wash bottles

7. *Wash bottles* for most work in semimicro analysis a 30 or 60 ml glass stoppered bottle is a suitable container for distilled water; the latter is handled with a reagent dropper. Alternatively, a bottle carrying its own dropper (Fig. 2.14*a* or *b*) may be used. A small conical flask (25 or 50 ml) may be used for hot water. For those who prefer wash bottles, various types are available (Fig. 2.19): *a* is a 100 or 250 ml flat-bottomed flask with a jet of 0.5–1 mm diameter, and is mouth-operated; *b* is a wash bottle with full-glass joints (Quickfit MF 25/3), *c* and *d* are polyethylene bottles, from which the wash solution can be dispensed by squeezing.

8. *Transferring of precipitates* in some cases precipitates can be transferred from semimicro test-tubes with a small spatula (two convenient types in nickel or monel metal are shown in Fig. 2.20). This operation is usually difficult, particularly for centrifuge tubes. Indeed, in semimicro analysis it is rarely necessary to transfer actual precipitates from one vessel to another.

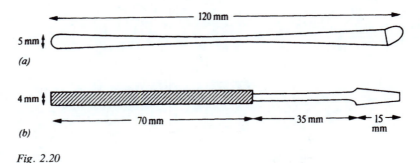

Fig. 2.20

If, for some reason, transfer of the precipitate is essential, a wash liquid or the reagent itself is added, the mixture vigorously stirred and the resulting suspension transferred to a reagent dropper, and the contents of the latter ejected into the other vessel; if required, the liquid is removed by centrifugation.

If the precipitate in a test-tube is to be treated with a reagent in an evaporating dish or crucible, the reagent is added first, the precipitate brought into suspension by agitation with a stirring rod, and the suspension is then poured into the open dish or crucible. The test-tube may be washed by holding it in an almost vertical (upside down) position with its mouth over the receptacle and directing a fine stream of solution or water from a capillary dropper onto the sides of the test-tube.

Heating solutions

9. *Heating of solutions*   solutions in semimicro centrifuge tubes cannot be heated over a free flame owing to the serious danger of 'bumping' (and consequent loss of part or all of the liquid) in such narrow tubes. The 'bumping' or spattering of hot solutions may often be dangerous and lead to serious burns if the solution contains strong acids or bases. Similar remarks apply to semimicro test-tubes. However, by heating the side of the test-tube (8 ml) and not the bottom alone with a micro burner, as indicated in Fig. 2.21, and withdrawing from the flame periodically and shaking gently, 'bumping' does not usually occur. The latter heating operation requires very careful manipulation and should not be attempted by beginners. The mouth of the test-tube must be pointed away from others nearby. The danger of 'bumping' may be considerably reduced by employing the anti-bumping device shown in Fig. 2.10; the tube should be about 1 cm longer than the test-tube to facilitate removal.

Water bath

On the whole it is better to resort to safer methods of heating. The simplest procedure is to employ a small water bath. This may consist of a 250 ml Pyrex beaker, three-quarters filled with water and covered with a stainless steel plate (Fig. 2.22) drilled with two holes to accommodate a test-tube and a centrifuge tube. It is a good plan to wind a thin rubber band about 5 mm from the top of

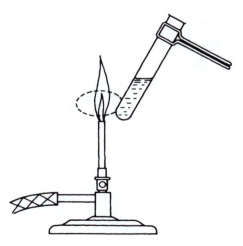

Fig. 2.21

Fig. 2.22

Evaporation

the tube; this will facilitate the removal of the hot tube from the water bath without burning the fingers and, furthermore, the rubber band can be used for attaching small pieces of folded paper containing notes of contents, etc.

A more elaborate arrangement which will, however, meet the requirements of several workers, is Barber's water bath rack (Fig. 2.23). The dimensions of a rack of smaller size are given in Fig. 2.24. This rack will accommodate four centrifuge tubes and four semimicro test-tubes. The apparatus is constructed of monel metal, stainless steel, a plastic material which is unaffected by water at 100°C, or of brass which is subsequently tinned. The brass may be tinned by boiling with 20% sodium hydroxide solution containing a few lumps of metallic tin.

10. *Evaporations* where rapid concentration of a liquid is required or where volatile gases must be expelled rapidly, the semimicro boiling tube (*c* in Fig. 2.11) may be employed. Useful holders, constructed of a light metal alloy, are shown in Fig. 2.25; *b* is to be preferred as the boiling tube cannot fall out by mere pressure on the holder at the point where it is usually held.[6] Slow evaporation may be achieved by heating in a test-tube, crucible, or beaker on a water bath.

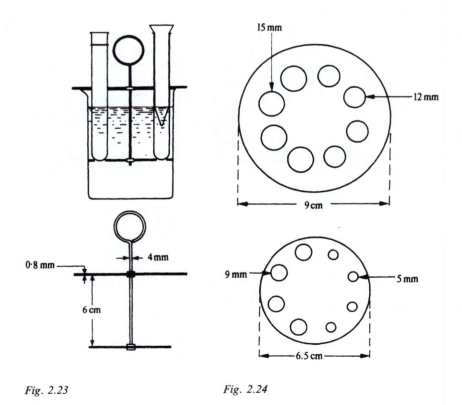

Fig. 2.23   *Fig. 2.24*

[6] A small wooden clothes-peg (spring type) may also be used for holding semimicro test-tubes.

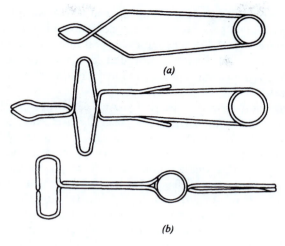

Fig. 2.25

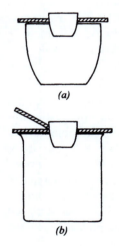

(a)

(b)

Fig. 2.26
Dissolution

If evaporation to dryness is required, a small casserole (*c.* 6 ml) or crucible (3–8 ml) may be employed. This may be placed in an air bath consisting of a 30 ml nickel crucible and supported thereon by a ring (Fig. 2.26*a*) and heated with a semimicro burner. Alternatively, a small Pyrex beaker (say, of 50 or 100 ml capacity) may be used with a silica or nichrome triangle to support the crucible or casserole (Fig. 2.26*b*): this device may also be used for evaporations in semimicro beakers.

Evaporation to dryness may also be accomplished by direct intermittent heating with a micro burner of a crucible (supported on a nichrome or silica triangle) or of a semimicro beaker (supported on a wire gauze). A little practice is required in order to achieve regular boiling by intermittent heating with a flame and also to avoid 'bumping' and spattering; too hot a flame should not be used. In many cases the flame may be removed whilst a little liquid remains; the heat capacity of the vessel usually suffices to complete the evaporation without further heating. If corrosive fumes are evolved, the operation should be conducted in the fume cupboard.

11. *Dissolving of precipitates* the reagent is added and the suspension is warmed, if necessary, on the water bath until the precipitate has dissolved. If only partial solution occurs, the suspension may be centrifuged.

Some precipitates may undergo an ageing process when set aside for longer times (e.g. overnight). This can result in an increased resistance against the reagent used for dissolution. In later chapters, when dealing with the particulars of reactions, such a tendency of a precipitate is always noted. It is advisable to arrange the timetable of work with these precipitates in such a way that dissolution can be attempted soon after precipitation.

Use of H_2S on a semimicro scale

12. *Precipitations with hydrogen sulphide* various automatic generators of the Kipp type are marketed. Owing to the highly poisonous and very obnoxious character of hydrogen sulphide, these generators are always kept in a

fume cupboard (draught chamber or hood). A wash bottle containing water should always be attached to the generator in order to remove acid spray (compare Fig. 2.6a). The tube dipping into the liquid in the wash bottle should preferably be a heavy-walled capillary; this will give a better control of the gas flow and will also help to prolong the life of the charge in the generator.

Precipitation may be carried out in a centrifuge tube, but a semimicro test-tube or 10 ml conical flask is generally preferred; for a large volume of solution, a 25 ml Erlenmeyer flask may be necessary. The delivery tube is drawn out to a thick-walled capillary (1–2 mm in diameter) and carries at the upper end a small rubber stopper which fits the semimicro test-tube[7] (Fig. 2.27) or conical flask. The same equipment with glass joints can be constructed of Quickfit MF 24/0/4 test tube or FE 25/0 Erlenmeyer flask, combined with MF 15/0B/SC delivery tube. The perfectly clean delivery tube is connected to the source of hydrogen sulphide as in Fig. 2.6, and a slow stream of gas is passed through the liquid for about 30 seconds in order to expel the air, the cork is pushed into position and the passage of hydrogen sulphide is continued until very few bubbles pass through the liquid; the vessel is shaken gently from time to time. The tap in the gas generator is then turned off; the delivery tube is disconnected and immediately rinsed with distilled water. If the liquid must be warmed, the stopper is loosened, the vessel placed in the water bath for a few minutes, and the gas introduced again.

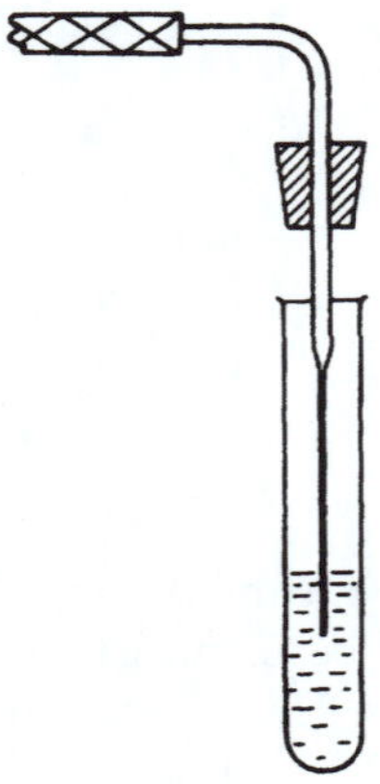

Fig. 2.27

Testing gases

13. *Identification of gases*   many anions (e.g. carbonate, sulphide, sulphite, thiosulphate, and hypochlorite) are usually identified by the volatile decomposition products obtained with the appropriate reagents. Suitable devices for this purpose are shown in Fig. 2.28. The simplest form (*a*) consists of a semimicro test-tube with the accompanying 'filter tube':[8] a strip of test paper (or of filter paper moistened with the necessary reagent) about 3–4 mm wide is suspended in the 'filter tube'. In those cases where spray is likely to affect the test paper, a loose plug of cotton wool should be placed at the narrow end of the 'filter tube'. Apparatus *b* is employed when the test reagent is a liquid. A short length of wide-bore rubber tubing is fitted over the mouth of a semimicro test-tube with about 1 cm protruding over the upper edge. The chemical reaction is started in the test-tube and a 'filter tube' containing a tightly-packed plug of cotton wool is inserted through the rubber collar, the filter plug again packed down with a 4 mm glass rod, and 0.5–1 ml of the reagent introduced. An alternative apparatus for liquid reagents is shown in *c*: it is a type of absorption pipette and is attached to the 4 ml test-tube by a rubber stopper. It is possible to use all-glass apparatus if available. The Quickfit version of *c* can be put together

---

[7] For a 3 ml centrifuge tube or a 4 ml test-tube, the top part of the rubber teat (bulb) from a dropper makes a satisfactory stopper: a small hole is made in the rubber bulb and the delivery tube is carefully pushed through it.

[8] Test-tubes (75 × 100 mm and 100 × 12 mm) together with the matching 'filter tubes' (55 × 7 mm and 80 × 9 mm) are standard products; they are inexpensive and therefore eminently suitable for routine work. The smaller size is satisfactory for most purposes.

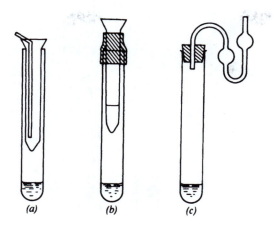

(a)   *(b)*   *(c)*

Fig. 2.28

from the MF 24/0 test-tube and CNB 10 joint; the two bulbs however have to be created by a glassblower. A drop or two of the liquid reagent is introduced into the absorption tube. The double bulb ensures that all the evolved gas reacts, and it also prevents the test reagent from being sucked back into the reaction mixture. All the above apparatus, *a–c*, may be warmed by placing in the hot water rack (e.g. Fig. 2.23). They will meet all normal requirements for testing gases evolved in reactions in qualitative analysis.

When the amounts of evolved gas are likely to be small, the apparatus of Fig. 2.29 may be used; all the evolved gas may be swept through the reagent by a stream of air introduced by a rubber bulb of about 100 ml capacity. The approximate dimensions of the essential part of the apparatus are given in *a*, whilst the complete assembly is depicted in *b*. The sample under test is placed in B, the test reagent is introduced into the 'filter tube' A over a tightly-packed plug of cotton wool (or other medium), and the acid or other liquid reagent is added through the 'filter tube' C. The rubber bulb is inserted into C, and by depressing it gently air is forced through the apparatus, thus sweeping out the gases through the test reagent in A. The apparatus may be warmed by placing it in a hot water bath. An all-glass apparatus can be constructed from a Quickfit two-neck flask FP 50/1/1A, equipped with an SF 4T31 conical semimicro filter funnel in the left-side neck and an MF 15/1 air inlet tube in the right neck.

Cleaning of apparatus

14. *Cleaning of apparatus* it is essential to keep all apparatus scrupulously clean if trustworthy results are to be obtained. All apparatus must be thoroughly cleaned with chromosulphuric acid (concentrated sulphuric acid containing about 100 g of potassium dichromate per litre) or with a brush and cleaning powder. **The cleaning agent is highly corrosive!** The apparatus is then rinsed several times with tap water, and repeatedly with distilled water. Special brushes (Fig. 2.30) are available for semimicro test-tubes and centrifuge tubes; the commercial 'pipe cleaners' are also satisfactory. The tubes may be allowed to drain in a special stand or else they

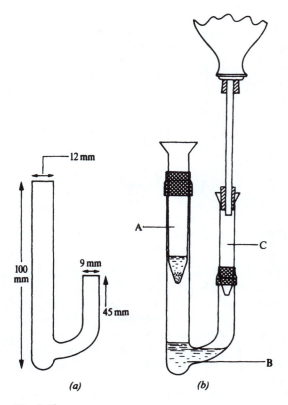

Fig. 2.29

may be inverted in a small beaker on the bottom of which there are several folds of filter paper or a filter paper pad to absorb the water. Larger apparatus may be allowed to drain on a clean linen towel or glass cloth. Droppers are best cleaned by first removing the rubber bulbs or teats and allowing distilled water to run through the tubes; the teats are cleaned by repeatedly filling them with distilled water and emptying them. When clean, they are allowed to dry on a linen glass cloth. At the end of the laboratory period, the clean apparatus is placed in a box with cover, so that it remains clean until required.

15. *Spot plates; drop-reaction paper* these are employed chiefly for confirmatory tests (see Section 2.6 for a full discussion).

Fig. 2.30

16. *Some practical hints*

(a) Upon commencing work, arrange the more common and frequently used apparatus in an orderly manner on your bench. Each item of apparatus should have a definite place so that it can be found readily when required. All apparatus should have been cleaned during the previous laboratory period.

(b) Weigh the sample for analysis (if a solid) to the nearest milligram: 50 mg is a suitable quantity.

(c) Read the laboratory directions carefully and be certain that you understand the purpose of each operation – addition of reagents, etc. Examine the label on the bottle before adding the reagent. Serious errors leading to a considerable loss of time and, possibly, personal injury may result from the use of the wrong reagent. Return each reagent bottle to its proper place immediately after use.

(d) When transferring a liquid reagent with a reagent dropper always hold the dropper just above the mouth of the vessel and allow the reagent to 'drop' into the vessel. Do not allow the dropper tip to touch anything outside the reagent bottle; the possible introduction of impurities is thus avoided. Similar remarks apply to the use of TK dropper bottles.

(e) Never dip your own dropper into a reagent. Pour a little of the reagent (e.g. a corrosive liquid) into a small clean vessel (test-tube, crucible, beaker, etc.) and introduce your dropper into this. Never return the reagent to the bottle; it is better to waste a little of the reagent than to take the risk of contaminating the whole supply.

(f) Do not introduce your spatula into a reagent bottle to remove a little solid. Pour or shake a little of the solid on to a clean, dry watch glass, and use this. Do not return the solid reagent to the stock bottle. Try to estimate your requirements and pour out only the amount necessary.

(g) All operations resulting in the production of fumes (acid vapours, volatile ammonium salts, etc.) or of poisonous or disagreeable gases (hydrogen sulphide, chlorine, sulphur dioxide, etc.) must be performed in the fume cupboard.

(h) Record your observations briefly in your note-book in ink immediately after each operation has been completed.

(i) Keep your droppers scrupulously clean. Never place them on the bench. Rinse the droppers several times with distilled water after use. At the end of each laboratory period, remove the rubber teat or cap and rinse it thoroughly.

(j) During the course of the work place dirty centrifuge tubes, test-tubes, etc., in a definite place, preferably in a beaker, and wash them at convenient intervals. This task can often be done while waiting for a solution to evaporate or for a precipitate to dissolve while being heated on a water bath.

(k) Adequately label all solutions and precipitates which must be carried over to the next laboratory period.

(l) When in difficulty, or if you suspect any apparatus (e.g. the centrifuge) is not functioning efficiently, consult the supervisor.

17. *Semimicro apparatus* the apparatus suggested for each workbench is listed below. (The liquid reagents recommended are given in Section 2.4(4).)

1 wash bottle, polythene (Fig. 2.19*d*)

1 beaker, Griffin form, 250 ml

1 hot water rack constructed of tinned copper (Figs. 2.23 and 2.24) or

1 stainless steel cover (Fig. 2.22) for water bath

1 beaker, 5 ml

1 beaker, 10 ml

1 flat-bottomed flask, 100 ml, and one 50 ml rubber bulb (for wash bottle, Fig. 2.19)

2 conical flasks, 10 ml

1 conical flask, 25 ml

6 test-tubes, 75×10 mm,[9] 4 ml, with rim

2 test-tubes, 75×10 mm,[9] 4 ml, without rim

2 'filter tubes', 55×7 mm[9] (Fig. 2.28)

1 gas absorption pipette (with 75×10 mm test-tube and rubber stopper, Fig. 2.28)

4 centrifuge tubes, 3 ml (Fig. 2.11*d*)

2 semimicro boiling tubes, 60×25 mm, 20 ml

1 wooden stand (to house test-tubes, 'filter tubes', gas absorption pipette, conical flasks, etc.)

2 medicine droppers, complete with rubber teats (bulbs)

2 reagent droppers (Fig. 2.13*a*)

2 capillary droppers (Fig. 2.13*b*)

1 stand for droppers

2 anti-bump tubes (Fig. 2.10)

1 crucible, porcelain, 3 ml (23×15 mm)

1 crucible, porcelain, 6 ml (28×20 mm)

1 crucible, porcelain, 8 ml (32×19 mm)

3 rubber stoppers (one 1×2 cm, two 0.5×1.5 cm, to fit 25 ml conical flask, test-tube, and gas absorption pipette)

30 cm glass tubing, 4 mm outside diameter (for H_2S apparatus)

10 cm rubber tubing, 3 mm (for H_2S apparatus)

5 cm rubber tubing, 5 mm (for 'filter tubes')

30 cm glass rod, 3 mm (for stirring rods, Fig. 2.12)

1 measuring cylinder, 5 ml

1 watch glass, 3.5 cm diameter

2 cobalt glasses, 3×3 cm

2 microscope slides

1 platinum wire (5 cm of 0.3 mm diameter)

1 forceps, 10 cm

1 semimicro spatula (Fig. 2.20*a* or *b*)

1 semimicro test-tube holder (Fig. 2.25)

1 semimicro test-tube brush (Fig. 2.30)

---

[9] Semimicro test-tubes (100×12.5 mm, 8 ml) and the appropriate 'filter tubes' (80×9 mm) are marketed, and these may find application in analysis. The smaller 4 ml test-tubes will generally suffice: their great advantage is that they can be used directly in a semimicro centrifuge in the buckets provided for the 3 ml centrifuge tubes.

1 pipe cleaner
1 spot plate (6 cavities)
1 wide-mouthed bottle, 25 ml filled with cotton wool
1 wide-mouthed bottle, 25 ml filled with strips (2 × 2 cm) of drop reaction
 paper
1 dropping bottle, labelled DISTILLED WATER
1 packet blue litmus paper
1 packet red litmus paper
1 triangular file (small)
1 tripod and wire gauze (for water bath)
1 retort stand and one iron ring 7.5 cm
1 wire gauze (for ring)
1 triangle, silica
1 triangle, nichrome
1 Bunsen burner
1 semimicro burner

2.5   Micro apparatus and microanalytical operations

The micro scale

In micro analysis the scale of operations is reduced by a factor of 100–200 as compared with macro analysis. Thus whereas in macro analysis the weights and volumes for analysis are 0.5–1 g and about 10 ml, and in semimicro analysis 50 mg and 1 ml respectively, in micro analysis the corresponding quantities are about 5 mg and 0.1 ml. Micro analysis is sometimes termed milligram analysis to indicate the order of weight of the sample employed. It must be pointed out that whilst the weight of the sample for analysis has been reduced, the ratio of weight to volume has been retained and in consequence the concentration of the individual ions, and other species, is maintained. A special technique must be used for handling such small quantities of materials. There is no sharp line of demarcation between semi-micro and micro analysis and much of the technique described for the former can, with suitable modifications to allow for the reduction in scale by about one-tenth, be utilized for the latter. Some of the modifications, involving comparatively simple apparatus, will be described. No attempt will be made to deal with operations centred round the microscope (magnification up to 250) as the specialized technique is outside the scope of this volume.[10]

Spot tests

The small amounts of material obtained after the usual systematic separations can be detected, in many cases, by what is commonly called **spot analysis**, i.e. analysis which utilizes spots of solutions (about 0.05 ml or smaller) or a fraction of a milligram of solids. Spot analysis has been developed as a result of the researches of numerous chemists: the names of Tananaeff, Krumholz, Wenger, van Niewenburg, Gutzeit and, particularly, Feigl and their collaborators must be mentioned in this connection.

---

[10] For a detailed account see: H. H. Emons, H. Keune and H. H. Seyfarth: *Chemical Microscopy* in G. Svehla (Ed.): *Wilson & Wilson's Comprehensive Analytical Chemistry* Vol. XVI, Elsevier, 1982.

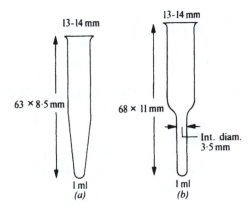

Fig. 2.31

In general, spot reactions are preferable to tests which depend upon the formation and recognition of crystals under the microscope, in that they are easier and quicker to carry out, less susceptible to slight variations of experimental conditions, and can be interpreted more readily. Incomplete schemes of qualitative inorganic analysis have been proposed which are based largely upon spot tests: these cannot, however, be regarded as entirely satisfactory as only very few spot tests are specific for particular ions and also the adoption of such schemes will not, in the long run, help in the development of micro qualitative analysis. It seems that the greatest potential for progress lies in the use of the common macro procedures (or simple modifications of them) to carry out preliminary separations followed by the utilization of spot tests after the group or other separation has been effected. Hence the following pages will contain an account of the methods which can be used for performing macro operations on a micro scale and also a discussion of the technique of spot analysis.

Equipment for micro work

Micro centrifuge tubes (Fig. 2.31) of 0.5–2 ml capacity replace test-tubes, beakers and flasks for most operations. Two types of centrifuge tubes are shown: *b* is particularly useful when very small amounts of precipitate are being handled. Centrifuge tubes are conveniently supported in a rack consisting of a wooden block provided with 6 to 12 holes, evenly spaced, of 1.5 cm diameter and 1.3 cm deep.

Solutions are separated from precipitates by centrifuging. Semimicro centrifuges (Section 2.4.5), either hand-operated or electrically-driven, can be used. Adaptors are provided inside the buckets (baskets) in order to accommodate micro centrifuge tubes with narrow open ends.

Precipitations are usually carried out in micro centrifuge tubes. After centrifuging, the precipitate collects in the bottom of the tube. The supernatant liquid may be removed either by a capillary dropper (Fig. 2.13) or by means of a **transfer capillary pipette**. The latter consists of a thin glass tube (internal diameter about 2 mm: this can be prepared from wider tubing) 20 to 25 cm in length with one end drawn out in a micro flame to a tip with a fine opening. The correct method of transferring the liquid to the

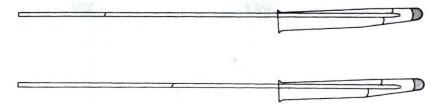

Fig. 2.32

capillary pipette will be evident from Fig. 2.32. The centrifuge cone is held in the right hand and the capillary pipette is pushed slowly towards the precipitate so that the point of the capillary remains just below the surface of the liquid. As the liquid rises in the pipette, the latter is gradually lowered, always keeping the tip just below the surface of the liquid until the entire solution is in the pipette and the tip is about 1 mm above the precipitate. The pipette is removed and the liquid blown or drained out into a clean dry centrifuge tube.

Another useful method of **transferring the centrifugate** to, say, another centrifuge tube will be evident upon reference to Fig. 2.33. The siphon is made of thermometer capillary and is attached to the capillary pipette by means of a short length of rubber tubing of 1 mm bore. The small hole A in the side of the tube permits perfect control of the vacuum; the side arm of the small test-tube is situated near the bottom so as to reduce spattering of the liquid in the centrifuge tube inside the test-tube when the vacuum is released. Only gentle suction is applied and the opening A is closed with the finger; upon removing the finger, the action of the capillary siphon ceases immediately. An all-glass apparatus can be put together from Quickfit BC/24/C/14T centrifuge tubes, an MF 24/3/8 test-tube equipped with an MF 28/3/125 inlet tube, which has to be bent by a glassblower.

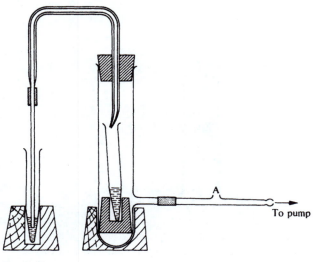

Fig. 2.33

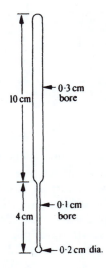

10 cm

0·3 cm bore

4 cm

0·1 cm bore

0·2 cm dia.

Fig. 2.34

For the **washing of precipitates** the wash solution is added directly to the precipitate in the centrifuge tube and stirred thoroughly either by a platinum wire or by means of a micro stirrer, such as is shown in Fig. 2.34; the latter can readily be made from thin glass rod. The mixture is then centrifuged, and the clear solution removed by a transfer capillary pipette as already described. It may be necessary to repeat this operation two or three times to ensure complete washing.

The **transfer of precipitates** is comparatively rare in micro qualitative analysis. Most of the operations are usually so designed that it is only necessary to transfer solutions. However, if transfer of a precipitate should be essential and the precipitate is crystalline, the latter may be sucked up by a dry dropper pipette and transferred to the appropriate vessel. If the precipitate is gelatinous, it may be transferred with the aid of a narrow glass, nickel, monel metal, or platinum spatula. The centrifuge tube must be of type *a* (Fig. 2.31) if most of the precipitate is to be removed.

The **heating of solutions** in centrifuge tubes is best carried out by supporting them in a suitable stand (compare Figs. 2.22 and 2.23) and heating on a water bath. When higher temperatures are required, as for evaporation, the liquid is transferred to a micro beaker or micro crucible; this is supported by means of a nichrome wire triangle as indicated in Fig. 2.35. Micro beakers may be heated by means of the device shown in Fig. 2.36, which is laid on a water bath.

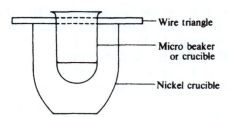

Wire triangle

Micro beaker or crucible

Nickel crucible

Fig. 2.35

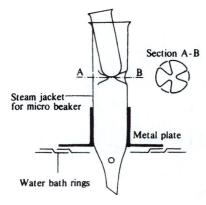

Section A-B

A B

Steam jacket for micro beaker

Metal plate

Water bath rings

Fig. 2.36

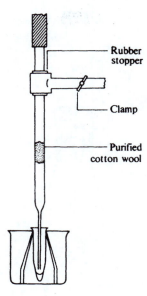

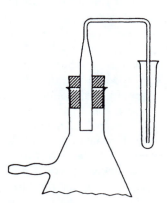

Fig. 2.37

Another valuable method for concentrating solutions or evaporating to dryness directly in a centrifuge tube consists of conducting the operation on a water bath in a stream of filtered air supplied through a capillary tube fixed just above the surface of the liquid. The experimental details will be evident by reference to Fig. 2.37.

Cleaning micro glassware

Micro centrifuge tubes are cleaned with a feather, 'pipe cleaner' or small test-tube brush (cf. Fig. 2.30). They are then filled with distilled water and emptied by suction as in Fig. 2.38. After suction has commenced and the liquid removed, the tube is filled several times with distilled water without removing the suction device. Dropper pipettes are cleaned by repeatedly filling and emptying them with distilled water, finally separating rubber bulb and glass tube, and rinsing both with distilled water from a wash

Fig. 2.38

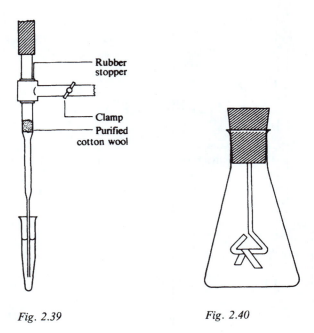

Fig. 2.39   *Fig. 2.40*

bottle. A transfer capillary pipette is cleaned by blowing a stream of water through it. An all-glass apparatus can be assembled from Quickfit FB 100/2 Buchner flask and MF 15/2 suction tube.

The **passage of hydrogen sulphide into a solution in a micro centrifuge tube** is carried out by leading the gas through a fine capillary tube in order not to blow the solution out of the tube. The delivery tube may be prepared by drawing out part of a length of glass tubing of 6 mm diameter to a capillary of 1–2 mm bore and 10–20 cm long. A plug of pure cotton wool is inserted into the wide part of the tubing, and then the capillary tube is drawn out by means of a micro burner to a finer tube of 0.3–0.5 mm bore and about 10 cm long. The complete arrangement is illustrated in Fig. 2.39. Such a fine capillary delivers a stream of very small bubbles of gas; large bubbles would throw the solution out of the micro centrifuge tube. The flow of the hydrogen sulphide must be commenced before introducing the point of the capillary into the centrifuge cone. If this is not done, the solution will rise in the capillary and when hydrogen sulphide is admitted, a precipitate will form in the capillary tube and clog it. The point of saturation of the solution is indicated by an increase in the size of the bubbles; this is usually after about two minutes.

The **identification of gases** obtained in the reactions for anions may be conducted in an Erlenmeyer (or conical) flask of 5 to 10 ml capacity; it is provided with a rubber stopper which carries at its lower end a 'wedge point' pin, made preferably of nickel or monel metal (Fig. 2.40). A small strip of impregnated test paper is placed on the loop of the pin. The stoppered flask, containing the reaction mixture and test paper, is placed on a water bath for five minutes. An improved apparatus consists of a small flask provided with a ground glass stopper onto which a glass hook is fused (see Fig. 2.52); the test paper is suspended from the glass hook. An all-glass apparatus is also available, it can be assembled from Quickfit FE 100/3 Erlenmeyer flask and SB 24/S Schöniger stopper. If the evolved gas is to be passed through a liquid reagent, the apparatus of Fig. 2.28*b* or *c*, proportionately reduced, may be employed.

Confirmatory tests for ions may be carried out either on drop reaction paper or upon a spot plate. The technique of spot tests is fully described in Section 2.6.

The following **micro apparatus**[11] will be found useful:

micro porcelain, silica, and platinum crucibles of 0.5 to 2 ml capacity
micro beakers (5 ml) and micro conical flasks (5 ml)
micro centrifuge tubes (0.5, 1.0 and 2.0 ml)
micro test-tubes ($40-50 \times 8$ mm)
micro volumetric flasks (1, 2, and 5 ml)
micro nickel, monel, or platinum spatula, $7-10$ cm in length and flattened at one end
micro burner
micro agate pestle and mortar
magnifying lens, $5 \times$ or $10 \times$
a pair of small forceps
a small platinum spoon, capacity $0.5-1$ ml, with handle fused into a glass tube; this may be used for fusions

2.6 Spot test analysis

The term 'spot reaction' is applied to micro and semimicro tests for compounds or for ions. In these chemical tests manipulation with drops (macro, semimicro, and micro) play an important part. Spot reactions may be carried out by any of the following processes:

(a) By bringing together one drop of the test solution and of the reagent on porous or non-porous surfaces (paper, glass, or porcelain).

(b) By placing a drop of the test solution on an appropriate medium (e.g. filter paper) impregnated with the necessary reagents.

(c) By subjecting a strip of reagent paper or a drop of the reagent to the action of the gases liberated from a drop of the test solution or from a minute quantity of the solid substance.

(d) By placing a drop of the reagent on a small quantity of the solid sample, including residues obtained by evaporation or ignition.

(e) By adding a drop of the reagent to a small volume (say, $0.5-2$ ml) of the test solution and then extracting the reaction products with organic solvents.

The actual 'spotting' is the fundamental operation in spot test analysis, but it is not always the only manipulation involved. Preliminary preparation

---

[11] All glass apparatus must be of borosilicate (e.g. Pyrex) glass.

is usually necessary to produce the correct reaction conditions. The preparation may involve some of the operations of macro analysis on a diminished scale (see Section 2.5), but it may also utilize certain operations and apparatus peculiar to spot test analysis. An account of the latter forms the subject matter of the present section.

Measures of sensitivity

Before dealing with the apparatus required for spot test reactions, it is necessary to define clearly the various terms which are employed to express the sensitivity of a test. The **limit of identification** is the smallest amount recognizable, and is usually expressed in **micrograms** (μg), one microgram being one-thousandth part of a milligram or one-millionth part of a gram:

$$1\,\mu g = 0.001\,mg = 10^{-6}\,g$$

Throughout this text the term **sensitivity** will be employed synonymously with limit of identification. The **concentration limit** is the greatest dilution in which the test gives positive results; it is expressed as a ratio of substance to solvent or solution. For these two terms to be comparable, a standard size drop must be used in performing the test. Throughout this book, unless otherwise stated, sensitivity will be expressed in terms of a standard drop of 0.05 ml.

Operations in spot test analysis

The removal and addition of drops of test and reagent solutions is most simply carried out by using glass tubing, about 20 cm long and 3 mm external diameter; drops from these tubes have an approximate volume of 0.05 ml. The capillary dropper (Fig. 2.13b) may also be employed. A useful glass pipette, about 20 cm long, may be made from 4 mm tubing and drawn out at one end in the flame (Fig. 2.41); the drawn-out ends of these pipettes may be of varying bores. Polyethylene micro pipettes with elastic bulb on the top are also suitable. A liberal supply of glass tubes and pipettes should always be kept at hand. They may be stored in a beaker about 10 cm high with the constricted end downwards and resting upon a pad of pure cotton wool; the beaker and pipettes can be protected against dust by covering with a sheet of polythene. Pipettes which are used frequently may be supported horizontally on a stand constructed of thin glass rod. After use, they should be immersed in beakers filled with distilled water: interchanges are thus prevented and subsequent thorough cleaning is facilitated.

Very small and even-sized drops can be obtained by means of **platinum wire loops**; the size of the loop can be varied and by calibrating the various loops (by weighing the drops delivered), the amount of liquid delivered from each loop is known fairly accurately. A number of loops are made by bending platinum wire of suitable thickness; the wires should be attached in the usual manner to lengths of glass rod or tubing to act as handles. They are kept in Pyrex test-tubes fitted with corks or rubber stoppers and labelled with particulars of the size of drop delivered. It must be pointed out that new smooth platinum wire allows liquids to drop off too readily, and hence it is essential to roughen it by dipping into chloroplatinic acid solution, followed by heating to glowing in a flame; this should be repeated several times. Micro burettes sometimes find application for the delivery of drops.

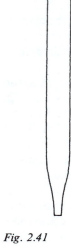

Fig. 2.41

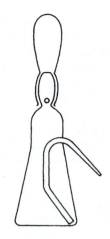

Fig. 2.42

Spot plates

Fig. 2.43

Fig. 2.44

Reagent solutions can be added from dropping bottles of 25–30 ml capacity (see Section 2.4.4). A stock bottle for water and for solutions which do not deteriorate on keeping, is shown in Fig. 2.42; it permits easy addition in drops. This stock bottle is constructed from a Pyrex flask into which a tube with a capillary end is fused; a small rubber bulb is placed over the drawn-out neck, which has a small hole to admit air.

Digestion of solid samples with acid or solvent may be performed in small crucibles heated on a metal hot plate, or in an air bath (Fig. 2.35), or in the glass apparatus illustrated in Fig. 2.43. The last-named is heated over a micro burner, and then rotated so that supernatant liquid or solution may be poured off drop-wise without danger or loss.

Spot tests may be performed in a number of ways: on a spot plate, in a micro crucible, test tube, or centrifuge tube, or on filter paper. Gas reactions are carried out in special apparatus.

The commercial **spot plates** are made from glazed porcelain or plastic and usually contain 6 to 12 depressions of equal size that hold 0.5 to 1 ml of liquid. It is advisable, however, to have several spot plates with depressions of different sizes. The white porcelain background enables very small colour changes to be seen in reactions that give coloured products; the colour changes are more readily perceived by comparison with blank tests in adjacent cavities of the spot plate. Where light-coloured or colourless precipitates or turbidities are formed, it is better to employ black spot plates. Transparent spot plates of glass are also available; these may be placed upon glossy paper of suitable colour. The drops of test solution and reagent brought together on a spot plate must always be mixed thoroughly; a glass stirrer (Fig. 2.34) or a platinum wire may be employed.

Traces of turbidity and of colour are also readily distinguished in micro test-tubes (50×8 mm) or in micro centrifuge tubes. As a general rule, these vessels are employed in testing dilute solutions so as to obtain a sufficient depth of colour. The liquid in a micro centrifuge tube or in a test-tube may be warmed in a special stand immersed in a water bath (see Figs. 2.22 and 2.23) or in the apparatus depicted in Fig. 2.44. The latter is constructed of thin aluminium or nickel wire; the tubes will slip through the openings and rest on their collars. The wire holder is arranged to fit over a small beaker, which can be filled with water at the appropriate temperature.

For heating to higher temperatures ($>100°C$) micro porcelain crucibles may be employed: they are immersed in an air bath (Fig. 2.35) and the latter heated by a micro burner, or they can be heated directly. Small silica watch glasses also find application for evaporations.

It must be emphasized that all glass and porcelain apparatus, including spot plates and crucibles, must be kept scrupulously clean. It is a good plan to wash all apparatus (particularly spot plates) immediately after use. Glassware and porcelain crucibles are best cleaned by immersion in a chromic acid–sulphuric acid mixture or in a mixture of concentrated sulphuric acid and hydrogen peroxide, followed by washing with a liberal quantity of distilled water, and drying. The use of chromic acid mixture is not recommended for spot plates. **These cleaning mixtures are highly corrosive!**

The great merit of glass and porcelain apparatus is that it can be employed with any strength of acid and base: it is also preferred when weakly coloured compounds (especially yellow) are produced or when the test depends upon slight colour differences. Filter paper, however, cannot be used with strongly acidic solutions for the latter cause it to tear, whilst strongly basic solutions produce a swelling of the paper. Nevertheless, for many purposes and especially for those dependent upon the application of capillary phenomena, spot reactions carried out on filter paper possess advantages over those in glass or porcelain: the tests generally have greater sensitivity and a permanent (or semi-permanent) record of the experiment is obtained.

Tests on filter paper

Spot reactions upon filter paper are usually performed with Whatman No. 3 filter paper. This paper may contain traces of iron and phosphate: spot reactions for these are better made with quantitative filter paper (Whatman No. 42 or, preferably, the hardened variety No. 542). The paper should be cut into strips 6×2 cm or 2×2 cm, and stored in petri dishes or in vessels with tightly-fitting stoppers.

Spot reactions on paper do not always involve interaction between a drop of test solution and one of the reagent. Sometimes the paper is impregnated with the reagent and the dry **impregnated reagent paper** is spotted with a drop of the solution. Special care must be taken in the choice of the impregnating reagent. Organic reagents, that are only slightly soluble in water but dissolve readily in alcohol or other organic solvents, find extensive application. Water-soluble salts of the alkali metals are frequently not very stable in paper. This difficulty can often be surmounted by the use of sparingly soluble salts of other metals. In this way the concentration of the reactive ion can be regulated automatically by the proper selection of the impregnating salt, and the specificity of the test can be greatly improved by thus restricting the number of possible reactions. Thus potassium xanthate (a selective reagent for molybdenum) has little value as an impregnating agent since it decomposes rapidly and is useless after a few days. When, however, cadmium xanthate is used, a paper is obtained which gives sensitive reactions only with copper and molybdenum and will keep for months. Similarly the colourless zinc hexacyanoferrate(II) offers parallel advantages as a source of hexacyanoferrate(II) ions: it provides a highly sensitive test for iron(II) ions. A further example is paper impregnated with zinc, cadmium, or antimony sulphides: such papers are stable, each with its maximum sulphide-ion concentration (controlled by its solubility product) and hence only those metallic sulphides are precipitated whose solubility products are sufficiently low. Antimony(III) sulphide paper precipitates only silver, copper, and mercury in the presence of lead, cadmium, tin, iron, nickel, cobalt and zinc. It might be expected that the use of 'insoluble' reagents would decrease the reaction rate. This retardation is not significant when paper is the medium because of the fine state of dispersion and the great surface area available. Reduction in sensitivity becomes appreciable only when the solubility products of the reagent and the reaction product approach the same order of magnitude. This is naturally avoided in the selection of reagents.

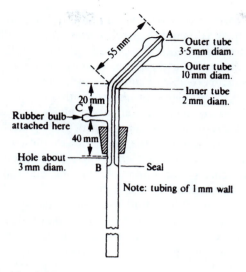

Fig. 2.45

Filter paper may be impregnated with reagents according to their solubility. For reagents that are **soluble** in water or in organic solvents, strips of filter paper are bathed in the solutions contained in beakers or in dishes. Care must be taken that the strips do not stick to the sides of the vessel or to one another, as this will prevent a uniform impregnation. The immersion should last about 20 minutes and the solution should be stirred frequently or the vessel gently rotated to produce a swirling of the solution. The strips are removed from the bath, allowed to drain, pinned to a string (stretched horizontally), and allowed to dry in the air. Uniform drying is of great importance.

Alternatively, the reagent may be sprayed on to the filter paper. The all-glass spray shown in Fig. 2.45 (not drawn to scale) gives excellent results. A rubber bulb is attached at C; the cork is fitted into a boiling tube or small flask charged with the impregnating solution. The paper is sprayed first on one side and then on the other. There are a number of spray units available commercially; plastic-made aerosol sprays (e.g. Fisons AET-100A or Unitary WBS-520-J) and full-glass devices (e.g. Fisons CIJ-700-010T atomiser, or Quickfit 7 CR spray).

For reagents that are **precipitated** on the paper, the strips are soaked rapidly and uniformly with the solution of one of the reactants, dried and then immersed similarly in a solution of the precipitant. The excess reagents are then removed by washing, and the strips dried. The best conditions (concentration of solutions, order in which applied, etc.) must be determined by experiment. In preparing highly impregnated paper, the precipitation should never be made with concentrated solutions as this may lead to an inhomogeneous precipitation and the reagent will tend to fall off the paper after it is washed and dried. It is essential to carry out the soaking and precipitation separately with dilute solutions and to dry the paper between the individual precipitations. Sometimes it is preferable to use a reagent in the gaseous form, e.g. hydrogen sulphide for sulphides and ammonia for hydroxides; there is then no danger of washing away the precipitate.

The reactions are carried out by adding a drop of the test solution from a capillary pipette, etc., to the centre of the horizontal reagent paper resting across a porcelain crucible or similar vessel; an unhindered capillary spreading follows and a circular spot results. With an impregnated reagent paper, the resulting change in colour may occur almost at once, or it may develop after the application of a further reagent. It is usually best not to place a drop of the test solution on the paper but to allow it to run slowly from a capillary tip (0.2–1 mm diameter) by touching the tip on the paper. The test drop then enters over a minute area, precipitation or adsorption of the reaction product occurs in the immediate surrounding region, where it remains fixed in the fibres whilst the clear liquid spreads radially outwards by capillarity. A concentration of the coloured product, which would otherwise be spread over the whole area originally wetted by the test drop, is obtained, thus rendering minute quantities distinctly visible. A greater sensitivity is thus obtained than by adding a free drop of the test solution. Contact with the fingers should be avoided in manipulation with drop-reaction papers; a corner of the strip should be held with a pair of clean forceps.

Separation of solids and liquids

The problem of **separating solid and liquid phases** either before or after taking a sample drop or two of the test solution frequently arises in spot test analysis. When there is a comparatively large volume of liquid and the solid matter is required, centrifugation in a micro centrifuge tube (Fig. 2.31) may be employed. Alternatively, a micro sintered glass filter tube (Fig. 2.46), placed in a test-tube of suitable size, may be subjected to centrifugation: this device simplifies the washing of a precipitate. If the solid is not required, the liquid may be collected in a capillary pipette by sucking through a small pad of purified cotton wool placed in the capillary end; upon removing the cotton wool and wiping the pipette, the liquid may be delivered clear and free from suspended matter.

Filter pipette

A useful **filter pipette** is shown in Fig. 2.47. It is constructed of tubing of 6 mm diameter. A rubber bulb is attached to the short arm A; the arm B is

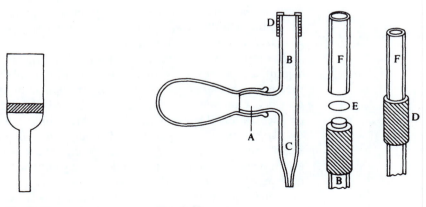

Fig. 2.46   Fig. 2.47

ground flat, whilst the arm C is drawn out to a fine capillary; a short piece of rubber tubing is fitted over the top of B. For filtering, a disc of filter paper of the same diameter as the outside diameter of the tube (cut out from filter paper by means of a sharp cork borer or by a hand punch) is placed on the flat ground surface of B, the tube F placed upon it and then held in position by sliding the rubber tubing just far enough over the paper to hold it when the tube F is removed. The filter pipette may be used either by placing a drop of the solution on the filter disc or by immersing the tube end B into the crucible, test-tube, or receptacle containing the solution to be filtered. The bulb is squeezed by the thumb and middle finger, and the dropper point closed with the forefinger thus allowing the solution to be drawn through the paper when the bulb is released. To release the drops of filtered liquid thus obtained, the filter pipette is inverted over the spot plate, in an inclined position with the bulb uppermost. Manipulation of the bulb again forces the liquid in the tip on to the spot plate. The precipitate on the paper can be withdrawn for any further treatment by simply sliding the rubber tubing D down over the arm B.

Filter stick

Another method involves the use of an **Emich filter stick** fitted through a rubber stopper into a thick-walled suction tube; the filtrate is collected in a micro test-tube (Fig. 2.48). The filter stick has a small pad of pure cotton wool above the constriction.

Filter pipettes and micro filtration devices are available commercially. Thus, the Whatman syringe filters No. 402/0803/06 (with 4 mm effective diameter) and 402/0805/06 (with 13 mm effective diameter) are suitable for micro- or semimicro-work respectively.

The apparatus illustrated in Fig. 2.49 may be employed when the filter paper (or drop reaction paper) must be **heated in steam**; the filter paper is

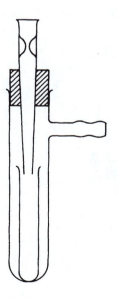

Fig. 2.48

Fig. 2.49

placed on the side arm support. By charging the flask with hydrogen sulphide solution, ammonia solution, chlorine or bromine water, the apparatus can be used for treating the filter paper with the respective gases or vapours.

Fusion and solution of a melt may be conducted either in a platinum wire loop or in a platinum spoon (0.5–1 ml capacity) attached to a heavy platinum wire and fused into a glass holder.

Gas reactions

Gas reactions may be performed in specially devised apparatus. Thus in testing for carbonates, sulphides, etc., it is required to absorb the gas liberated in a drop of water or reagent solution. The apparatus is shown in Fig. 2.50, and consists of a micro test-tube of about 1 ml capacity, which can be closed with a small ground glass stopper fused to a glass knob. The reagent and test solution or test solid are placed in the bottom of the tube, and a drop of the reagent for the gas is suspended on the knob of the stopper. The gas is evolved in the tube, if necessary, by gentle warming, and is absorbed by the reagent on the knob. Since the apparatus is closed, no gas can escape, and if sufficient time is allowed it is absorbed quantitatively by the reagent. A drop of water may replace the reagent on the stopper; the gas is dissolved, the drop may be washed on to a spot plate or into a micro crucible and treated with the reagent. The apparatus, shown in Fig. 2.51, which is closed by a rubber stopper, is sometimes preferable, particularly when minute quantities of gas are concerned; the glass tube, blown into a small bulb at the lower end, may be raised or lowered at will, whilst the change of colour and reaction products may be rendered more easily visible by filling the bulb with gypsum or magnesia powder. In some reactions, e.g. in testing for ammonia, it may be desirable to suspend a small strip of reagent paper from a glass hook fused to the stopper as in Fig. 2.52. When a particular gas has to be identified in the presence of other gases, the apparatus shown in Fig. 2.53 should be used; here the stopper for the micro test-tube consists of a small glass funnel on top of

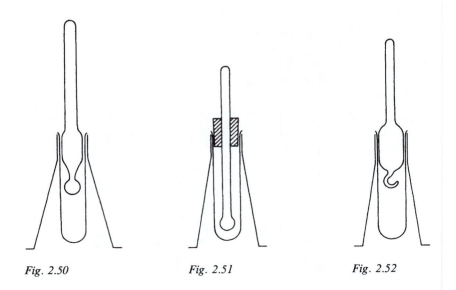

Fig. 2.50   Fig. 2.51   Fig. 2.52

Fig. 2.53   *Fig. 2.54*   *Fig. 2.55*

which the impregnated filter paper is laid in order to absorb the gas. The impregnated filter paper permits the passage of other gases and only retains the gas to be tested by the formation of a non-volatile compound that can be identified by means of a spot test. Another useful apparatus is shown in Fig. 2.54; it consists of a micro test-tube into which is placed a loosely-fitting glass tube narrowed at both ends. The lower capillary end is filled to a height of about 1 mm with a suitable reagent solution; if the gas liberated forms a coloured compound with the reagent, it can easily be seen in the capillary.

Where **high temperatures** are necessary for the evolution of the gas, a simple hard glass tube supported in a circular hole in a Teflon plate (Fig. 2.55) may be used. The open end of the tube should be covered by a small piece of reagent paper kept in position by means of a glass cap.

Distillation

Micro distillation is sometimes required, e.g. in the chromyl chloride test for a chloride (see Section 4.14). The apparatus depicted in Fig. 2.56 is suitable for the distillation of very small quantities of a mixture. A micro crucible or a micro centrifuge tube may be employed as a receiver.

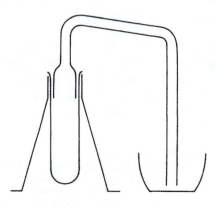

Fig. 2.56

In Chapters 3, 4 and 6, the experimental details are given for the detection of a number of ions by spot tests. The sensitivities given are, as a general rule, for a solution containing only the ion in question. It must be remembered that this is the most favourable case, and that in actual practice the presence of other ions usually necessitates a modification of the procedure which, more often than not, involves a loss of sensitivity. Almost without exception each test is subject to interference from the presence of other ions, and the possibility of these interferences occurring must be taken into consideration when a test is applied. Furthermore, the sensitivities when determined upon drop-reaction paper will depend upon the type of paper used. The figures given in the text have been obtained largely with the Schleicher and Schuell spot paper: substantially similar results are given by the equivalent Whatman papers.

It is important to draw attention to the difference between the terms 'specific' and 'selective' when used in connection with reagents or reactions. Reactions (and reagents), which under the experimental conditions employed are indicative of one substance (or ion) only, are designated as specific, whilst those reactions (and reagents) which are characteristic of a comparatively small number of substances are classified as selective. Hence we may describe reactions (or reagents) as having varying degrees of selectivity; however, a reaction (or reagent) can be only specific or non-specific.

2.7 Detection of metals by flame atomic spectrometric methods

Most modern laboratories are nowadays equipped with flame atomic spectrometers, which are routinely used for the quantitative determination of metals in solution. If such equipment is available, detection of metals can be carried out much faster than with the usual wet tests, especially if separations are involved. Usually 1–2 ml solution is consumed during one test; however in most cases the tests are so sensitive, that a portion of the original sample solution can be diluted 10–100 fold, leaving enough material for separations. Note that flame atomic spectrometric tests do not provide information about the oxidation state of the metal (e.g. they cannot differentiate between divalent and trivalent iron, trivalent or hexavalent chromium, etc.).[12]

Safety

It is important to emphasize some aspects of **safety**. Cylinders of fuel and oxidant gases must be fastened to a wall or laboratory bench. The exhaust gases have to be channelled through an adequate ventilation system and released into the atmosphere. In some cases (e.g. when employing nitrous oxide) the exhaust gases are toxic. Special care has to be exercised when lighting and extinguishing the flame.

---

[12] In the present text only brief outlines of these methods can be presented. For more detailed discussion of the theoretical principles, instrumentation and practice of atomic spectrometry, modern texts of instrumental analysis should be consulted, e.g. Daniel C. Harris: *Quantitative Chemical Analysis*, 3rd edn, W. H. Freeman and Co, 1991, chapters 20 and 21.

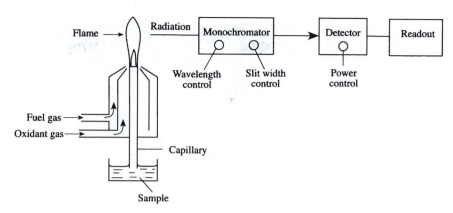

Fig. 2.57

There are two principal methods available, which require somewhat different instrumentation; these will be described below. Historically, atomic emission instruments were designed first; one can find such older instruments in many laboratories. Nowadays atomic absorption spectrophotometers are more popular, these however can be used also in emission mode.

Flame atomic emission (FAES)

In **flame atomic emission spectrometry (FAES)** the sample solution is sprayed into a flame, and by means of a monochromator the emitted radiation is analysed. As the emission spectrum is characteristic for each element, by identifying the wavelengths of the lines in the spectrum, the presence or absence of a metal can thus be established.

Spectrometer

The typical layout of a flame atomic emission spectrometer is shown in Fig. 2.57. A flame is produced and maintained by mixing a **fuel gas** (usually acetylene, but hydrogen or propane–butane may also be used) and an **oxidant gas** (air, oxygen or nitrous oxide N_2O). The gas flow rates (pressures) have to be regulated accurately with precision valves and rotameters. The **nebuliser unit** disperses the sample solution into very fine droplets and transports these through the burner into the flame. Two different types of burners can be employed. A **total consumption type burner**, as shown on the figure, using turbulent flow, uses up all the sample, thus resulting in higher sensitivities. A **pre-mix type burner**, like the one shown in Fig. 2.58 for atomic absorption (but with a tubular, Meker-type burner head) uses laminar flow and produces a quiet, steady flame. Larger droplets of the spray settle out in the mixing chamber, and only 10–25% of the sample solution reaches the burner, hence sensitivities are lower.

Monochromator

The radiation emitted by the flame first enters the **monochromator** through an **entrance slit**. By placing a **collimator** lens or mirror at the focal length distance from the entrance slit the now parallel rays are directed onto a prism or a grating. The rays undergo dispersion, leaving the prism or grating at different angles, according to their wavelengths. A **focusing** lens or mirror then focuses these radiations onto a plane, thus creating separate spectrum lines (i.e. images of the entrance slit). By rotating the prism or grating, and using a fixed **exit slit**, each individual spectrum line can be projected onto the radiation **detector**. The signal created by the detector is

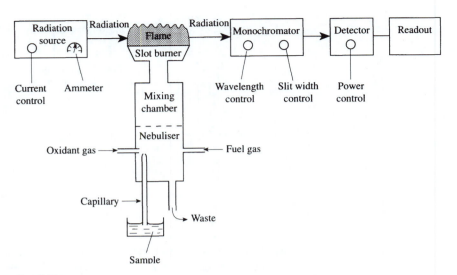

Fig. 2.58

then amplified and fed into a **readout** device, where radiation intensities are displayed.

Flame atomic emission spectrometers have separate controls for maintaining gas pressures, selecting wavelengths and adjusting slit widths (optimal values depend on the wavelength examined; figures are given in the instrument handbook). The power control allows the adjustment of (relative) radiation intensities (readouts).

Operations in FAES

To **operate** a flame atomic emission instrument, consult the instrument handbook or seek assistance from an experienced operator. After switching on the instrument, the flow rate of the oxidant gas has to be adjusted to a suitable value, followed by regulation of the fuel gas. The mixture is then ignited (usually with an electric spark). If nitrous oxide is employed as the oxidant gas, the flame has to be started first with air, followed by a gradual switch to N_2O. Once the flame has been established, spray pure water into the flame, adjust slit widths and wavelength. The newest instruments allow digital adjustment with an accuracy of one-hundredth of a nanometer (hence all wavelength data given in this text are quoted with this accuracy). Small changes of temperature, however, cause fluctuations in refractive indices and therefore true wavelengths may slightly differ from the nominal ones. It is advisable therefore to scan wavelengths within a few nanometers around the nominal figure and take measurements when the emission signal is highest. Now spray a standard solution, containing the suspected metal in known concentration, and, using the power control, adjust to a given emission reading (say 50%). Spray water again until the emission reading falls back to near zero, and then spray the test solution into the flame and note the wavelength and emission reading.

Emission spectrometric characteristics of the various metals, with further relevant information, will be described among the reactions of the cations.

Plasma torches

The most up-to-date atomic emission instruments employ **high-temperature plasma torches** instead of flames as radiation sources. Their sensitivities are much higher than those of the flame instruments and emissions are less prone to chemical interferences. They can also be used for the detection and determination of certain metallic (Th, Ce) and non-metallic elements (like B, P, S, Se, Te). The discussion of such instruments is beyond the scope of this book.

Flame atomic absorption (FAAS)

In **flame atomic absorption spectrometry (FAAS)** monochromatic radiation, characteristic for the metal in question, is passed through a long, thin flame, into which the sample solution is sprayed (see Fig. 2.58). A flame is produced in a pre-mix type burner with a 10 cm long slit. The burner head has to be aligned carefully so that the radiation should pass exactly over this slit. The role of the flame in this case is to produce a population of ground-state atoms, capable of absorbing the radiation coming from the **radiation source**. The latter, in most cases, is a **hollow-cathode lamp**. Emission is produced by electric discharge between a thimble-like cathode, made from the pure metal (or sometimes an alloy), and a tungsten anode. The glass lamp is filled with low-pressure neon or argon gas, and has a

Spectrometer

small silica window through which the radiation is directed onto the flame. Under these circumstances extremely narrow spectral lines are emitted, which are then partly absorbed by the ground-state atoms in the flame. Steady emission can be controlled by adjusting an optimal lamp current with the aid of a variable resistor and an ammeter. For each element to be detected, a different radiation source has to be used; this could be a disadvantage if solutions containing a number of metals have to be analysed. Hollow cathode lamps are expensive and have limited lifetimes. On the other hand, the sensitivities are usually higher than achievable with FAES.

Monochromator

The role of the monochromator is mainly to isolate a selected line from the few spectral lines generated by the radiation source, hence relatively wide entrance slits can be used. All the other components and controls of these instruments have the same roles as described under FAES, though with some instruments there are separate power controls for adjusting zero and 100% transmittances.

Operations in FAAS

To **operate** flame atomic absorption instruments an instrument handbook or an experienced operator should again be consulted. After switching on the instrument, the hollow-cathode lamp has to be turned on. By increasing the voltage on the electrodes of the lamp (usually to 600–1000 V), the lamp 'strikes', which can be detected from the ammeter, which now displays a measurable current (at the same time the voltage between the electrodes falls to about 200–400 V). The lamp current has to be adjusted with the variable resistor to an optimal value (these are different for each lamp; the instrument manual will recommend values). The flame has now to be ignited in a similar manner to FAES, and the wavelength and slit width must also be adjusted. This is followed by adjusting the detector output. First block the radiation from entering the monochromator and set zero transmittance with the appropriate control. Remove the block, and spray pure solvent into the flame and adjust to 100% transmittance (or zero absorbance) with the power control. Finally, while spraying the test solution into the flame, the absorbance of the solution can be measured, and compared to a standard solution containing the suspected analyte.

Absorption spectrometric characteristics of the various metals, with further relevant information, will be described among the reactions of the cations.

Instead of flames, atomic absorption spectrometers sometimes employ graphite furnaces or (relatively speaking) cold quartz tubes as atomisers; these devices are not normally required for the purpose of qualitative analysis (unless the volume of sample available is very small), and will not be discussed here.

Chapter 3

Reactions of the cations

3.1 Classification of cations (metal ions) into analytical groups

For the purpose of systematic qualitative analysis, cations are classified into five groups on the basis of their behaviour against some reagents. By the systematic use of these so-called group reagents we can decide about the presence or absence of groups of cations, and can also separate these groups for further examination. The reactions of cations will be dealt with here according to the order defined by this group system. Apart from being the traditional way of presenting the material, it makes the study of these reactions easier because ions of analogous behaviour are dealt with within one group.

The group reagents used for the classification of most common cations are hydrochloric acid, hydrogen sulphide, ammonium sulphide, and ammonium carbonate. Classification is based on whether a cation reacts with these reagents by the formation of precipitates or not. It can therefore be said that **classification** of the most common **cations is based on the differences of solubilities of their chlorides, sulphides, and carbonates.**

The five groups
The five groups of cations and the characteristics of these groups are as follows:

Group I cations of this group form precipitates with dilute hydrochloric acid. Ions of this group are **lead**, **mercury(I)**, and **silver**.

Group II cations of this group do not react with hydrochloric acid, but form precipitates with hydrogen sulphide in dilute mineral acid medium. Ions of this group are **mercury(II)**, **copper**, **bismuth**, **cadmium**, **arsenic(III)**, **arsenic(V)**, **antimony(III)**, **antimony(V)**, **tin(II)**, and **tin(IV)**. The first four form the sub-group IIa and the last six the sub-group IIb. While sulphides of cations in Group IIa are insoluble in ammonium polysulphide, those of cations in Group IIb are soluble.

Group III cations of this group do not react either with dilute hydrochloric acid, or with hydrogen sulphide in dilute mineral acid medium. However they form precipitates with ammonium sulphide in neutral or ammoniacal media. Cations of this group are **cobalt(II)**, **nickel(II)**, **iron(II)**, **iron(III)**, **chromium(III)**, **aluminium**, **zinc**, and **manganese(II)**.

Group IV cations of this group do not react with the reagents of Groups I, II, and III. They form precipitates with ammonium carbonate in the presence of ammonium chloride in neutral or slightly acidic media. Cations of this group are **calcium**, **strontium**, and **barium**.

Group V Common cations, which do not react with reagents of the previous groups, form the last group of cations, which includes **magnesium**, **sodium**, **potassium**, **ammonium**, **lithium**, and **hydrogen ions**.

This group system of cations can be extended to include less common ions as well. Classification of these ions, together with their reactions, will be given in Chapter 6.

3.2   Notes on the study of the reactions of ions

When studying the reactions of ions, experimental techniques described in Chapter 2 should be applied. The reactions can be studied both in macro and semimicro scale, and the majority of the reactions can be applied as a spot test as well. Hints on the preparation of reagents are given in the appendix of this book. The reagents are listed there in alphabetical order, with notes on their stability. Most reagents are poisonous to some extent, and should therefore be handled with care. Those reagents which are exceptionally poisonous or hazardous must be specially labelled and must be used with utmost care. In the list of reactions these reagents will be marked as **(POISON)** or **(HAZARD)**. One should not use these reagents when working alone in a laboratory; the supervisor or a colleague should always be notified before using them.

The concentration of reagents is in most cases chosen to be molar, meaning that it is easy to calculate the relative volumes of the reactant and the reagent needed to complete the reaction. It is not advisable to add the calculated amount of reagent at once to the solution (cf. Chapter 2), but the final amount should be equal to or more than the equivalent. In some cases it is impossible or impractical to prepare a M reagent; thus 0.5 or even 0.1M reagents have to be used sometimes. It is easy to predict the volume of a particular reagent needed to complete the reaction from the concentrations. Acids and bases are applied mostly in 2M concentrations in order to avoid unnecessary dilution of the mixture.

How to take notes   **Taking notes** when studying these reactions is absolutely necessary. A logical, clear way of making notes is essential. Although it would be wrong to copy the text of this book, it is important to note (*a*) the reagent, and any special experimental circumstances applied when performing the test, (*b*) the changes observed, and (*c*) the equation of the reaction or some other explanation of what had happened. A useful way of making notes is the following. The left page of an open notebook is divided into two equal sections by a vertical line. The left column can be headed 'TEST' and should contain a brief description of the test itself, including the reagent and experimental circumstances. The second column (still on the left-hand page), headed 'OBSERVATION' should contain the visible change which occurred when carrying out the test. Finally, the entire right-hand page should be reserved for 'EXPLANATION', where the reaction equation can be entered. A typical page of a notebook containing some of the reactions of lead(II) ions is shown on Table 3.1. It is also advisable to summarize reactions within one group

Table 3.1   **One page of a laboratory notebook**

Test	Observation	Explanation
Group 1 Pb^{2+}		
1. HCl	white ppt.	$Pb^{2+} + 2Cl^- \rightarrow PbCl_2\downarrow$
+NH$_3$	no change	no ammine complexes (but $Pb(OH)_2\downarrow$)
+hot water	dissolves	33.4 g $PbCl_2$ dissolves per litre at 100°C
2. H$_2$S(+HCl)	black ppt.	$Pb^{2+} + H_2S \rightarrow PbS\downarrow + 2H^+$
+conc. HNO$_3$	white ppt.	$3PbS\downarrow + 8HNO_3 \rightarrow 3Pb^{2+} + 2NO\uparrow + 4H_2O$ $+3S\downarrow + 6NO_3^-$
+boiling	white ppt. (different)	$S\downarrow + 2HNO_3 \rightarrow SO_4^{2-} + 2H^+ + 2NO\uparrow$ $Pb^{2+} + SO_4^{2-} \rightarrow PbSO_4\downarrow$
3. NH$_3$	white ppt.	$Pb^{2+} + 2NH_3 + 2H_2O \rightarrow Pb(OH)_2\downarrow + 2NH_4^+$
+excess	no change	Pb^{2+} does not form ammine complexes
4. NaOH	white ppt.	$Pb^{2+} + 2OH^- \rightarrow Pb(OH)_2\downarrow$
+excess	dissolves	$Pb(OH)_2\downarrow + 2OH^- \rightleftharpoons [Pb(OH)_4]^{2-}$ $Pb(OH)_2$: amphoteric
5. KI	yellow ppt.	$Pb^{2+} + 2I^- \rightarrow PbI_2\downarrow$
+excess (6M)	dissolves	$PbI_2\downarrow + 2I^- \rightleftharpoons [PbI_4]^{2-}$

in the form of tables, as shown on Table 3.2 for the first group of cations.

The column 'TEST' should be made up before making the actual experiments. The 'OBSERVATION' column should be filled when actually making the experiments, while the 'EXPLANATION' page should be made up after leaving the laboratory. Finally, the reaction tables should be made up when the reactions of the particular group have been studied and explained. This systematic way of study enables us to devote precious laboratory time entirely to experiments and, by dealing with one particular reaction altogether four times, helps to learn the subject.

3.3   First group of cations: lead(II), mercury(I), and silver(I)

Group reagent   dilute (2M) hydrochloric acid.

Pb(II), Hg(I), Ag(I)

Group reaction   white precipitates of lead chloride PbCl$_2$, mercury(I) chloride Hg$_2$Cl$_2$, and silver chloride AgCl.

Cations of the first group form insoluble chlorides. Lead chloride, however, is slightly soluble in water and therefore lead is never completely precipitated by adding dilute hydrochloric acid to a sample; the rest of the lead ions are quantitatively precipitated with hydrogen sulphide in an acidic medium together with the cations of the second group.

Nitrates of these cations are very soluble. Among sulphates, lead sulphate is practically insoluble, but silver sulphate dissolves to a much greater extent. The solubility of mercury(I) sulphate is somewhere between. Bromides and iodides are also insoluble, though precipitation of lead halides is incomplete, and they dissolve quite easily in hot water. Sulphides are insoluble. Acetates are more soluble, though silver acetate might be precipitated

Table 3.2 **Tabulating reactions of Group I cations**

	Pb^{2+}	Hg_2^{2+}	Ag^+
HCl	white $PbCl_2\downarrow$	white $Hg_2Cl_2\downarrow$	white $AgCl\downarrow$
+NH$_3$	no change	black $Hg\downarrow + HgNH_2Cl\downarrow$	dissolves $[Ag(NH_3)_2]^+$
+hot water	dissolves	no change	no change
H$_2$S(+HCl)	black $PbS\downarrow$	black $Hg\downarrow + HgS\downarrow$	black $Ag_2S\downarrow$
+ccHNO$_3$, boiling	white $PbSO_4\downarrow$	white $Hg_2(NO_3)_2S\downarrow$	dissolves Ag^+
NH$_3$, small amounts	white $Pb(OH)_2\downarrow$	black $Hg\downarrow + HgO \cdot HgNH_2NO_3\downarrow$	brown $Ag_2O\downarrow$
+excess	no change	no change	dissolves $[Ag(NH_3)_2]^+$
NaOH, small amounts	white $Pb(OH)_2\downarrow$	black, $Hg_2O\downarrow$	brown, $Ag_2O\downarrow$
+excess	dissolves $[Pb(OH)_4]^{2-}$	no change	no change
KI, small amounts	yellow $PbI_2\downarrow$	green $Hg_2I_2\downarrow$	yellow $AgI\downarrow$
+excess	no change	grey $Hg\downarrow + [HgI_4]^{2-}$	no change
K$_2$CrO$_4$	yellow $PbCrO_4\downarrow$	red $Hg_2CrO_4\downarrow$	red $Ag_2CrO_4\downarrow$
+NH$_3$	no change	black $Hg\downarrow + HgNH_2NO_3\downarrow$	dissolves $[Ag(NH_3)_2]^+$
KCN, small amounts	white $Pb(CN)_2\downarrow$	black $Hg\downarrow + Hg(CN)_2$	white $AgCN\downarrow$
+excess	no change	no change	dissolves $[Ag(CN)_2]^-$
Na$_2$CO$_3$	white $PbO \cdot PbCO_3\downarrow$	yellowish-white $Hg_2CO_3\downarrow$	yellowish-white $Ag_2CO_3\downarrow$
+boiling	no change	black $Hg\downarrow + HgO\downarrow$	brown $Ag_2O\downarrow$
Na$_2$HPO$_4$	white $Pb_3(PO_4)_2\downarrow$	white $Hg_2HPO_4\downarrow$	yellow $Ag_3PO_4\downarrow$
Specific reaction	K_2CrO_4 yellow $PbCrO_4\downarrow$	Diphenyl carbazide violet colour	p-dimethylamino-benzylidene-rhodanine (+HNO$_3$) violet colour

from more concentrated solutions. Hydroxides and carbonates are precipitated with an equivalent amount of the reagent, an excess however might produce varied results. There are differences in their behaviour towards ammonia as well.

3.4 Lead, Pb (A_r: 207.19)

Lead is a bluish-grey metal with a high density (11.48 g ml^{-1} at room temperature). It readily dissolves in moderately concentrated nitric acid (8M), and nitrogen oxide is formed also:

$$3Pb + 8HNO_3 \rightarrow 3Pb^{2+} + 6NO_3^- + 2NO\uparrow + 4H_2O$$

The colourless nitrogen oxide gas, when mixed with air, is oxidized to red nitrogen dioxide:

$$2NO\uparrow(\text{colourless}) + O_2\uparrow \rightarrow 2NO_2\uparrow(\text{red})$$

With concentrated nitric acid a protective film of lead nitrate is formed on the surface of the metal and prevents further dissolution. Dilute hydrochloric or sulphuric acid have little effect owing to the formation of insoluble lead chloride or sulphate on the surface.

Reactions of lead(II) ions   A solution of lead nitrate (0.25M) or lead acetate (0.25M) can be used for the study of these reactions.

1. *Dilute hydrochloric acid (or soluble chlorides)*   a white precipitate in cold solution if the acid is not too dilute:

$$Pb^{2+} + 2Cl^- \rightleftharpoons PbCl_2\downarrow$$

The precipitate is soluble in hot water (33.4 g l^{-1} at 100°C while only 9.9 g l^{-1} at 20°C), but separates again in long, needle-like crystals when cooling. It is also soluble in concentrated hydrochloric acid or concentrated potassium chloride when the tetrachloroplumbate(II) ion is formed:

$$PbCl_2\downarrow + 2Cl^- \rightarrow [PbCl_4]^{2-}$$

If the precipitate is washed by decantation and dilute ammonia is added, no visible change occurs [difference from mercury(I) or silver ions], though a precipitate-exchange reaction takes place and lead hydroxide is formed:

$$PbCl_2\downarrow + 2NH_3 + 2H_2O \rightarrow Pb(OH)_2\downarrow + 2NH_4^+ + 2Cl^-$$

2. *Hydrogen sulphide (gas or saturated aqueous solution) in neutral or dilute acid medium*   black precipitate of lead sulphide:

$$Pb^{2+} + H_2S \rightarrow PbS\downarrow + 2H^+$$

Precipitation is incomplete if strong mineral acids are present in more than 2M concentration. Because hydrogen ions are formed in the above reaction, it is advisable to buffer the mixture with sodium acetate.

When hydrogen sulphide gas is introduced into a mixture which contains a precipitate of white lead chloride, the latter is converted into (black) lead sulphide in a precipitate-exchange reaction:

$$PbCl_2\downarrow + H_2S \rightarrow PbS\downarrow + 2H^+ + 2Cl^-$$

If the test is carried out in the presence of larger amounts of chloride [potassium chloride (saturated)], initially a red precipitate of lead sulpho-chloride is formed when introducing hydrogen sulphide gas:

$$2Pb^{2+} + H_2S + 2Cl^- \rightarrow Pb_2SCl_2\downarrow + 2H^+$$

This however decomposes on dilution or on further addition of hydrogen sulphide and a black precipitate of lead sulphide is formed:

$$Pb_2SCl_2\downarrow \rightarrow PbS\downarrow + PbCl_2\downarrow$$

$$Pb_2SCl_2 + H_2S \rightarrow 2PbS\downarrow + 2Cl^- + 2H^+$$

Lead sulphide decomposes when concentrated nitric acid is added, and white, finely divided elemental sulphur is precipitated:

$$3PbS\downarrow + 8HNO_3 \rightarrow 3Pb^{2+} + 6NO_3^- + 3S\downarrow + 2NO\uparrow + 4H_2O$$

If the mixture is boiled, sulphur is oxidized by nitric acid to sulphate, which immediately forms white lead sulphate precipitate with the lead ions in the solution:

$$S\downarrow + 2HNO_3 \rightarrow SO_4^{2-} + 2H^+ + 2NO\uparrow$$

$$Pb^{2+} + SO_4^{2-} \rightarrow PbSO_4\downarrow$$

On boiling lead sulphide with hydrogen peroxide (3%), the black precipitate turns white owing to the formation of lead sulphate:

$$PbS\downarrow + 4H_2O_2 \rightarrow PbSO_4\downarrow + 4H_2O$$

The great insolubility of lead sulphide in water $(4.9 \times 10^{-11} \, \mathrm{g\,l^{-1}})$ explains why hydrogen sulphide is such a sensitive reagent for the detection of lead, and why it can be detected in the filtrate from the separation of the sparingly soluble lead chloride in dilute hydrochloric acid.

Hydrogen sulphide is a highly poisonous gas, and all operations with the gas must be conducted in the fume cupboard. Every precaution must be observed to prevent the escape of hydrogen sulphide into the air of the laboratory.

NH_3
$\downarrow$
white $Pb(OH)_2$

3. *Ammonia solution*   white precipitate of lead hydroxide

$$Pb^{2+} + 2NH_3 + 2H_2O \rightarrow Pb(OH)_2\downarrow + 2NH_4^+$$

The precipitate is insoluble in excess reagent.

$NaOH$
$\downarrow$
white $Pb(OH)_2$

Dissolves in excess

4. *Sodium hydroxide*   white precipitate of lead hydroxide

$$Pb^{2+} + 2OH^- \rightarrow Pb(OH)_2\downarrow$$

The precipitate dissolves in excess reagent, when tetrahydroxoplumbate(II) ions are formed:

$$Pb(OH)_2\downarrow + 2OH^- \rightarrow [Pb(OH)_4]^{2-}$$

Thus, lead hydroxide has an amphoteric character.

Hydrogen peroxide or ammonium peroxodisulphate, when added to a solution of tetrahydroxoplumbate(II), forms a black precipitate of lead dioxide by oxidizing bivalent lead to the tetravalent state:

$$[Pb(OH)_4]^{2-} + H_2O_2 \rightarrow PbO_2\downarrow + 2H_2O + 2OH^-$$

$$[Pb(OH)_4]^{2-} + S_2O_8^{2-} \rightarrow PbO_2\downarrow + 2H_2O + 2SO_4^{2-}$$

H_2SO_4
$\downarrow$
white $PbSO_4$

5. *Dilute sulphuric acid (or soluble sulphates)*   white precipitate of lead sulphate:

$$Pb^{2+} + SO_4^{2-} \rightarrow PbSO_4\downarrow$$

The precipitate is insoluble in excess reagent. Hot, concentrated sulphuric acid dissolves the precipitate owing to formation of lead hydrogen sulphate:

$$PbSO_4\downarrow + H_2SO_4 \rightarrow Pb^{2+} + 2HSO_4^-$$

Solubility is much lower in the presence of ethanol.

Lead sulphate is soluble in more concentrated solutions of ammonium acetate (6M) or ammonium tartrate (6M) in the presence of ammonia, when tetraacetoplumbate(II) and ditartratoplumbate(II) ions are formed:

$$PbSO_4\downarrow + 4CH_3COO^- \rightarrow [Pb(CH_3COO)_4]^{2-} + SO_4^{2-}$$

$$PbSO_4\downarrow + 2C_4H_4O_6^{2-} \rightarrow [Pb(C_4H_4O_6)_2]^{2-} + SO_4^{2-}$$

The stabilities of these complexes are not very great; chromate ions, for example, can precipitate lead chromate from their solution.

When boiled with sodium carbonate, the lead sulphate is transformed into lead carbonate in a precipitate-exchange reaction:

$$PbSO_4\downarrow + CO_3^{2-} \rightarrow PbCO_3\downarrow + SO_4^{2-}$$

On washing the precipitate by decantation with hot water, sulphate ions can be removed and the precipitate will dissolve in dilute nitric acid

$$PbCO_3\downarrow + 2H^+ \rightarrow Pb^{2+} + H_2O + CO_2\uparrow$$

K$_2$CrO$_4$
↓
yellow PbCrO$_4$

6. *Potassium chromate in neutral, acetic acid or ammonia solution*   yellow precipitate of lead chromate

$$Pb^{2+} + CrO_4^{2-} \rightarrow PbCrO_4\downarrow$$

Nitric acid or sodium hydroxide dissolve the precipitate:

$$2PbCrO_4\downarrow + 2H^+ \rightleftarrows 2Pb^{2+} + Cr_2O_7^{2-} + H_2O$$

$$PbCrO_4\downarrow + 4OH^- \rightleftarrows [Pb(OH)_4]^{2-} + CrO_4^{2-}$$

Both reactions are reversible; by buffering the solution with ammonia or acetic acid respectively, lead chromate reprecipitates.

KI
↓
yellow PbI$_2$

7. *Potassium iodide*   yellow precipitate of lead iodide

$$Pb^{2+} + 2I^- \rightarrow PbI_2\downarrow$$

The precipitate is moderately soluble in boiling water to yield a colourless solution, from which it separates as golden yellow plates on cooling.

An excess of a more concentrated (6M) solution of the reagent dissolves the precipitate and tetraiodoplumbate(II) ions are formed:

$$PbI_2\downarrow + 2I^- \rightleftarrows [PbI_4]^{2-}$$

The action is reversible; on diluting with water the precipitate reappears.

Na$_2$SO$_3$
↓
white PbSO$_3$

8. *Sodium sulphite in neutral solution*   white precipitate of lead sulphite

$$Pb^{2+} + SO_3^{2-} \rightarrow PbSO_3\downarrow$$

The precipitate is less soluble than lead sulphate, though it can be dissolved by both dilute nitric acid and sodium hydroxide

$$PbSO_3\downarrow + 2H^+ \rightarrow Pb^{2+} + H_2O + SO_2\uparrow$$

$$PbSO_3\downarrow + 4OH^- \rightarrow [Pb(OH)_4]^{2-} + SO_3^{2-}$$

Na$_2$CO$_3$
↓
white ppt with mixed composition

9. *Sodium carbonate*   white precipitate of a mixture of lead carbonate and lead hydroxide

$$2Pb^{2+} + 2CO_3^{2-} + H_2O \rightarrow Pb(OH)_2\downarrow + PbCO_3\downarrow + CO_2\uparrow$$

On boiling, no visible change takes place [difference from mercury(I) and silver(I) ions]. The precipitate dissolves in dilute nitric acid, and even in acetic acid, and CO_2 gas is liberated:

$$Pb(OH)_2\downarrow + PbCO_3\downarrow + 4H^+ \rightarrow 2Pb^{2+} + 3H_2O + CO_2\uparrow$$

Na_2HPO_4
↓
white $Pb_3(PO_4)_2$

10. *Disodium hydrogen phosphate*   white precipitate of lead phosphate

$$3Pb^{2+} + 2HPO_4^{2-} \rightleftharpoons Pb_3(PO_4)_2\downarrow + 2H^+$$

The reaction is reversible; strong acids (nitric acid) dissolve the precipitate. The precipitate is also soluble in sodium hydroxide.

KCN
↓
white $Pb(CN)_2$

11. *Potassium cyanide* (**POISON**)   white precipitate of lead cyanide

$$Pb^{2+} + 2CN^- \rightarrow Pb(CN)_2\downarrow$$

which is insoluble in the excess of the reagent. This reaction can be used to distinguish lead(II) ions from mercury(I) and silver(I), which react quite differently.

12. *Di-(4-dimethylaminodiphenyl)methane (or 'tetrabase') (0.5%)*

$$(CH_3)_2N - \underset{}{\bigcirc} - CH_2 - \underset{}{\bigcirc} - N(CH_3)_2$$

a blue oxidation product {hydrol: $-CH_2 \rightarrow -CH(OH)$} is formed under the conditions given below.

Tetrabase
↓
blue colour

Place 1 ml test solution in a 5 ml centrifuge tube, add 1 ml 2M potassium hydroxide and 0.5–1 ml 3% hydrogen peroxide solution. Allow to stand for 5 minutes. Separate the precipitate by centrifugation, and wash once with cold water. Add 2 ml reagent, shake and centrifuge. The supernatant liquid is coloured blue.

The ions of bismuth, cerium, manganese, thallium, cobalt, and nickel give a similar reaction: iron and large quantities of copper interfere.

Concentration limit 1 in 10000.

13. *Gallocyanine (1%)*

$$\underset{OH}{\overset{COOH}{\bigcirc}} \quad N(CH_3)_3 \cdot HCl$$

deep-violet precipitate of unknown composition. The test is applicable to finely divided lead sulphate precipitated on filter paper.

Gallocyanine
↓
violet ppt

Place a drop of the test solution upon drop-reaction paper, followed by a drop each of 1% aqueous pyridine and the gallocyanine reagent (blue). Remove the excess of the reagent by placing several filter papers beneath the drop-reaction paper and adding drops of the pyridine solution to the spot until the wash liquid percolating through is colourless; move the filter papers to a fresh position after each addition of pyridine. A deep violet spot is produced.

Sensitivity 1–6 µg Pb.

Concentration limit 1 in 50 000.

In the presence of silver, bismuth, cadmium, or copper, proceed as follows. Transfer a drop of the test solution to a drop-reaction paper and add a drop of M sulphuric acid to fix the lead as lead sulphate. Remove the soluble sulphates of the other metals by washing with about 3 drops of M sulphuric acid, followed by a little 96% ethanol. Dry the paper on a water bath, and then apply the test as detailed above.

14. *Diphenylthiocarbazone (or Dithizone) (0.005%)*

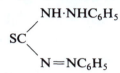

brick-red complex salt in neutral, ammoniacal, alkaline, or alkalicyanide solution.

Dithizone
↓
red complex

Place 1 ml of the neutral or faintly alkaline solution in a micro test-tube, introduce a few small crystals of potassium cyanide (**POISON**), and then 2 drops of the reagent. Shake for 30 seconds. The green colour of the reagent changes to red.

Sensitivity 0.1 µg Pb (in neutral solution).

Concentration limit 1 in 1 250 000.

Heavy metals (silver, mercury, copper, cadmium, antimony, nickel, and zinc, etc.) interfere, but this effect may be eliminated by conducting the reaction in the presence of much alkali cyanide: excess of alkali hydroxide is also required for zinc. The reaction is extremely sensitive, but it is not very selective.

Atomic spectrometry

15. *Atomic spectrometric tests*

(a) *Flame emission spectrometry*   lead cannot be easily detected with the air/acetylene flame; however, using a nitrous oxide/acetylene flame good emission readings are obtainable at 405.78 nm, when 100 ppm lead can easily be identified.

(b) *Flame atomic absorption spectrometry*   using an air/acetylene flame and a lead hollow-cathode lamp, 5 ppm lead can be easily detected at 283.31 nm.

(c) *Lead standard, 1000 ppm*   dissolve 1.5985 g analytical grade lead nitrate $Pb(NO_3)_2$ in water and dilute to 1 litre.

Dry tests

16. *Dry tests*

(a) *Blowpipe test*   when a lead salt is heated with alkali carbonate upon charcoal, a malleable bead of lead (which is soft and will mark paper) surrounded with a yellow incrustation of lead monoxide is obtained.

(b) *Flame test*   pale blue (inconclusive).

3.5   Mercury, Hg (A_r: 200.59) – mercury(I)

Mercury is a silver-white, liquid metal at ordinary temperatures and has a density of 13.534 g ml^{-1} at 25°C. It is unaffected when treated with hydrochloric or dilute sulphuric acid (2M), but reacts readily with nitric acid. Cold, medium concentrated (8M) nitric acid with an excess of mercury yields mercury(I) ions:

$$6Hg + 8HNO_3 \rightarrow 3Hg_2^{2+} + 2NO\uparrow + 6NO_3^- + 4H_2O$$

With an excess of hot concentrated nitric acid, mercury(II) ions are formed:

$$3Hg + 8HNO_3 \rightarrow 3Hg^{2+} + 2NO\uparrow + 6NO_3^- + 4H_2O$$

Hot, concentrated sulphuric acid also dissolves mercury. The product is mercury(I) ion if mercury is in excess

$$2Hg + 2H_2SO_4 \rightarrow Hg_2^{2+} + SO_4^{2-} + SO_2\uparrow + 2H_2O$$

while if the acid is in excess, mercury(II) ions are formed:

$$Hg + 2H_2SO_4 \rightarrow Hg^{2+} + SO_4^{2-} + SO_2\uparrow + 2H_2O$$

The two ions, mercury(I) and mercury(II), behave quite differently with reagents used in qualitative analysis, and hence belong to two different analytical groups. Mercury(I) ions belong to the first group of cations, their reactions will therefore be treated here. Mercury(II) ions, on the other hand, are in the second cation group; their reactions will be dealt with later, together with the other members of that group.

Hg(I) reactions

Reactions of mercury(I) ions A solution of mercury(I) nitrate (0.05M) can be used for the study of these reactions. **All compounds of mercury are toxic!**

HCl
↓
white Hg_2Cl_2

1. _Dilute hydrochloric acid or soluble chlorides_ white precipitate of mercury(I) chloride (calomel)

$$Hg_2^{2+} + 2Cl^- \rightarrow Hg_2Cl_2\downarrow$$

The precipitate is insoluble in dilute acids.

Ammonia solution converts the precipitate into a mixture of mercury(II) amidochloride and mercury metal, which are both insoluble.

$$Hg_2Cl_2 + 2NH_3 \rightarrow Hg\downarrow + Hg(NH_2)Cl\downarrow + NH_4^+ + Cl^-$$

The reaction involves disproportionation: mercury(I) is converted partly to mercury(II) and partly to mercury metal. This reaction can be used to differentiate mercury(I) ions from lead(II) and silver(I).

The mercury(II) amidochloride is a white precipitate, but the finely divided mercury makes it shiny black. The name calomel, coming from Greek ($\kappa\alpha\lambda ov\ \mu\varepsilon\lambda\alpha\sigma$ = nice black) refers to this characteristic of the originally white mercury(I) chloride precipitate.

Mercury(I) chloride dissolves in aqua regia, forming undissociated but soluble mercury(II) chloride:

$$3Hg_2Cl_2\downarrow + 2HNO_3 + 6HCl \rightarrow 6HgCl_2 + 2NO\uparrow + 4H_2O$$

H_2S
↓
black mixed ppt

2. _Hydrogen sulphide (gas or saturated aqueous solution) in neutral or dilute acid medium_ black precipitate, which is a mixture of mercury(II) sulphide and mercury metal

$$Hg_2^{2+} + H_2S \rightarrow Hg\downarrow + HgS\downarrow + 2H^+$$

Owing to the extremely low solubility product of mercury(II) sulphide, the reaction is very sensitive.

Sodium sulphide (colourless) dissolves the mercury(II) sulphide (but leaves mercury metal) and a disulphomercurate(II) complex is formed:

$$HgS + S^{2-} \rightarrow [HgS_2]^{2-}$$

After removing the mercury metal by filtration, black mercury(II) sulphide can again be precipitated by acidification with dilute mineral acids:

$$[HgS_2]^{2-} + 2H^+ \rightarrow HgS\downarrow + H_2S\uparrow$$

Sodium disulphide (yellow) dissolves both mercury and mercury(II) sulphide:

$$HgS\downarrow + Hg\downarrow + 3S_2^{2-} \rightarrow 2[HgS_2]^{2-} + S_3^{2-}$$

Aqua regia dissolves the precipitate, yielding undissociated mercury(II) chloride and sulphur:

$$12HCl + 4HNO_3 + 3Hg\downarrow + 3HgS\downarrow = 6HgCl_2 + 3S\downarrow + 4NO\uparrow + 8H_2O$$

When heated with aqua regia, sulphur is oxidized to sulphuric acid and the solution becomes clear:

$$S\downarrow + 6HCl + 2HNO_3 \rightarrow SO_4^{2-} + 6Cl^- + 8H^+ + 2NO\uparrow$$

NH_3
↓
black mixed ppt

3. *Ammonia solution*   black precipitate which is a mixture of mercury metal and basic mercury(II) amidonitrate (which itself is a white precipitate)

$$2Hg_2^{2+} + NO_3^- + 4NH_3 + H_2O \rightarrow HgO \cdot Hg\!\!\begin{array}{c} \diagup NH_2 \\ \diagdown NO_3 \end{array}\!\!\downarrow + 2Hg\downarrow + 3NH_4^+$$

This reaction can be used to differentiate between mercury(I) and mercury(II) ions.

NaOH
↓
black Hg_2O

4. *Sodium hydroxide*   black precipitate of mercury(I) oxide

$$Hg_2^{2+} + 2OH^- \rightarrow Hg_2O\downarrow + H_2O$$

The precipitate is insoluble in excess reagent, but dissolves readily in dilute nitric acid.

When boiling, the colour of the precipitate turns to grey, owing to disproportionation, when mercury(II) oxide and mercury metal are formed:

$$Hg_2O\downarrow \rightarrow HgO\downarrow + Hg\downarrow$$

K_2CrO_4
↓
red Hg_2CrO_4

5. *Potassium chromate in hot solution*   red crystalline precipitate of mercury(I) chromate

$$Hg_2^{2+} + CrO_4^{2-} \rightarrow Hg_2CrO_4\downarrow$$

If the test is carried out cold, a brown amorphous precipitate is formed with an undefined composition. When heated the precipitate turns to red crystalline mercury(I) chromate.

Sodium hydroxide turns the precipitate into black mercury(I) oxide:

$$Hg_2CrO_4\downarrow + 2OH^- \rightarrow Hg_2O\downarrow + CrO_4^{2-} + H_2O$$

KI
↓
green Hg_2I_2

6. *Potassium iodide, added slowly in cold solution* green precipitate of mercury(I) iodide

$$Hg_2^{2+} + 2I^- \rightarrow Hg_2I_2\downarrow$$

If excess reagent is added, a disproportionation reaction takes place, soluble tetraiodomercurate(II) ions and a black precipitate of finely divided mercury being formed:

$$Hg_2I_2\downarrow + 2I^- \rightarrow [HgI_4]^{2-} + Hg\downarrow$$

When boiling the mercury(I) iodide precipitate with water, disproportionation again takes place, and a mixture of red mercury(II) iodide precipitate and finely distributed black mercury is formed:

$$Hg_2I_2\downarrow \rightarrow HgI_2\downarrow + Hg\downarrow$$

Na_2CO_3
↓
yellow Hg_2CO_3
↓
turns slowly black

7. *Sodium carbonate in cold solution* yellow precipitate of mercury(I) carbonate:

$$Hg_2^{2+} + CO_3^{2-} \rightarrow Hg_2CO_3\downarrow$$

The precipitate turns slowly to blackish grey, as mercury(II) oxide and mercury are formed:

$$Hg_2CO_3\downarrow \rightarrow HgO\downarrow + Hg\downarrow + CO_2\uparrow$$

The decomposition can be speeded up by heating the mixture.

Na_2HPO_4
↓
white Hg_2HPO_4

8. *Disodium hydrogen phosphate* white precipitate of mercury(I) hydrogen phosphate:

$$Hg_2^{2+} + HPO_4^{2-} \rightarrow Hg_2HPO_4\downarrow$$

KCN
↓
black Hg

9. *Potassium cyanide* (**POISON**) produces mercury(II) cyanide solution and a precipitate of mercury:

$$Hg_2^{2+} + 2CN^- \rightarrow Hg\downarrow + Hg(CN)_2$$

Mercury(II) cyanide, though soluble, is virtually undissociated.

$SnCl_2$
↓
black Hg

10. *Tin(II) chloride* reduces mercury(I) ions to mercury metal, which appears in the form of a greyish-black precipitate:

$$Hg_2^{2+} + Sn^{2+} \rightarrow 2Hg\downarrow + Sn^{4+}$$

Mercury(II) ions react in a similar way.

Cu
↓
Hg deposit

11. *Polished copper sheet or copper coin* if a drop of mercury(I) nitrate is placed on a polished copper surface, a deposit of mercury metal is formed:

$$Cu + Hg_2^{2+} \rightarrow Cu^{2+} + 2Hg\downarrow$$

Rinsing, drying, and rubbing the surface with a dry cloth, a glittery, silverish spot is obtained. Heating the spot in a Bunsen-flame evaporates the mercury, and the red copper surface becomes visible again. Mercury(II) solutions react in a similar way.

Al
↓
amalgam and
white Al(OH)$_3$

12. *Aluminium sheet*   if a drop of mercury(I) nitrate is placed on a clean aluminium surface, aluminium amalgam is formed and aluminium ions pass into solution:

$$3Hg_2^{2+} + 2Al \rightarrow 2Al^{3+} + 6Hg\downarrow$$

The aluminium which is dissolved in the amalgam is oxidized rapidly by the oxygen of the air and a voluminous precipitate of aluminium hydroxide is formed. The remaining mercury amalgamates a further batch of aluminium, which again is oxidized, thus considerable amounts of aluminium are corroded.

Diphenylcarbazide
↓
violet complex

13. *1,5-Diphenylcarbazide (1% in alcohol)*

$$C{=}O \begin{cases} NH-NH-C_6H_5 \\ NH-NH-C_6H_5 \end{cases}$$

forms a violet-coloured compound with mercury(I) or mercury(II) ions, the composition of which is not quite understood. In the presence of 0.2M nitric acid the test is selective for mercury. Under these circumstances the sensitivity is 1 μg Hg_2^{2+} or Hg^{2+} with a concentration limit of 1 in 5×10^4.
Drop test   Impregnate a piece of filter paper with the freshly prepared reagent. Add 1 drop 0.4M nitric acid, and on top of the latter one drop of the test solution. In the presence of mercury a violet colour is observable. The test is most sensitive if the filter paper is left to dry at room temperature.

Atomic spectrometry

14. *Atomic spectrometric tests*   about 20 ppm mercury can be detected by emission, using a nitrous oxide/acetylene flame at 253.65 nm. By absorption, however, mercury cannot be reliably and easily detected with ordinary equipment. The methods used in environmental analysis for the determination of mercury require special devices and training; the discussion of these is beyond the scope of this book.
Mercury standard, 1000 ppm   dissolve 1.3534 g of analytical grade mercury(II) chloride $HgCl_2$ in water and dilute to 1 litre. To avoid adsorption by glass walls, keep the solution in a polythene bottle.

Dry test

15. *Dry test*   all compounds of mercury when heated with a large excess (7–8 times the bulk) of anhydrous sodium carbonate in a small dry test-tube yield a grey mirror, consisting of fine drops of mercury, in the upper part of the tube. The globules coalesce when they are rubbed with a glass rod.
   MERCURY VAPOUR IS EXTREMELY POISONOUS and not more than 0.1 gram of the substance should be used in the test.

3.6   Silver, Ag (A_r: 107.868)

Silver is a white, malleable, and ductile metal. It has a high density (10.5 g ml^{-1}) and melts at 960.5°C. It is insoluble in hydrochloric, dilute sulphuric (M) or dilute nitric (2M) acid. In more concentrated nitric acid (8M) (a) or in hot, concentrated sulphuric acid (b) it dissolves:

$$6Ag + 8HNO_3 \rightarrow 6Ag^+ + 2NO\uparrow + 6NO_3^- + 4H_2O \quad\quad (a)$$

$$2Ag + 2H_2SO_4 \rightarrow 2Ag^+ + SO_4^{2-} + SO_2\uparrow + 2H_2O \quad\quad (b)$$

Silver forms a monovalent ion in solution, which is colourless. Silver(II) compounds are unstable, but play an important role in silver-catalysed oxidation–reduction processes. Silver nitrate is readily soluble in water, silver acetate, nitrite and sulphate are less soluble, while all the other silver compounds are practically insoluble. Silver complexes are however soluble. Silver halides are sensitive to light; these characteristics are widely utilized in photography.

Ag$^+$ reactions

Reactions of silver(I) ions   A solution of silver nitrate (0.1M) can be used to study these reactions.

HCl
↓
white AgCl

1. *Dilute hydrochloric acid (or soluble chlorides)*   white precipitate of silver chloride

$$Ag^+ + Cl^- \rightarrow AgCl\downarrow$$

With concentrated hydrochloric acid precipitation does not occur. After decanting the liquid from over the precipitate, it can be dissolved in concentrated hydrochloric acid, when a dichloroargentate complex is formed:

$$AgCl\downarrow + Cl^- \rightleftarrows [AgCl_2]^-$$

On dilution with water, the equilibrium shifts back to the left and the precipitate reappears.

Dilute ammonia solution dissolves the precipitate to form the diammine-argentate complex ion:

$$AgCl\downarrow + 2NH_3 \rightleftarrows [Ag(NH_3)_2]^+ + Cl^-$$

Dilute nitric acid or hydrochloric acid neutralizes the excess ammonia, and the precipitate reappears because the equilibrium is shifted back towards the left.

The solution obtained after dissolving in ammonia should be acidified quickly with 2M nitric acid and disposed of. Otherwise, when set aside, a precipitate of silver nitride Ag_3N ('fulminating silver') is formed slowly, which **explodes readily even in a wet form!**

Potassium cyanide (**POISON**) dissolves the precipitate with formation of the dicyanoargentate complex:

$$AgCl\downarrow + 2CN^- \rightarrow [Ag(CN)_2]^- + Cl^-$$

The safest way to study this reaction is as follows: decant the liquid from the precipitate, and wash it 2–3 times with water by decantation. Then apply the reagent.

Sodium thiosulphate dissolves the precipitate with the formation of a dithiosulphatoargentate complex:

$$AgCl\downarrow + 2S_2O_3^{2-} \rightarrow [Ag(S_2O_3)_2]^{3-} + Cl^-$$

This reaction takes place when fixing photographic negatives or positive prints after development.

Sunlight or ultraviolet irradiation decomposes the silver chloride precipitate, which turns to greyish or black owing to the formation of silver metal:

$$2AgCl\downarrow \xrightarrow{(h\nu)} 2Ag\downarrow + Cl_2\uparrow$$

The reaction is slow and the actual reaction mechanism is very complicated. Other silver halides show similar behaviour. Photography is based on these reactions. In the camera these processes are only initiated; the photographic material has to be 'developed' to complete the reaction. Greyish or black silver particles appear on places irradiated by light; a 'negative' image of the object is therefore obtained. The excess of silver halide has to be removed (to make the developed negative insensitive to light) by fixation.

H_2S
↓
black Ag_2S

2. *Hydrogen sulphide (gas or saturated aqueous solution) in neutral or acidic medium*   black precipitate of silver sulphide

$$2Ag^+ + H_2S \rightarrow Ag_2S\downarrow + 2H^+$$

Hot concentrated nitric acid decomposes the silver sulphide, and sulphur remains in the form of a white precipitate:

$$3Ag_2S\downarrow + 8HNO_3 \rightarrow S\downarrow + 2NO\uparrow + 6Ag^+ + 6NO_3^- + 4H_2O$$

If the mixture is heated with concentrated nitric acid for a considerable time, sulphur is oxidized to sulphate and the precipitate disappears:

$$S\downarrow + 2HNO_3 \rightarrow SO_4^{2-} + 2NO\uparrow + 2H^+$$

The precipitate is insoluble in ammonium sulphide, ammonium polysulphide, ammonia, potassium cyanide, or sodium thiosulphate. Silver sulphide can be precipitated from solutions containing diammine-, dicyanato- or dithiosulphato-argentate complexes with hydrogen sulphide.

NH_3
↓
brown Ag_2O

3. *Ammonia solution*   brown precipitate of silver oxide

$$2Ag^+ + 2NH_3 + H_2O \rightarrow Ag_2O\downarrow + 2NH_4^+$$

The reaction reaches an equilibrium and therefore precipitation is incomplete at any stage. (If ammonium nitrate is present in the original solution or the solution is strongly acidic no precipitation occurs.) The precipitate dissolves in excess of the reagent, and diammineargentate complex ions are formed:

Dissolves in excess

$$Ag_2O\downarrow + 4NH_3 + H_2O \rightarrow 2[Ag(NH_3)_2]^+ + 2OH^-$$

The solution obtained after dissolving in ammonia should be discarded quickly to avoid a serious **explosion!** (Cf. Section 3.6, reaction 1.)

NaOH
↓
brown Ag_2O

4. *Sodium hydroxide*   brown precipitate of silver oxide:

$$2Ag^+ + 2OH^- \rightarrow Ag_2O\downarrow + H_2O$$

A well-washed suspension of the precipitate shows a slight alkaline reaction owing to the hydrolysis equilibrium:

$$Ag_2O\downarrow + H_2O \rightleftarrows 2Ag(OH)_2\downarrow \rightleftarrows 2Ag^+ + 2OH^-$$

The precipitate is insoluble in excess reagent.

The precipitate dissolves in ammonia solution (a) and in nitric acid (b):

$$Ag_2O\downarrow + 4NH_3 + H_2O \rightarrow 2[Ag(NH_3)_2]^+ + 2OH^- \qquad \text{(a)}$$

$$Ag_2O\downarrow + 2H^+ \rightarrow 2Ag^+ + H_2O \qquad \text{(b)}$$

KI
↓
yellow Ag$_2$I

5. *Potassium iodide*   yellow precipitate of silver iodide:

$$Ag^+ + I^- \rightarrow AgI\downarrow$$

The precipitate is insoluble in dilute or concentrated ammonia, but dissolves readily in potassium cyanide (**POISON**) (a) and in sodium thiosulphate (b):

$$AgI + 2CN^- \rightarrow [Ag(CN)_2]^- + I^- \qquad \text{(a)}$$

$$AgI + 2S_2O_3^{2-} \rightarrow [Ag(S_2O_3)_2]^{3-} + I^- \qquad \text{(b)}$$

K$_2$CrO$_4$
↓
red Ag$_2$CrO$_4$

6. *Potassium chromate in neutral solution*   red precipitate of silver chromate:

$$2Ag^+ + CrO_4^{2-} \rightarrow Ag_2CrO_4\downarrow$$

Spot test   place a drop of the test solution on a watch glass or on a spot plate, add a drop of ammonium carbonate solution and stir (this renders any mercury(I) or lead ions unreactive by precipitation as the highly insoluble carbonates). Remove one drop of the clear liquid and place it on drop-reaction paper together with a drop of the potassium chromate reagent. A red ring of silver chromate is obtained.

The reaction can be used for a microscopic test when a piece of potassium chromate crystal has to be dropped into the test solution. The formation of needle-like red crystals of silver chromate can be observed distinctly.

The precipitate is soluble in dilute nitric acid (a) and in ammonia solution (b):

$$2Ag_2CrO_4\downarrow + 2H^+ \rightleftarrows 4Ag^+ + Cr_2O_7^{2-} + H_2O \qquad \text{(a)}$$

$$Ag_2CrO_4\downarrow + 4NH_3 \rightarrow 2[Ag(NH_3)_2]^+ + CrO_4^{2-} \qquad \text{(b)}$$

The acidified solution turns to orange because of the formation of dichromate.

KCN
↓
white AgCN

7. *Potassium cyanide* (**POISON**) when added dropwise to a neutral solution of silver nitrate: white precipitate of silver cyanide:

$$Ag^+ + CN^- \rightarrow AgCN\downarrow$$

Dissolves in excess

When potassium cyanide is added in excess, the precipitate disappears owing to the formation of dicyanoargentate ions:

$$AgCN\downarrow + CN^- \rightarrow [Ag(CN)_2]^-$$

Na$_2$CO$_3$
↓
white Ag$_2$CO$_3$

8. *Sodium carbonate*   yellowish-white precipitate of silver carbonate:

$$2Ag^+ + CO_3^{2-} \rightarrow Ag_2CO_3\downarrow$$

Turns brown

When heating, the precipitate decomposes and brown silver oxide precipitate is formed:

$$Ag_2CO_3\downarrow \rightarrow Ag_2O\downarrow + CO_2\uparrow$$

Nitric acid and ammonia solution dissolve the precipitate.

Na$_2$HPO$_4$
↓
yellow Ag$_3$PO$_4$

9. *Disodium hydrogen phosphate in neutral solution* yellow precipitate of silver phosphate:

$$3Ag^+ + HPO_4^{2-} \rightarrow Ag_3PO_4\downarrow + H^+$$

Nitric acid and ammonia solution dissolve the precipitate.

Silver mirror test

10. *Hydrazine sulphate (saturated)* when added to a solution of diammine-argentate ions, forms finely divided silver metal, while gaseous nitrogen is evolved:

$$4[Ag(NH_3)_2]^+ + H_2N-NH_2 \cdot H_2SO_4 \rightarrow$$

$$\rightarrow 4Ag\downarrow + N_2\uparrow + 6NH_4^+ + 2NH_3 + SO_4^{2-}$$

If the vessel in which the reaction is carried out is clean, silver adheres to the glass walls forming an attractive mirror.

Procedure fill two-thirds of a test-tube with chromosulphuric acid (concentrated), and set aside overnight (note that chromosulphuric acid is **highly corrosive**). Next day empty the test-tube, rinse cautiously with running cold water, then with distilled water. Into this test-tube pour 2 ml silver nitrate (0.1M) and 2 ml distilled water. Then add dilute ammonia (2M) dropwise, mixing the solution vigorously by shaking, until the last traces of the silver oxide precipitate disappear. Then add 2 ml saturated hydrazine sulphate solution and shake the mixture vigorously. The silver mirror forms within a few seconds. **The solution should be discarded after the test** (cf. reaction 1). The silver mirror can be removed most easily by dissolving it in nitric acid (8M).

Rhodanine
↓
violet ppt

11. *4-Dimethylaminobenzylidenerhodanine (in short: rhodanine reagent, 0.3% solution in acetone)* reddish-violet precipitate in slightly acid solutions:

Mercury, copper, gold, platinum, and palladium salts form similar compounds and therefore interfere.

Spot test to 1 drop of the test solution add 1 drop nitric acid (2M), then 1 drop of the reagent. A red-violet precipitate or stain is formed if silver ions are present. Alternatively, the test may be performed on a spot plate, or in a semimicro test tube; in the latter case the excess of the reagent is extracted with diethyl ether or amyl alcohol, when violet specks of the silver complex will be visible under the yellow solvent layer.

In the presence of mercury, gold, platinum, or palladium first add 1 drop potassium cyanide (10%, **HIGHLY POISONOUS**) solution to the test solution then follow the procedure given above.

To detect silver in a mixture of lead chloride, mercury(I) chloride, and silver chloride (Group I), the mixture is treated with 10% potassium cyanide solution whereby mercury(II) cyanide, mercury, and dicyanoargentate

[Ag(CN)$_2$]$^-$ are formed: after filtration (or centrifugation), a little of the clear filtrate is treated on a spot plate with a drop of the reagent and 2 drops nitric acid (2M). A red colouration is formed in the presence of Ag in weakly acid solution.

Atomic spectrometry

12. *Atomic spectrometric tests*

(a) *Flame atomic emission spectrometry*   silver cannot be easily detected with the air/acetylene flame; however, using a nitrous oxide/acetylene flame good emission readings are obtainable at 328.07 nm, when 10 ppm silver can be identified.

(b) *Flame atomic absorption spectrometry*   using an air/acetylene flame and a silver hollow-cathode lamp 1 ppm silver can be easily detected at 328.07 nm.

(c) *Silver standard, 1000 ppm*   dissolve 1.5747 g analytical grade silver nitrate AgNO$_3$ in water and dilute to 1 litre.

Dry test

13. *Dry test (blowpipe test)*   when a silver salt is heated with alkali carbonate on charcoal, a white malleable bead without an incrustation of the oxide results; this is readily soluble in nitric acid. The solution is immediately precipitated by dilute hydrochloric acid, but not by very dilute sulphuric acid (difference from lead).

3.7   Second group of cations: mercury(II), lead(II), bismuth(III), copper(II), cadmium(II), arsenic(III) and (V), antimony(III) and (V), and tin(II) and (IV)

Second group of cations
Hg(II), Pb(II), Bi(III), Cu(II), Cd(II), As(III), As(V), Sb(III), Sb(V), Sn(II), Sn(IV)

Group reagent   hydrogen sulphide (gas or saturated aqueous solution).

Group reaction   precipitates of different colours; mercury(II) sulphide HgS (black), lead(II) sulphide PbS (black), copper(II) sulphide CuS (black), cadmium sulphide CdS (yellow), bismuth(III) sulphide Bi$_2$S$_3$ (brown), arsenic(III) sulphide As$_2$S$_3$ (yellow), arsenic(V) sulphide As$_2$S$_5$ (yellow), antimony(III) sulphide Sb$_2$S$_3$ (orange), antimony(V) sulphide Sb$_2$S$_5$ (orange), tin(II) sulphide SnS (brown), and tin(IV) sulphide SnS$_2$ (yellow).

Cations of the second group are traditionally divided into two sub-groups; the copper sub-group and the arsenic sub-group. The basis of this division is the solubility of the sulphide precipitates in ammonium polysulphide. While sulphides of the copper sub-group are insoluble in this reagent, those of the arsenic sub-group do dissolve with the formation of thiosalts.

Cu sub-group

The copper sub-group consists of mercury(II), lead(II), bismuth(III), copper(II), and cadmium(II). Although the bulk of lead(II) ions are precipitated with dilute hydrochloric acid together with other ions of Group I, this precipitation is rather incomplete owing to the relatively high solubility of lead(II) chloride. In the course of systematic analysis therefore lead ions will still be present when the precipitation of the second group of cations is the task. Reactions of lead(II) ions were already described with those of the cations of the first group (see Section 3.4).

The chlorides, nitrates, and sulphates of the cations of the copper sub-group are quite soluble in water. The sulphides, hydroxides, and carbonates are insoluble. Some of the cations of the copper sub-group (mercury(II), copper(II), and cadmium(II)) tend to form complexes (ammonia, cyanide ions, etc.).

As sub-group

The arsenic sub-group consists of the ions arsenic(III), arsenic(V), antimony(III), antimony(V), tin(II) and tin(IV). These ions have amphoteric character: their oxides form salts both with acids and bases. Thus, arsenic(III) oxide can be dissolved in hydrochloric acid (6M), and arsenic(III) cations are formed:

$$As_2O_3 + 6HCl \rightarrow 2As^{3+} + 6Cl^- + 3H_2O$$

At the same time arsenic(III) oxide dissolves in sodium hydroxide (2M), forming arsenite anions:

$$As_2O_3 + 6OH^- \rightarrow 2AsO_3^{3-} + 3H_2O$$

The dissolution of sulphides in ammonium polysulphide can be regarded as the formation of thiosalts from anhydrous thioacids. Thus the dissolution of arsenic(III) sulphide (anhydrous thioacid) in ammonium sulphide (anhydrous thiobase) yields the formation of ammonium and thioarsenite ions (ammonium thioarsenite: a thiosalt):

$$As_2S_3\downarrow + 3S^{2-} \rightarrow 2AsS_3^{3-}$$

All the sulphides of the arsenic sub-group dissolve in (colourless) ammonium sulphide except tin(II) sulphide: to dissolve the latter, ammonium polysulphide is needed, which acts partly as an oxidizing agent, thiostannate ions being formed:

$$SnS\downarrow + S_2^{2-} \rightarrow SnS_3^{2-}$$

Note that while tin is bivalent in the tin(II) sulphide precipitate, it is tetravalent in the thiostannate ion.

Arsenic(III), antimony(III), and tin(II) ions can be oxidized to arsenic(V), antimony(V), and tin(IV) ions respectively. On the other hand, the latter three can be reduced by proper reducing agents. The oxidation-reduction potentials of the arsenic(V)–arsenic(III) and antimony(V)–antimony(III) systems vary with pH, therefore the oxidation or reduction of the relevant ions can be assisted by choosing an appropriate pH for the reaction.

3.8   Mercury, Hg (A_r: 200.59) – mercury(II)

The most important physical and chemical properties of the metal were described in Section 3.5.

Hg(II) reactions

Reactions of mercury(II) ions   The reactions of mercury(II) ions can be studied with a dilute solution of mercury(II) nitrate (0.05M). **All compounds of mercury are toxic!**

H_2S
↓
black HgS

1. *Hydrogen sulphide (gas or saturated aqueous solution)*   in the presence of dilute (M) hydrochloric acid, initially a white precipitate of mercury(II) chlorosulphide forms, which reacts with further amounts of hydrogen

sulphide and finally a black precipitate of mercury(II) sulphide is formed:

$$3Hg^{2+} + 2Cl^- + 2H_2S \rightarrow Hg_3S_2Cl_2\downarrow + 4H^+$$

$$Hg_3S_2Cl_2\downarrow + H_2S \rightarrow 3HgS\downarrow + 2H^+ + 2Cl^-$$

Mercury(II) sulphide is one of the least soluble precipitates known ($K_s = 4 \times 10^{-54}$).

The precipitate is insoluble in water, hot dilute nitric acid, alkali hydroxides, or (colourless) ammonium sulphide.

Sodium sulphide (2M) dissolves the precipitate when the disulphomercurate(II) complex ion is formed:

$$HgS\downarrow + S^{2-} \rightarrow [HgS_2]^{2-}$$

Adding ammonium chloride to the solution, mercury(II) sulphide precipitates again.

Aqua regia dissolves the precipitate:

$$3HgS\downarrow + 6HCl + 2HNO_3 \rightarrow 3HgCl_2 + 3S\downarrow + 2NO\uparrow + 4H_2O$$

Mercury(II) chloride is practically undissociated under these circumstances. Sulphur remains as a white precipitate, which however dissolves readily if the solution is heated, to form sulphuric acid:

$$2HNO_3 + S\downarrow \rightarrow SO_4^{2-} + 2H^+ + 2NO\uparrow$$

NH₃
↓
white ppt of mixed composition

2. *Ammonia solution* white precipitate with a mixed composition; essentially it consists of mercury(II) oxide and mercury(II) amidonitrate:

$$2Hg^{2+} + NO_3^- + 4NH_3 + H_2O \rightarrow HgO\cdot Hg(NH_2)NO_3\downarrow + 3NH_4^+$$

The salt, like most of the mercury compounds, sublimes at atmospheric pressure.

NaOH
↓
yellow HgO

3. *Sodium hydroxide when added in small amounts* brownish-red precipitate with varying composition; if added in stoichiometric amounts the precipitate turns to yellow when mercury(II) oxide is formed:

$$Hg^{2+} + 2OH^- \rightarrow HgO\downarrow + H_2O$$

The precipitate is insoluble in excess sodium hydroxide. Acids dissolve the precipitate readily.

This reaction is characteristic for mercury(II) ions, and can be used to differentiate mercury(II) from mercury(I).

KI
↓
red HgI₂

4. *Potassium iodide when added slowly to the solution* red precipitate of mercury(II) iodide:

$$Hg^{2+} + 2I^- \rightarrow HgI_2\downarrow$$

Dissolves in excess

The precipitate dissolves in excess reagent, when colourless tetraiodomercurate(II) ions are formed:

$$HgI_2 + 2I^- \rightarrow [HgI_4]^{2-}$$

An alkaline solution of potassium tetraiodomercurate(II) serves as a selective and sensitive reagent for ammonium ions (Nessler's reagent, cf. Section 3.38, reaction 2).

KCN no effect

5. *Potassium cyanide* (**POISON**) does not cause any change in dilute solutions (difference from other ions of the copper sub-group).

SnCl$_2$
↓
white Hg$_2$Cl$_2$

6. *Tin(II) chloride* when added in moderate amounts: white, silky precipitate of mercury(I) chloride (calomel):

$$2Hg^{2+} + Sn^{2+} + 2Cl^- \rightarrow Hg_2Cl_2\downarrow + Sn^{4+}$$

This reaction is widely used to remove the excess of tin(II) ions, used for prior reduction, in oxidation-reduction titrations.

With excess, black Hg

If more reagent is added, mercury(I) chloride is further reduced and black precipitate of mercury is formed:

$$Hg_2Cl_2\downarrow + Sn^{2+} \rightarrow 2Hg\downarrow + Sn^{4+} + 2Cl^-$$

Cu
↓
Hg deposit

7. *Copper sheet or coin* reduces mercury(II) ions to the metal:

$$Cu + Hg^{2+} \rightarrow Cu^{2+} + Hg\downarrow$$

For practical hints for the test see Section 3.5, reaction 11.

8. *Diphenylcarbazide* reacts with mercury(II) ions in a similar way to mercury(I). For details see Section 3.5, reaction 13.

Co(SCN)$_2$
↓
blue ppt

9. *Cobalt(II) thiocyanate test* to the test solution add an equal volume of the reagent (about 10%, freshly prepared), and stir the wall of the vessel with a glass rod. A deep-blue crystalline precipitate of cobalt tetrathiocyanatomercurate(II) is formed:

$$Hg^{2+} + Co^{2+} + 4SCN^- \rightarrow Co[Hg(SCN)_4]\downarrow$$

Drop test place a drop of the test solution on a spot plate, add a small crystal of ammonium thiocyanate followed by a little solid cobalt(II) acetate. A blue colour is produced in the presence of mercury(II) ions.
Sensitivity 0.5 µg Hg^{2+}.
Concentration limit 1 in 10^5.

10. *Atomic spectrometric tests* these were discussed in Section 3.5, reaction 14.

11. *Dry test* all mercury compounds, irrespective of their valency state, form mercury metal when heated with excess anhydrous sodium carbonate. For practical hints see Section 3.5, reaction 15.

3.9 Bismuth, Bi (A_r: 208.98)

Bismuth is a brittle, crystalline, reddish-white metal. It melts at 271.5°C. It is insoluble in hydrochloric acid because of its standard potential (0.2 V), but dissolves in oxidizing acids such as concentrated nitric acid (a), aqua regia (b), or hot, concentrated sulphuric acid (c).

$$2Bi + 8HNO_3 \rightarrow 2Bi^{3+} + 6NO_3^- + 2NO\uparrow + 4H_2O \tag{a}$$

$$Bi + 3HCl + HNO_3 \rightarrow Bi^{3+} + 3Cl^- + NO\uparrow + 2H_2O \tag{b}$$

$$2Bi + 6H_2SO_4 \rightarrow 2Bi^{3+} + 3SO_4^{2-} + 3SO_2\uparrow + 6H_2O \tag{c}$$

Bismuth forms tervalent and pentavalent ions. Tervalent bismuth ion Bi^{3+} is the most common. The hydroxide, $Bi(OH)_3$, is a weak base; bismuth salts therefore hydrolyse readily, when the following process occurs:

$$Bi^{3+} + H_2O \rightleftarrows BiO^+ + 2H^+$$

The bismuthyl ion, BiO^+, forms insoluble salts, like bismuthyl chloride, $BiOCl$, with most ions. If we want to keep bismuth ions in solution, we must acidify the solution, when the above equilibrium shifts towards the left.

Pentavalent bismuth forms the bismuthate BiO_3^- ion. Most of its salts are insoluble in water.

Bi(III) reactions

Reactions of bismuth(III) ions These reactions can be studied with a 0.2M solution of bismuth(III) nitrate, which contains about 3–4% nitric acid.

H_2S
↓
black Bi_2S_3

1. *Hydrogen sulphide (gas or saturated aqueous solution)* black precipitate of bismuth sulphide:

$$2Bi^{3+} + 3H_2S \rightarrow Bi_2S_3\downarrow + 6H^+$$

The precipitate is insoluble in cold, dilute acid and in ammonium sulphide. Boiling concentrated hydrochloric acid dissolves the precipitate, when hydrogen sulphide gas is liberated:

$$Bi_2S_3\downarrow + 6HCl \rightarrow 2Bi^{3+} + 6Cl^- + 3H_2S\uparrow$$

Hot dilute nitric acid dissolves bismuth sulphide, leaving behind sulphur in the form of a white precipitate:

$$Bi_2S_3\downarrow + 8H^+ + 2NO_3^- \rightarrow 2Bi^{3+} + 3S\downarrow + 2NO\uparrow + 4H_2O$$

NH_3
↓
white basic salt

2. *Ammonia solution* white basic salt of variable composition. The approximate chemical reaction is:

$$Bi^{3+} + NO_3^- + 2NH_3 + 2H_2O \rightarrow Bi(OH)_2NO_3\downarrow + 2NH_4^+$$

The precipitate is insoluble in excess reagent (distinction from copper or cadmium).

NaOH
↓
white $Bi(OH)_3$

3. *Sodium hydroxide* white precipitate of bismuth(III) hydroxide:

$$Bi^{3+} + 3OH^- \rightarrow Bi(OH)_3\downarrow$$

The precipitate is very slightly soluble in excess reagent in cold solution, 2–3 mg bismuth dissolves per 100 ml sodium hydroxide (2M). The precipitate is soluble in acids:

$$Bi(OH)_3\downarrow + 3H^+ \rightarrow Bi^{3+} + 3H_2O$$

When boiled, the precipitate loses water and turns yellowish-white:

$$Bi(OH)_3\downarrow \rightarrow BiO\cdot OH\downarrow + H_2O$$

Both the hydrated and the dehydrated precipitate can be oxidized by 4–6 drops of concentrated hydrogen peroxide when yellowish-brown bismuthate ions are formed:

$$BiO \cdot OH\downarrow + H_2O_2 \rightarrow BiO_3^- + H^+ + H_2O$$

KI
↓
black BiI₃

4. *Potassium iodide when added dropwise*   black precipitate of bismuth(III) iodide:

$$Bi^{3+} + 3I^- \rightarrow BiI_3\downarrow$$

Dissolves in excess as
orange complex

The precipitate dissolves readily in excess reagent, when orange-coloured tetraiodobismuthate ions are formed:

$$BiI_3\downarrow + I^- \rightleftarrows [BiI_4]^-$$

When diluted with water, the above reaction is reversed and black bismuth iodide is reprecipitated. Heating the precipitate with water, it turns orange, owing to the formation of bismuthyl iodide:

$$BiI_3\downarrow + H_2O \rightarrow BiOI\downarrow + 2H^+ + 2I^-$$

Sn(II) in
alkaline medium
↓
black Bi

5. *Sodium tetrahydroxostannate(II) (0.125M, freshly prepared)*   in cold solution bismuth(III) ions are reduced to bismuth metal, which separates in the form of a black precipitate. First the sodium hydroxide present in the reagent reacts with bismuth(III) ions (cf. reaction 3), bismuth(III) hydroxide is then reduced by tetrahydroxostannate(II) ions to form bismuth metal and hexahydroxostannate(IV) ions:

$$Bi^{3+} + 3OH^- \rightarrow Bi(OH)_3\downarrow$$

$$2Bi(OH)_3\downarrow + 3[Sn(OH)_4]^{2-} \rightarrow 2Bi\downarrow + 3[Sn(OH)_6]^{2-}$$

The reagent must be freshly prepared, and the test must be carried out in cold solution.

Test by induced reaction   in the absence of bismuth(III) ions, the reaction between tetrahydroxoplumbate(II) ions (cf. Section 3.4, reaction 4) and tetrahydroxostannate(II) is slow:

$$[Pb(OH)_4]^{2-} + [Sn(OH)_4]^{2-} \xrightarrow{\text{(slow)}} Pb\downarrow + [Sn(OH)_6]^{2-} + 2OH^-$$

With dilute solutions (using 0.25M lead nitrate and 0.125M sodium tetrahydroxostannate(II) reagent), the formation of the black precipitate of lead metal is not observable within an hour. In the presence of bismuth the reaction is accelerated; at the same time bismuth is also precipitated:

$$[Pb(OH)_4]^{2-} + [Sn(OH)_4]^{2-} \xrightarrow{\text{(fast)}} Pb\downarrow + [Sn(OH)_6]^{2-} + 2OH^-$$

$$2Bi(OH)_3\downarrow + 3[Sn(OH)_4]^{2-} \xrightarrow{\text{(fast)}} 2Bi\downarrow + 3[Sn(OH)_6]^{2-}$$

Induced reactions

Such reactions are called **induced reactions** (that is, bismuth induces the reduction of lead). They are distinguished from catalytic processes by the fact that the inductor is used up itself in the course of the reaction (and not regenerated, as in a catalytic process). Induced reactions are quite common among oxidation-reduction processes.

Silver, copper, and mercury interfere with the reaction. Copper can be deactivated by the addition of some potassium cyanide.

H$_2$O
↓
white basic salt

6. *Water* when a solution of a bismuth salt is poured into a large volume of water, a white precipitate of the corresponding basic salt is produced. This is soluble in dilute mineral acids, but is insoluble in tartaric acid (distinction from antimony) and in alkali hydroxides (distinction from tin).

$$Bi^{3+} + NO_3^- + H_2O \rightarrow BiO(NO_3)\downarrow + 2H^+$$

$$Bi^{3+} + Cl^- + H_2O \rightarrow BiO \cdot Cl\downarrow + 2H^+$$

Pyrogallol
↓
yellow ppt

7. *Pyrogallol (10%, freshly prepared) reagent* when added in slight excess to a hot, faintly acid solution of bismuth ions produces a yellow precipitate of bismuth pyrogallate:

$$Bi^{3+} + C_6H_3(OH)_3 \rightarrow Bi(C_6H_3O_3)\downarrow + 3H^+$$

It is best to neutralize the test solution first by ammonia against litmus paper, then add some drops of dilute nitric acid, and then the reagent. The test is a very sensitive one. Antimony interferes and should be absent.

Specific test with cinchonine + KI

8. *Cinchonine–potassium iodide reagent (1%)* orange-red colouration or precipitate in dilute acid solution.
Test on filter paper moisten a piece of drop-reaction paper with the reagent and place a drop of the slightly acid test solution upon it. An orange-red spot is obtained.
   Sensitivity 0.15 μg Bi.
   Concentration limit 1 in 350 000.
The test may also be carried out on a spot plate.
   Lead, copper, and mercury salts interfere because they react with the iodide. Nevertheless, bismuth may be detected in the presence of salts of these metals as they diffuse at different rates through the capillaries of the paper, and are fixed in distinct zones. When a drop of the test solution containing bismuth, lead, copper, and mercury ions is placed upon absorbent paper impregnated with the reagent, four zones can be observed: (i) a white central ring, containing the mercury; (ii) an orange ring, due to bismuth; (iii) a yellow ring of lead iodide; and (iv) a brown ring of iodine liberated by the reaction with copper. The thicknesses of the rings will depend upon the relative concentrations of the various metals.

Test with 8-hydroxyquinoline + KI

9. *8-Hydroxyquinoline (5%) and potassium iodide (6M) in acidic medium* red precipitate of 8-hydroxyquinoline-tetraiodobismuthate

$$Bi^{3+} + C_9H_7ON + H^+ + 4I^- \rightarrow C_9H_7ON \cdot HBiI_4\downarrow$$

If other halide ions are absent, the reaction is characteristic for bismuth.

Atomic spectrometry

10. *Atomic spectrometric tests*
   (a) *Flame atomic emission spectrometry* about 10 ppm bismuth can be detected in an air/acetylene flame at 223.06 nm; the test becomes more sensitive in a nitrous oxide/acetylene flame.
   (b) *Flame atomic absorption spectrometry* using an air/acetylene flame, 1 ppm bismuth can be identified at 223.06 nm, with a bismuth hollow-cathode lamp.

(c) *Bismuth standard, 1000 ppm* dissolve 1.0000 g analytical grade bismuth metal in 50 ml concentrated nitric acid, and dilute the solution to 1 litre.

Dry test

11. *Dry test (blowpipe test)* when a bismuth compound is heated on charcoal with sodium carbonate in the blowpipe flame, a brittle bead of metal, surrounded by a yellow incrustation of the oxide, is obtained.

3.10 Copper, Cu (A_r: 63.55)

Copper is a light-red metal which is soft, malleable, and ductile. It melts at 1038°C. Because of its positive standard electrode potential (+0.34 V for the Cu/Cu^{2+} couple), it is insoluble in hydrochloric acid and in dilute sulphuric acid, although in the presence of oxygen some dissolution might take place. Moderately concentrated nitric acid (8M) dissolves copper readily:

$$3Cu + 8HNO_3 \rightarrow 3Cu^{2+} + 6NO_3^- + 2NO\uparrow + 4H_2O$$

Hot, concentrated sulphuric acid also dissolves copper:

$$Cu + 2H_2SO_4 \rightarrow Cu^{2+} + SO_4^{2-} + SO_2\uparrow + 2H_2O$$

Copper is readily dissolved in aqua regia:

$$3Cu + 6HCl + 2HNO_3 \rightarrow 3Cu^{2+} + 6Cl^- + 2NO\uparrow + 4H_2O$$

There are two series of copper compounds. Copper(I) compounds are derived from the red copper(I) oxide Cu_2O and contain the copper(I) ion Cu^+. These compounds are colourless. Most copper(I) salts are insoluble in water, and their behaviour generally resembles that of the silver(I) compounds. They are readily oxidized to copper(II) compounds, which are derivable from the black copper(II) oxide, CuO. Copper(II) compounds contain the copper(II) ion, Cu^{2+}. Copper(II) salts are generally blue both in solid, hydrated form and in dilute aqueous solution; the colour is characteristic really for the tetraaquocuprate(II) ion $[Cu(H_2O)_4]^{2+}$ only: The limit of visibility of the colour of the tetraaquocuprate(II) complex (i.e. the colour of copper(II) ions in aqueous solutions) is 500 µg in a concentration limit of 1 in 10^4. Anhydrous copper(II) salts, like anhydrous copper(II) sulphate $CuSO_4$, are white (or slightly yellow). In aqueous solutions the blue tetraaquo complex ion is always present; for the sake of simplicity however in this text it will be denoted as the simple copper(II) ion Cu^{2+}.

In analytical practice only the copper(II) ion is important; therefore only its reactions are described.

Cu(II) reactions

Reactions of copper(II) ions These reactions can be studied with a 0.25M solution of copper(II) sulphate.

H_2S
↓
black CuS

1. *Hydrogen sulphide (gas or saturated aqueous solution)* black precipitate of copper(II) sulphide:

$$Cu^{2+} + H_2S \rightarrow CuS\downarrow + 2H^+$$

$K_s(CuS; 25\,°C) = 10^{-44}$.

Sensitivity 1 µg Cu^{2+}.

Concentration limit 1 in 5×10^6.

The solution must be acidic (M in hydrochloric acid) in order to obtain a crystalline, easily filterable precipitate. In the absence of acid, or in very slightly acid solutions, a colloidal, brownish-black precipitate or colouration is obtained. By adding some acid and boiling, coagulation can be achieved.

The precipitate is insoluble in boiling dilute (M) sulphuric acid (distinction from cadmium), in sodium hydroxide, sodium sulphide and ammonium sulphide. It is only very slightly soluble in polysulphides.

Hot, concentrated nitric acid dissolves the copper(II) sulphide, leaving behind sulphur as a white precipitate:

$$3CuS\downarrow + 8HNO_3 \rightarrow 3Cu^{2+} + 6NO_3^- + 3S\downarrow + 2NO\uparrow + 4H_2O$$

When boiled for longer, sulphur is oxidized to sulphuric acid and a clear, blue solution is obtained:

$$S\downarrow + 2HNO_3 \rightarrow 2H^+ + SO_4^{2-} + 2NO\uparrow$$

Potassium cyanide (**POISON**) dissolves the precipitate, when colourless tetracyanocuprate(I) ions and disulphide ions are formed:

$$2CuS\downarrow + 8CN^- \rightarrow 2[Cu(CN)_4]^{3-} + S_2^{2-}$$

Note that this is an oxidation-reduction process (copper is reduced, sulphur is oxidized) coupled with a formation of a complex.

When exposed to air, in the moist state, copper(II) sulphide tends to oxidize to copper(II) sulphate:

$$CuS\downarrow + 2O_2 \rightarrow CuSO_4$$

and therefore becomes water soluble. A considerable amount of heat is liberated during this process. A filter paper with copper(II) sulphide precipitate on it should never be thrown into a waste container, with paper or other inflammable substances in it. Instead the precipitate should be washed away first with running water.

NH_3
↓
blue basic salt

2. *Ammonia solution when added sparingly* blue precipitate of a basic salt (basic copper sulphate):

$$2Cu^{2+} + SO_4^{2-} + 2NH_3 + 2H_2O \rightarrow Cu(OH)_2 \cdot CuSO_4\downarrow + 2NH_4^+$$

Dissolves in excess as blue tetrammine complex

which is soluble in excess reagent, when a deep blue colouration is obtained owing to the formation of tetramminocuprate(II) complex ions:

$$Cu(OH)_2 \cdot CuSO_4\downarrow + 8NH_3 \rightarrow 2[Cu(NH_3)_4]^{2+} + SO_4^{2-} + 2OH^-$$

If the solution contains ammonium salts (or it was highly acidic and larger amounts of ammonia were used up for its neutralization), precipitation does not occur at all, but the blue colour is formed right away.

The reaction is characteristic for copper(II) ions in the absence of nickel.

3. *Sodium hydroxide in cold solution* blue precipitate of copper(II) hydroxide:

$$Cu^{2+} + 2OH^- \rightarrow Cu(OH)_2\downarrow$$

The precipitate is insoluble in excess reagent.

When heated, the precipitate is converted to black copper(II) oxide by dehydration:

$$Cu(OH)_2\downarrow \rightarrow CuO\downarrow + H_2O$$

In the presence of a solution of tartaric acid or of citric acid, copper(II) hydroxide is not precipitated by solutions of caustic alkalis, but the solution is coloured an intense blue. If the alkaline solution is treated with certain reducing agents, such as hydroxylamine, hydrazine, glucose, and acetaldehyde, yellow copper(I) hydroxide is precipitated from the warm solution. It is converted into red copper(I) oxide Cu_2O on boiling. The alkaline solution of copper(II) salt containing tartaric acid is usually known as Fehling's solution; it contains the complex ion $[Cu(C_4H_2O_6)_2]^{2-}$.

4. *Potassium iodide* precipitates copper(I) iodide, which is white, but the solution is intensely brown because of the formation of tri-iodide ions (iodine):

$$2Cu^{2+} + 5I^- \rightarrow 2CuI\downarrow + I_3^-$$

Adding an excess of sodium thiosulphate to the solution, tri-iodide ions are reduced to colourless iodide ions and the white colour of the precipitate becomes visible. The reduction with thiosulphate yields tetrathionate ions:

$$I_3^- + 2S_2O_3^{2-} \rightarrow 3I^- + S_4O_6^{2-}$$

These reactions are used in quantitative analysis for the iodometric determination of copper.

5. *Potassium cyanide* (**POISON**) when added sparingly forms first a yellow precipitate of copper(II) cyanide:

$$Cu^{2+} + 2CN^- \rightarrow Cu(CN)_2\downarrow$$

The precipitate quickly decomposes into white copper(I) cyanide and cyanogen (HIGHLY **POISONOUS** GAS):

$$2Cu(CN_2)\downarrow \rightarrow 2CuCN\downarrow + (CN)_2\uparrow$$

Excess reagent dissolves the precipitate, and the colourless tetracyanocuprate(I) complex is formed:

$$CuCN\downarrow + 3CN^- \rightarrow [Cu(CN)_4]^{3-}$$

The complex is so stable (i.e. the concentration of copper(I) ions is so low) that hydrogen sulphide cannot precipitate copper(I) sulphide from this solution (distinction from cadmium, cf. Section 3.11, reactions 1 and 4).

6. *Potassium thiocyanate* black precipitate of copper(II) thiocyanate:

$$Cu^{2+} + 2SCN^- \rightarrow Cu(SCN)_2\downarrow$$

Turns slowly white

The precipitate decomposes slowly to form white copper(I) thiocyanate and thiocyanogen is formed:

$$2Cu(SCN)_2\downarrow \rightarrow 2CuSCN\downarrow + (SCN)_2\uparrow$$

Thiocyanogen decomposes rapidly in aqueous solutions.

Copper(II) thiocyanate can be transformed to copper(I) thiocyanate immediately by adding a suitable reducing agent. A saturated solution of sulphur dioxide is the most suitable agent:

$$2Cu(SCN)_2\downarrow + SO_2 + 2H_2O \rightarrow 2CuSCN\downarrow + 2SCN^- + SO_4^{2-} + 4H^+$$

Fe
↓
Cu

7. *Iron* if a clean iron nail or a blade of a penknife is immersed in a solution of a copper salt, a red deposit of copper is obtained:

$$Cu^{2+} + Fe \rightarrow Fe^{2+} + Cu$$

and an equivalent amount of iron dissolves. The electrode potential of copper (more precisely of the copper–copper(II) system) is more positive than that of iron (or the iron–iron(II) system).

Specific test with cupron

8. *Benzoin-α-oxime (or cupron) (5% in alcohol)*

$$(C_6H_5 \cdot CHOH \cdot C(=NOH) \cdot C_6H_5)$$

forms a green precipitate of copper(II) benzoinoxime $Cu(C_{14}H_{11}O_2N)$, insoluble in dilute ammonia. In the presence of metallic salts which are precipitated by ammonia, their precipitation can be prevented by the addition of sodium potassium tartrate (10%). The reagent is specific for copper in ammoniacal tartrate solution. Large amounts of ammonium salts interfere and should be removed by evaporation and heating to glowing; the residue is then dissolved in a little dilute hydrochloric acid.

Treat some drop-reaction paper with a drop of the weakly acid test solution and a drop of the reagent, and then hold it over ammonia vapour. A green colouration is obtained.

Sensitivity 0.1 µg Cu.

Concentration limit 1 in 5×10^5.

If other ions, precipitable by ammonia solution, are present, a drop of potassium sodium tartrate solution (10%) is placed on the paper before the reagent is added.

9. *Salicylaldehyde oxime (salicylaldoxime) (1%)*

Salicylaldoxime
↓
green ppt

forms a greenish-yellow precipitate of copper salicylaldoxime $Cu(C_7H_6O_2N)_2$ in acetic acid solution, soluble in mineral acids. Only palladium and gold interfere giving $Pd(C_7H_6O_2N)_2$ and metallic gold respectively in acetic acid solution; they should therefore be absent.

Place a drop of the test solution, which has been neutralized and then acidified with acetic acid, in a micro test-tube and add a drop of the reagent. A yellow-green precipitate or opalescence (according to the amount of copper present) is obtained.

Sensitivity 0.5 µg Cu.
Concentration limit 1 in 10^5.

10. *Dithio oxamide (or rubeanic acid) (0.5%)*

$$\underset{\underset{S}{\|}}{H_2N-C}-\underset{\underset{S}{\|}}{C}-NH_2$$

Rubeanic acid
↓
black ppt

black precipitate of copper rubeanate $Cu[C(\!=\!NH)S]_2$ from ammoniacal or weakly acid solution. The precipitate is formed in the presence of alkali tartrates, but not in alkali-cyanide solutions. Only nickel and cobalt ions react under similar conditions yielding blue and brown precipitates respectively. Copper may, however, be detected in the presence of these elements by utilizing the capillary separation method on filter paper. Mercury(I) should be absent as it gives a black stain with ammonia.

Place a drop of the neutral test solution upon drop-reaction paper, expose it to ammonia vapour and add a drop of the reagent. A black or greenish-black spot is produced.

Sensitivity 0.01 µg Cu.
Concentration limit 1 in 2×10^6.

Traces of copper in distilled water give a positive reaction, hence a blank test must be carried out with the distilled water.

In the presence of nickel, proceed as follows. Impregnate drop-reaction paper with the reagent and add a drop of the test solution acidified with acetic acid, (2M). Two zones or circles are formed: the central olive-green or black ring is due to copper and the outer blue-violet ring to nickel.

Sensitivity 0.05 µg Cu in the presence of 20 000 times that amount of nickel.
Concentration limit 1 in 10^6.

In the presence of cobalt, the central green or black ring, due to copper, is surrounded by a yellow-brown ring of cobalt rubeanate.

Sensitivity 0.25 µg Cu in the presence of 20 000 times that amount of cobalt.
Concentration limit 1 in 2×10^5.

Catalytic test

11. *Catalytic test*   iron(III) salts react with thiosulphate according to the equations:

$$Fe^{3+} + 2S_2O_3^{2-} \rightarrow [Fe(S_2O_3)_2]^- \tag{a}$$

$$[Fe(S_2O_3)_2]^- + Fe^{3+} \rightarrow 2Fe^{2+} + S_4O_6^{2-} \tag{b}$$

Reaction (a) is fairly rapid; reaction (b) is a slow one, but is enormously accelerated by traces of copper. If the reaction is carried out in the presence of a thiocyanate, which serves as an indicator for the presence of iron(III) ions and also retards reaction (b), then the reaction velocity, which is proportional to the time taken for complete decolourization, may be employed for detecting minute amounts of copper(II) ions. Tungsten and, to a lesser extent, selenium cause a catalytic acceleration similar to that of copper: they should therefore be absent.

Upon adjacent cavities of a spot plate, place a drop of the test solution and a drop of distilled water. Add to each 1 drop iron(III) thiocyanate (0.05M) and 3 drops sodium thiosulphate (0.5M). The decolourization of the copper-free solution is complete in 1.5–2 minutes: if the test solution contains 1 μg copper, the decolourization is instantaneous. For smaller amounts of copper, the difference in times between the two tests is still appreciable.

Sensitivity 0.2 μg Cu^{2+}.

Concentration limit 1 in 2×10^6.

12. *Atomic spectrometric tests*

(a) *Flame atomic emission spectrometry* about 10 ppm copper can be identified in an air/acetylene flame at 324.75 nm. The test is more sensitive with a nitrous oxide/acetylene flame.

(b) *Flame atomic absorption spectrometry* with an air/acetylene flame, 1 ppm copper can be easily identified at 324.75 nm. In the absence of transition metals the test is more sensitive. A copper or brass hollow-cathode lamp can be used.

(c) *Copper standard, 1000 ppm* dissolve 3.9294 g analytical grade copper sulphate pentahydrate $CuSO_4 \cdot 5H_2O$ in water and dilute to 1 litre.

13. *Dry tests*

(a) *Blowpipe test* when copper compounds are heated with alkali carbonate upon charcoal, red metallic copper is obtained but no oxide is visible.

(b) *Borax bead* green while hot, and blue when cold after heating in an oxidizing flame; red in a reducing flame, best obtained by the addition of a trace of tin.

(c) *Flame test* green, especially in the presence of halides, e.g. by moistening with concentrated hydrochloric acid before heating.

3.11 Cadmium, Cd (A_r: 112.40)

Cadmium is a silver-white, malleable and ductile metal. It melts at 321°C. It dissolves slowly in dilute acids with the evolution of hydrogen (owing to its negative electrode potential):

$$Cd + 2H^+ \rightarrow Cd^{2+} + H_2\uparrow$$

Cadmium forms bivalent ions which are colourless. Cadmium chloride, nitrate, and sulphate are soluble in water; the sulphide is insoluble with a characteristic yellow colour. **All compounds of cadmium are toxic!**

Reactions of cadmium(II) ions These reactions can be studied most conveniently with a 0.25M solution of cadmium sulphate.

1. *Hydrogen sulphide (gas or saturated aqueous solution)* yellow precipitate of cadmium sulphide:

$$Cd^{2+} + H_2S \rightarrow CdS\downarrow + 2H^+$$

The reaction is reversible; if the concentration of strong acid in the solution is above 0.5M, precipitation is incomplete. Concentrated acids dissolve the precipitate for the same reason. The precipitate is insoluble in potassium cyanide (**POISON**); this distinguishes cadmium ions from copper.

NH₃
↓
white Cd(OH)₂

2. Ammonia solution when added dropwise   white precipitate of cadmium(II) hydroxide:

$$Cd^{2+} + 2NH_3 + 2H_2O \rightleftharpoons Cd(OH)_2\downarrow + 2NH_4^+$$

The precipitate dissolves in acid when the equilibrium shifts towards the left.

An excess of the reagent dissolves the precipitate, when tetrammine-cadmiate(II) ions are formed:

$$Cd(OH)_2\downarrow + 4NH_3 \rightarrow [Cd(NH_3)_4]^{2+} + 2OH^-$$

The complex is colourless.

NaOH
↓
white Cd(OH)₂

3. Sodium hydroxide   white precipitate of cadmium(II) hydroxide:

$$Cd^{2+} + 2OH^- \rightleftharpoons Cd(OH)_2\downarrow$$

The precipitate is insoluble in excess reagent; its colour and composition remain unchanged when boiled. Dilute acids dissolve the precipitate by shifting the equilibrium to the left.

KCN
↓
white Cd(CN)₂

4. Potassium cyanide (**POISON**)   white precipitate of cadmium cyanide, when added slowly to the solution:

$$Cd^{2+} + 2CN^- \rightarrow Cd(CN)_2\downarrow$$

Dissolves in excess

An excess of the reagent dissolves the precipitate, with the formation of tetracyanocadmiate(II) ions:

$$Cd(CN)_2\downarrow + 2CN^- \rightarrow [Cd(CN)_4]^{2-}$$

The colourless complex is not very stable; when hydrogen sulphide gas is introduced, cadmium sulphide is precipitated:

$$[Cd(CN)_4]^{2-} + H_2S \rightarrow CdS\downarrow + 2H^+ + 4CN^-$$

The marked differences in the stabilities of the copper and cadmium tetra-cyanato complexes serves as the basis for the separation of copper and cadmium ions.

5. Potassium iodide   forms no precipitate (distinction from copper).

Specific test with bis(*p*-nitrophenyl) carbazide

6. Bis(p-nitrophenyl) carbazide (0.1%)

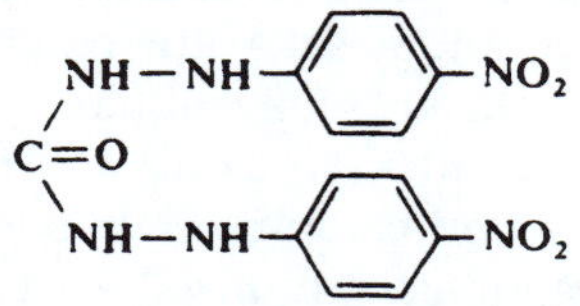

forms a brown-coloured product with cadmium hydroxide, which turns greenish-blue with formaldehyde.

Place a drop of the acid, neutral or ammoniacal test solution on a spot plate and mix it with 1 drop sodium hydroxide (2M) solution and 1 drop potassium cyanide (10%) solution. Introduce 1 drop of the reagent and

2 drops formaldehyde solution (40%). A brown precipitate is formed, which very rapidly becomes greenish-blue. The reagent alone is red in alkaline solution and is coloured violet with formaldehyde, hence it is advisable to compare the colour produced in a blank test with pure water when searching for minute amounts of cadmium.

Sensitivity 0.8 μg Cd.

Concentration limit 1 in 60 000.

In the presence of considerable amounts of copper, 3 drops each of the potassium cyanide and formaldehyde solution should be used; the sensitivity is 4 μg Cd in the presence of 400 times that amount of copper.

Specific test with 'Cadion 2B'

7. *4-Nitronaphthalene-diazoamino-azo-benzene ('Cadion 2B') (0.02%)*

Cadmium hydroxide forms a red-coloured lake with the reagent, which contrasts with the blue tint of the latter.

Place a drop of the reagent upon drop-reaction paper, add one drop of the test solution (which should be slightly acidified with acetic acid (2M) containing a little sodium potassium tartrate), and then one drop potassium hydroxide (2M). A bright-pink spot, surrounded by a blue circle, is produced.

Sensitivity 0.025 μg Cd.

Atomic spectrometry

8. *Atomic spectrometric tests*

(a) *Flame atomic emission spectrometry* emission in an air/acetylene flame is low, but 10 ppm cadmium can be identified in a nitrous oxide/acetylene flame at 326.11 nm.

(b) *Flame atomic absorption spectrometry* with an air/acetylene flame 1 ppm cadmium can be easily tested and even 0.1 ppm identified at 228.80 nm, using a cadmium hollow-cathode lamp.

(c) *Cadmium standard, 1000 ppm* dissolve 2.1930 g analytical grade cadmium sulphate $CdSO_4 \cdot 8/3H_2O$ in water and dilute to 1 litre.

Dry tests

9. *Dry tests*

(a) *Blowpipe test* all cadmium compounds when heated with alkali carbonate on charcoal give a brown incrustation of cadmium oxide CdO.

(b) *Ignition test* cadmium salts are reduced by sodium oxalate to elemental cadmium, which is usually obtained as a metallic mirror surrounded by a little brown cadmium oxide. Upon heating with sulphur, the metal is converted into yellow cadmium sulphide.

Place a little of the cadmium salt mixed with an equal weight of sodium oxalate in a small ignition tube, and heat. A mirror of metallic cadmium with brown edges is produced. Allow to cool, add a little flowers of sulphur and heat again. The metallic mirror is gradually converted into the orange-coloured sulphide, which becomes yellow after cooling. Do not confuse this with the yellow sublimate of sulphur.

3.12   Arsenic, As (A_r: 74.92) – arsenic(III)

Arsenic is a steel-grey, brittle solid with a metallic lustre. It sublimes on heating, and a characteristic, garlic-like odour is apparent; on heating in a free supply of air, arsenic burns with a blue flame yielding white fumes of arsenic(III) oxide. The element is insoluble in hydrochloric acid and in dilute sulphuric acid: it dissolves readily in dilute nitric acid yielding arsenite ions and in concentrated nitric acid, aqua regia or sodium hypochlorite solution forming arsenate:

$$As + 4H^+ + NO_3^- \rightarrow As^{3+} + NO\uparrow + 2H_2O$$

$$3As + 5HNO_3(conc) + 2H_2O \rightarrow 3AsO_4^{3-} + 5NO\uparrow + 9H^+$$

$$2As + 5OCl^- + 3H_2O \rightarrow 2AsO_4^{3-} + 5Cl^- + 6H^+$$

Two series of compounds of arsenic are common: that of arsenic(III) and arsenic(V). Arsenic(III) compounds can be derived from the amphoteric arsenic trioxide As_2O_3, which yields salts both with strong acids (e.g. arsenic(III) chloride, $AsCl_3$), and with strong bases (e.g. sodium arsenite, Na_3AsO_3). In strongly acidic solutions, therefore, the arsenic(III) ion As^{3+} is stable. In strongly basic solutions, the arsenite ions AsO_3^{3-} are predominant. Arsenic(V) compounds are derived from arsenic pentoxide, As_2O_5. This is the anhydride of arsenic acid, H_3AsO_4, which forms salts such as sodium arsenate Na_3AsO_4. Arsenic(V) therefore exists in solutions predominantly as the arsenate AsO_4^{3-} ion. **All compounds of arsenic are toxic!**

As(III) reactions

Reactions of arsenic(III) ions   A 0.1M solution of arsenic(III) oxide, As_2O_3, or sodium arsenite, Na_3AsO_3, can be used for these experiments. Arsenic(III) oxide does not dissolve in cold water, but by boiling the mixture for 30 minutes, dissolution is complete. The mixture can be cooled without the danger of precipitating the oxide.

H_2S
$\downarrow$
yellow As_2S_3

1. *Hydrogen sulphide (gas or saturated aqueous solution)*   yellow precipitate of arsenic(III) sulphide:

$$2As^{3+} + 3H_2S \rightarrow As_2S_3\downarrow + 6H^+$$

The solution must be strongly acidic; if there is not enough acid present a yellow colouration is visible only, owing to the formation of colloidal As_2S_3. The precipitate is insoluble in concentrated hydrochloric acid (distinction and method of separation from Sb_2S_3 and SnS_2), but dissolves in hot concentrated nitric acid:

$$3As_2S_3 + 28HNO_3 + 4H_2O \rightarrow 6AsO_4^{3-} + 9SO_4^{2-} + 36H^+ + 28NO\uparrow$$

It is also readily soluble in solutions of alkali hydroxides, and ammonia:

$$As_2S_3 + 6OH^- \rightarrow AsO_3^{3-} + AsS_3^{3-} + 3H_2O$$

Ammonium sulphide also dissolves the precipitate:

$$As_2S_3 + 3S^{2-} \rightarrow 2AsS_3^{3-}$$

In both cases thioarsenite (AsS_3^{3-}) ions are formed. On reacidifying, these both decompose, when arsenic(III) sulphide and hydrogen sulphide are formed:

$$2AsS_3^{3-} + 6H^+ \rightarrow As_2S_3\downarrow + 3H_2S\uparrow$$

Yellow ammonium sulphide (ammonium polysulphide) $(NH_4)_2S_2$ dissolves the precipitate, and thioarsenate AsS_4^{3-} ions are formed:

$$As_2S_3\downarrow + 4S_2^{2-} \rightarrow 2AsS_4^{3-} + S_3^{2-}$$

Upon acidifying this solution, yellow arsenic(V) sulphide is precipitated. It is, however, contaminated with sulphur because of the decomposition of the excess polysulphide reagent:

$$2AsS_4^{3-} + 6H^+ \rightarrow As_2S_5\downarrow + 3H_2S\uparrow$$

$$S_2^{2-} + 2H^+ \rightarrow H_2S\uparrow + S\downarrow$$

AgNO₃
↓
yellow Ag₃AsO₃

2. *Silver nitrate* yellow precipitate of silver arsenite in **neutral** solution (distinction from arsenates):

$$AsO_3^{3-} + 3Ag^+ \rightarrow Ag_3AsO_3\downarrow$$

The precipitate is soluble both in nitric acid and ammonia.

3. *Magnesia mixture (a solution containing MgCl₂, NH₄Cl, and a little NH₃)* no precipitate (distinction from arsenate).
A similar result is obtained with the magnesium nitrate reagent (a solution containing $Mg(NO_3)_2$, NH_4NO_3, and a little NH_3).

CuSO₄
↓
green ppt

4. *Copper sulphate solution* green precipitate of copper arsenite (Scheele's green), variously formulated as $CuHAsO_3$ and $Cu_3(AsO_3)_2 \cdot xH_2O$, from neutral solutions, soluble in acids, and also in ammonia solution forming a blue solution. The precipitate also dissolves in sodium hydroxide solution; upon boiling, copper(I) oxide is precipitated.

Iodine
↓
redox reaction

5. *Potassium tri-iodide (solution of iodine in potassium iodide)* oxidizes arsenite ions and is decolourized:

$$AsO_3^{3-} + I_3^- + H_2O \rightleftharpoons AsO_4^{3-} + 3I^- + 2H^+$$

The reaction is reversible, and an equilibrium is reached. If the hydrogen ions formed in this reaction are removed by adding sodium hydrogen carbonate as a buffer, the reaction goes to completion.

Bettendorf's test

6. *Tin(II) chloride solution and concentrated hydrochloric acid (Bettendorff's Test)* a few drops of the arsenite solution are added to 2 ml concentrated hydrochloric acid and 0.5 ml saturated tin(II) chloride solution, and the solution is gently warmed; the solution becomes dark brown and finally black, due to the separation of elemental arsenic:

$$2As^{3+} + 3Sn^{2+} \rightarrow 2As\downarrow + 3Sn^{4+}$$

If the test is made with the sulphide precipitated in acid solution, only mercury will interfere; by converting the arsenic into magnesium ammonium arsenate and heating to redness, the pyroarsenate $Mg_2As_2O_7$ remains and any mercury salts present are volatilized. This forms the basis of a delicate test for arsenic.

Mix a drop of the test solution in a micro crucible with 1–2 drops concentrated ammonia solution, 2 drops of hydrogen peroxide (3%), and 2 drops M magnesium sulphate solution. Evaporate slowly and finally heat until fuming ceases. Treat the residue with 1–2 drops of a solution of tin(II) chloride in concentrated hydrochloric acid, and warm slightly. A brown or black precipitate or colouration is obtained.

Sensitivity 1 µg As.

Concentration limit 1 in 50 000.

Atomic spectrometry

7. *Atomic spectrometric tests*

(a) *Flame atomic emission spectrometry* with an air/acetylene flame 10 ppm arsenic can be identified at 235.00 nm. The test can be made more sensitive using a nitrous oxide/acetylene flame.

(b) *Flame atomic absorption spectroscopy* using an arsenic hollow-cathode lamp and an air/acetylene flame, 1 ppm can be identified at 193.70 nm. Results are subject to background interferences, therefore it is better to use a nitrous oxide/acetylene flame.

(c) *Arsenic standard, 1 000 ppm* dissolve 1.3203 g analytical grade arsenic(III) oxide As_2O_3 in 50 ml concentrated hydrochloric acid under gentle warming. Dilute with water to 1 litre.

3.13   Arsenic, As (A_r: 74.92) – arsenic(V)

The properties of arsenic were summarized in Section 3.12.

As(V) reactions

Reactions of arsenate ions A 0.1M solution of disodium hydrogen arsenate $Na_2HAsO_4 \cdot 7H_2O$ can be used for the study of these reactions. The solution should contain some dilute hydrochloric acid. **All compounds of arsenic are toxic!**

H_2S
↓
yellow As_2S_3 + S

1. *Hydrogen sulphide (gas or saturated aqueous solution)* no immediate precipitate in the presence of dilute hydrochloric acid. If the passage of the gas is continued, a mixture of arsenic(III) sulphide, As_2S_3, and sulphur is slowly precipitated. Precipitation is more rapid in hot solution.

$$AsO_4^{3-} + H_2S \rightarrow AsO_3^{3-} + S\downarrow + H_2O$$

$$2AsO_3^{3-} + 3H_2S + 6H^+ \rightarrow As_2S_3\downarrow + 6H_2O$$

If a large excess of concentrated hydrochloric acid is present and hydrogen sulphide is passed rapidly into the cold solution, yellow arsenic pentasulphide As_2S_5 is precipitated:

$$2AsO_4^{3-} + 5H_2S + 6H^+ \rightarrow As_2S_5\downarrow + 8H_2O$$

Arsenic pentasulphide, like the trisulphide, is readily soluble in alkali hydroxides or ammonia, ammonium sulphide, ammonium polysulphide, and in sodium or ammonium carbonate:

$$As_2S_5\downarrow + 6OH^- \rightarrow AsS_4^{3-} + AsO_3S^{3-} + 3H_2O$$

$$As_2S_5\downarrow + 3S^{2-} \rightarrow 2AsS_4^{3-}$$

$$As_2S_5\downarrow + 6S_2^{2-} \rightarrow 2AsS_4^{3-} + 3S_3^{2-}$$

$$As_2S_5\downarrow + 3CO_3^{2-} \rightarrow AsS_4^{3-} + AsO_3S^{3-} + 3CO_2$$

Upon acidifying these solutions with hydrochloric acid, arsenic pentasulphide is reprecipitated:

$$2AsS_4^{3-} + 6H^+ \rightarrow As_2S_5\downarrow + 3H_2S\uparrow$$

For the rapid precipitation of arsenic from solutions of arsenates without using a large excess of hydrochloric acid, sulphur dioxide may be passed into the slightly acid solution in order to reduce the arsenic to the tervalent state and then the excess of sulphur dioxide is boiled off; on conducting hydrogen sulphide into the warm reduced solution, immediate precipitation of arsenic trisulphide occurs.

The precipitation can be greatly accelerated by the addition of small amounts of an iodide, say, 1 ml of a 0.05M solution, and a little concentrated hydrochloric acid. The iodide acts as a catalyst.

AgNO₃
↓
brown Ag₃AsO₄

2. *Silver nitrate solution* brownish-red precipitate of silver arsenate Ag_3AsO_4 from neutral solutions (distinction from arsenite and phosphate which yield yellow precipitates), soluble in acids and in ammonia solution, but insoluble in acetic acid.

$$AsO_4^{3-} + 3Ag^+ \rightarrow Ag_3AsO_4\downarrow$$

This reaction may be adapted as a sensitive test for arsenic in the following manner. The test is applicable only in the absence of chromates, hexacyanoferrate(II) and (III) ions, which also give coloured silver salts insoluble in acetic acid.

Place a drop of the test solution in a micro crucible, add a few drops of concentrated ammonia solution and of 3% hydrogen peroxide, and warm. Acidify with acetic acid and add 2 drops silver nitrate solution. A brownish-red precipitate or colouration appears.

Sensitivity 6 μg As.

Concentration limit 1 in 8000.

Magnesia mixture
↓
white ppt

3. *Magnesia mixture (see Section 3.12, reaction 3)* white, crystalline precipitate of magnesium ammonium arsenate $Mg(NH_4)AsO_4 \cdot 6H_2O$ from neutral or ammoniacal solution (distinction from arsenite):

$$AsO_4^{3-} + Mg^{2+} + NH_4^+ \rightarrow MgNH_4AsO_4\downarrow$$

For some purposes (e.g. the detection of arsenate in the presence of phosphate), it is better to use the magnesium nitrate reagent (a solution containing $Mg(NO_3)_2$, NH_4Cl, and a little NH_3).

Upon treating the white precipitate with silver nitrate solution containing a few drops of acetic acid, red silver arsenate is formed (distinction from phosphate):

$$MgNH_4AsO_4\downarrow + 3Ag^+ \rightarrow Ag_3AsO_4\downarrow + Mg^{2+} + NH_4^+$$

$(NH_4)_2MoO_4$
↓
yellow ppt

4. *Ammonium molybdate solution* when the reagent and nitric acid are added in considerable excess to a solution of an arsenate, a yellow crystalline precipitate of ammonium arsenomolybdate, $(NH_4)_3AsMo_{12}O_{40}$, is obtained on boiling (distinction from arsenites which give no precipitate, and from phosphates which yield a precipitate in the cold or upon gentle warming). The precipitate is insoluble in nitric acid, but dissolves in ammonia solution and in solutions of caustic alkalis.

$$AsO_4^{3-} + 12MoO_4^{2-} + 3NH_4^+ + 24H^+ \rightarrow (NH_4)_3AsMo_{12}O_{40}\downarrow + 12H_2O$$

The precipitate in fact contains trimolybdate $(Mo_3O_{10}^{2-})$ ions; each replacing one oxygen in AsO_4^{3-}. The composition of the precipitate should be written as $(NH_4)_3[As(Mo_3O_{10})_4]$.

KI
↓
brown I_2

In excess
↓
I_3^-

5. *Potassium iodide solution* in the presence of concentrated hydrochloric acid, iodine is precipitated; upon shaking the mixture with $1-2$ ml of chloroform, the latter is coloured violet by the iodine. The reaction may be used for the detection of arsenate in the presence of arsenite; oxidizing agents must be absent. Excess of the reagent dissolves iodine to form tri-iodide ions.

$$AsO_4^{3-} + 2H^+ + 2I^- \rightleftharpoons AsO_3^{3-} + I_2\downarrow + H_2O$$

$$I_2 + I^- \rightarrow I_3^-$$

The reaction is reversible (cf. Section 3.12, reaction 5); a large amount of acid must be present to complete this reaction.

5. *Atomic spectrometric tests* these have been described in Section 3.12, reaction 7.

3.14 Special tests for small amounts of arsenic

These tests are applicable to all arsenic compounds, and play an important role in forensic analysis.

Marsh's test

1. *Marsh's test* this test, which must be carried out in the fume cupboard, is based upon the fact that all soluble compounds of arsenic are reduced by 'nascent' hydrogen in acid solution to arsine, AsH_3, a colourless, **extremely poisonous gas** with a garlic-like odour. If the gas, mixed with hydrogen, is conducted through a heated glass tube, it is decomposed into hydrogen and metallic arsenic, which is deposited as a brownish-black 'mirror' just beyond the heated part of the tube.

On igniting the mixed gases, composed of hydrogen and arsine (after all the air has been expelled from the apparatus), they burn with a dark blue flame and white fumes of arsenic(III) oxide are evolved; if the inside of a small porcelain dish is pressed down upon the flame, a black deposit of arsenic is obtained on the cool surface, and the deposit is readily soluble in sodium hypochlorite or bleaching powder solution (distinction from antimony).

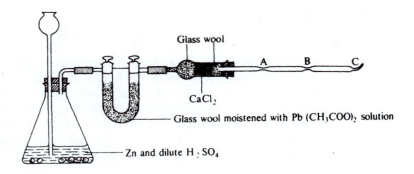

Fig. 3.1

The following reactions take place during these operations:

$$As^{3+} + 3Zn + 3H^+ \rightarrow AsH_3\uparrow + 3Zn^{2+}$$

$$AsO_4^{3-} + 4Zn + 11H^+ \rightarrow AsH_3\uparrow + 4Zn^{2+} + 4H_2O$$

$$4AsH_3\uparrow \rightarrow (\text{heat}) \rightarrow 4As\downarrow + 6H_2\uparrow$$

$$2AsH_3\uparrow + 3O_2 \rightarrow As_2O_3 + 3H_2O$$

$$2As\downarrow + 5OCl^- + 3H_2O \rightarrow 2AsO_4^{3-} + 5Cl^- + 6H^+$$

The Marsh test is best carried out as follows. The apparatus is fitted up as shown in Fig. 3.1. A conical flask of about 125 ml capacity is fitted with a two-holed rubber stopper carrying a thistle funnel reaching nearly to the bottom of the flask and a 5–7 mm right-angle tube; the latter is attached by a short piece of 'pressure' tubing to a U-tube filled with glass wool moistened with lead acetate solution to absorb any hydrogen sulphide evolved (this may be dispensed with, if desired, as its efficacy has been questioned), then to a small tube containing anhydrous calcium chloride of about 8 mesh, then to a hard glass tube, about 25 cm long and 7 mm diameter, constricted twice near the middle to about 2 mm diameter, the distance between the constrictions being 6–8 cm. The drying tubes and the tube ABC are securely supported by means of clamps. All reagents must be arsenic-free. Place 15–20 grams of arsenic-free zinc in the flask and add dilute sulphuric acid (3M) until hydrogen is vigorously evolved. The purity of the reagents is tested by passing the gases, by means of a delivery tube attached by a short piece of rubber tubing to the end of C, through silver nitrate solution (0.1M) for several minutes; the absence of a black precipitate or suspension proves that appreciable quantities of arsenic are not present.

The black precipitate is silver:

$$AsH_3\uparrow + 6Ag^+ \rightarrow As^{3+} + 3H^+ + 6Ag\downarrow$$

The solution containing the arsenic compound is then added in small amounts at a time to the contents of the flask. If much arsenic is present, there will be an almost immediate blackening of the silver nitrate solution.

Disconnect the rubber tube at C. Heat the tube at A to just below the softening point; a mirror of arsenic is deposited in the cooler, less constricted portion of the tube. A second flame may be applied at B to ensure complete decomposition (arsine is extremely poisonous). When a satisfactory mirror has been obtained, remove the flames at A and B and apply a light at C. Hold a cold porcelain dish in the flame, and test the solubility of the black or brownish deposit in sodium hypochlorite solution.

Smaller amounts of arsenic may be present in the silver nitrate solution as arsenious acid and can be detected by the usual tests, e.g. by hydrogen sulphide after removing the excess of silver nitrate with dilute hydrochloric acid, or by neutralizing and adding further silver nitrate solution, if necessary.

The original Marsh test involved burning and deposition of the arsenic upon a cold surface. Nowadays the mirror test is usually applied. The silver nitrate reaction (sometimes known as Hofmann's test) is very useful as a confirmatory test.

Gutzeit's test

2. *Gutzeit's test*   this is essentially a modification of Marsh's test, the chief difference being that only a test-tube is required and the arsine is detected by means of silver nitrate or mercury(II) chloride. Place 1–2 g arsenic-free zinc in a test-tube, add 5–7 ml dilute sulphuric acid, loosely plug the tube with purified cotton wool and then place a piece of filter paper moistened with 20% silver nitrate solution on top of the tube (Fig. 3.2). It may be necessary to warm the tube gently to produce a regular evolution of hydrogen. At the end of a definite period, say 2 minutes, remove the filter paper and examine the part that covered the test-tube; usually a light-brown spot is obtained owing to the traces of arsenic present in the reagents. Remove the cotton-wool plug, add 1 ml of the solution to be tested, replace the cotton wool and silver nitrate paper, the latter displaced so that a fresh portion is exposed. After 2 minutes, assuming that the rate of evolution of gas is approximately the same, remove the filter paper and examine the two spots. If much arsenic is present, the second spot (due to metallic silver), will appear black.

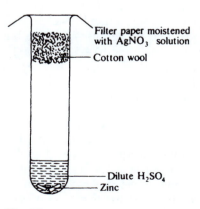

Fig. 3.2

Hydrogen sulphide, phosphine PH_3 and stibine SbH_3 give a similar reaction. They may be removed by means of a purified cotton-wool plug impregnated with copper(I) chloride.

The use of mercury(II) chloride paper, prepared by immersing filter paper in a 5% solution of mercury(II) chloride in alcohol (**POISON**) and drying in the atmosphere out of contact with direct sunlight, constitutes an improvement. This is turned yellow by a little arsine and reddish-brown by larger quantities. Filter paper, impregnated with a 0.33M aqueous solution of 'gold chloride' [sodium tetrachloroaurate(III)], $NaAuCl_4 \cdot 2H_2O$, may also be employed when a dark-red to blue-red stain is produced. A blank test must be performed with the reagent in all cases. In this reaction gold is formed:

$$AsH_3 + 2Au^{3+} \rightarrow As^{3+} + 2Au\downarrow + 3H^+$$

Fig. 3.3

The test may be performed on the semimicro scale with the aid of the apparatus shown in Fig. 3.3. Place 10 drops of the test solution in the semi-micro test-tube, add a few granules of arsenic-free zinc and 1 ml dilute sulphuric acid. Insert a loose wad of pure cotton wool moistened with lead nitrate solution in the funnel, and on top of this place a disc of drop-reaction paper impregnated with 20% silver nitrate solution; the paper may be held in position by a watch glass or microscope slide. Warm the test-tube gently (if necessary) on a water bath to accelerate the reaction, and allow to stand. Examine the silver nitrate paper after about 5 minutes. A grey spot will be obtained; this is occasionally yellow, due to the complex $Ag_3As \cdot 3AgNO_3$.

For minute quantities of arsenic, it is convenient to use the apparatus depicted in Fig. 2.53. Mix a drop of the test solution with a few grains of zinc and a few drops of dilute sulphuric acid in the micro test-tube. Insert the funnel with a flat rim, and place a small piece of drop-reaction paper moistened with 20% silver nitrate solution on the flat surface. A grey stain will be obtained.

Sensitivity 1 µg As.

Concentration limit 1 in 50 000.

A more sensitive test is provided by drop-reaction paper impregnated with gold chloride reagent (a 0.33M solution of $NaAuCl_4 \cdot 2H_2O$). Perform the test as described in the previous paragraph: a blue to blue-red stain of metallic gold is obtained after standing for 10–15 minutes. It is essential to perform a blank test with the reagents to confirm that they are arsenic-free.

Sensitivity 0.05 µg As.

Concentration limit 1 in 100 000.

Fleitmann's test

3. *Fleitmann's test* this test depends upon the fact that nascent hydrogen generated in alkaline solution, e.g. from aluminium or zinc and sodium hydroxide solution, reduces arsenic(III) compounds to arsine, but does not affect antimony compounds. A method of distinguishing arsenic and antimony compounds is thus provided. Arsenates must first be reduced to the tervalent state before applying the test. The *modus operandi* is as for the Gutzeit test, except that zinc or aluminium and sodium hydroxide solution

replace zinc and dilute sulphuric acid. It is necessary to warm the solutions. A black stain of silver is produced by the action of the arsine.

The apparatus of Fig. 3.3 may be used on the semimicro scale. Place 1 ml test solution in the test-tube, add some pure aluminium turnings, and 1 ml 2M potassium hydroxide solution. Gentle warming is usually necessary. A yellow or grey stain is produced after several minutes.

Dry tests

4. *Dry tests*

(a) *Blowpipe test*   arsenic compounds when heated upon charcoal with sodium carbonate give a white incrustation of arsenic(III) oxide, and an odour of garlic is apparent while hot.

(b) *Black mirror*   when heated with excess of potassium cyanide (**POISON**) and of anhydrous sodium carbonate in a dry bulb tube, a black mirror of arsenic, soluble in sodium hypochlorite solution, is produced in the cooler part of the tube.

3.15   Antimony, Sb (A_r: 121.75) – antimony(III)

Antimony is a lustrous, silver-white metal, which melts at 630°C. It is insoluble in hydrochloric acid, and in dilute sulphuric acid. In hot, concentrated sulphuric acid it dissolves slowly forming antimony(III) ions:

$$2Sb + 3H_2SO_4 + 6H^+ \rightarrow 2Sb^{3+} + 3SO_2\uparrow + 6H_2O$$

Nitric acid oxidizes antimony to an insoluble product, which can be regarded as a mixture of Sb_2O_3 and Sb_2O_5. These anhydrides, in turn, can be dissolved in tartaric acid. A mixture of nitric acid and tartaric acid dissolves antimony easily.

Aqua regia dissolves antimony, when antimony(III) ions are formed:

$$Sb + HNO_3 + 3HCl \rightarrow Sb^{3+} + 3Cl^- + NO\uparrow + 2H_2O$$

Two series of salts are known, with antimony(III) and antimony(V) ions in them; these are derived from the oxides Sb_2O_3 and Sb_2O_5.

Antimony(III) compounds are easily dissolved in acids, when the ion Sb^{3+} is stable. If the solution is made alkaline, or the concentration of hydrogen ions is decreased by dilution, hydrolysis occurs when antimonyl, SbO^+, ions are formed:

$$Sb^{3+} + H_2O \rightleftharpoons SbO^+ + 2H^+$$

Antimony(V) compounds contain the antimonate ion, SbO_4^{3-}. Their characteristics are similar to the corresponding arsenic compounds.

Sb(III) reactions

Reactions of antimony(III) ions   A 0.2M solution of antimony(III) chloride, $SbCl_3$, can be used to study these reactions. This can be prepared either by dissolving the solid antimony(III) chloride or antimony(III) oxide, Sb_2O_3, in dilute hydrochloric acid.

H_2S
↓
orange Sb_2S_3

1. *Hydrogen sulphide (gas or saturated aqueous solution)*   orange-red precipitate of antimony trisulphide, Sb_2S_3, from mildly acidic solutions. The precipitate is soluble in warm concentrated hydrochloric acid [distinction

and method of separation from arsenic(III) sulphide and mercury(II) sulphide], in ammonium polysulphide (forming a thioantimonate), and in alkali hydroxide solutions (forming antimonite and thioantimonite).

$$2Sb^{3+} + 3H_2S \rightarrow Sb_2S_3\downarrow + 6H^+$$

$$Sb_2S_3\downarrow + 6HCl \rightarrow 2Sb^{3+} + 6Cl^- + 3H_2S\uparrow$$

$$Sb_2S_3\downarrow + 4S_2^{2-} \rightarrow 2SbS_4^{3-} + S_3^{2-}$$

$$2Sb_2S_3\downarrow + 4OH^- \rightarrow SbO_2^- + 3SbS_2^- + 2H_2O$$

Upon acidification of the thioantimonate solution with hydrochloric acid, antimony pentasulphide is precipitated initially but usually decomposes partially into the trisulphide and sulphur:

$$2SbS_4^{3-} + 6H^+ \rightarrow Sb_2S_5\downarrow + 3H_2S\uparrow$$

$$Sb_2S_5\downarrow \rightarrow Sb_2S_3\downarrow + 2S\downarrow$$

Acidification of the antimonite-thioantimonite mixture leads to the precipitation of the trisulphide:

$$SbO_2^- + 3SbS_2^- + 4H^+ \rightarrow 2Sb_2S_3\downarrow + 2H_2O$$

H_2O
$\downarrow$
white basic salt

2. *Water* when the solution is poured into water, a white precipitate of antimonyl chloride $SbO \cdot Cl$ is formed, soluble in hydrochloric acid and in tartaric acid solution (difference from bismuth). With a large excess of water the hydrated oxide $Sb_2O_3 \cdot xH_2O$ is produced.

NaOH
$\downarrow$
white Sb_2O_3

3. *Sodium hydroxide or ammonia solution* white precipitate of the hydrated antimony(III) oxide $Sb_2O_3 \cdot xH_2O$ soluble in concentrated (5M) solutions of caustic alkalis forming antimonites.

$$2Sb^{3+} + 6OH^- \rightarrow Sb_2O_3\downarrow + 3H_2O$$

$$Sb_2O_3\downarrow + 2OH^- \rightarrow 2SbO_2^- + H_2O$$

Zn
$\downarrow$
black Sb

4. *Zinc* a black precipitate of antimony is produced. If a little of the antimony trichloride solution is poured upon platinum foil and a fragment of metallic zinc is placed on the foil, a black stain of antimony is formed upon the platinum; the stain (or deposit) should be dissolved in a little warm dilute nitric acid and hydrogen sulphide passed into the solution after dilution: an orange precipitate of antimony trisulphide will be obtained.

$$2Sb^{3+} + 3Zn\downarrow \rightarrow 2Sb\downarrow + 3Zn^{2+}$$

Some stibine SbH_3 may be evolved when zinc is used; it is preferable to employ tin.

$$2Sb^{3+} + 3Sn\downarrow \rightarrow 2Sb\downarrow + 3Sn^{2+}$$

A modification of the above test is to place a drop of the solution containing antimony upon a silver foil and to touch the foil through the drop with a piece of tin or zinc. A black spot consisting of antimony is formed.

Fe
↓
black Sb

5. *Iron wire* black precipitate of antimony. This may be confirmed as described in reaction 4.

$$2Sb^{3+} + 3Fe \rightarrow 2Sb\downarrow + 3Fe^{2+}$$

KI
↓
yellow colour

6. *Potassium iodide solution*   yellow colouration owing to the formation of a complex salt.

$$Sb^{3+} + 6I^- \rightarrow [SbI_6]^{3-}$$

Rhodamine-B
↓
violet colour

7. *Rhodamine-B (or tetraethylrhodamine) reagent*

violet or blue colouration with pentavalent antimony. Trivalent antimony does not respond to this test, hence it must be oxidized with potassium or sodium nitrite in the presence of strong hydrochloric acid. In Group IIB $SbCl_3$ is always formed together with $SnCl_4$ when the precipitate is treated with hydrochloric acid: by oxidizing Sb(III) to Sb(V) with a little solid nitrite, an excellent means of testing for Sb in the presence of a large excess of Sn is available. Mercury, gold, thallium, molybdates, vanadates, and tungstates in solution give similar colour reactions.

The test solution should be made strongly acidic by adding hydrochloric acid (6M). Antimony should be oxidized to Sb(V) with the addition of a little solid sodium or potassium nitrite (avoiding a large excess). Place 1 ml reagent on a spot plate and add 1 ml of the test solution. The bright-red colour of the reagent changes to blue.

Sensitivity 0.5 µg Sb and applicable in the presence of 12 500 times that amount of Sn.

Concentration limit 1 in 100 000.

Molybdenum blue test

8. *Phosphomolybdic acid reagent* $(H_3[PMo_{12}O_{40}])$ 'molybdenum blue' is produced by antimony(III) salts. Of the ions in Group II, only tin(II) interferes with the test. The test solution may consist of the filtered solution obtained by treating the Group IIB precipitate with hydrochloric acid: the antimony is present as Sb^{3+} and the tin as Sb^{4+}, which has no effect upon the reagent.

Place a drop of the test solution upon filter paper which has been impregnated with the phosphomolybdic acid reagent and hold the paper in steam. A blue colouration appears within a few minutes.

Sensitivity 0.2 µg Sb.

Concentration limit 1 in 250 000.

Alternatively, place 1 ml test solution in a semimicro test-tube, add 0.5–1 ml reagent, and heat for a short time. The reagent is reduced to a blue compound, which can be extracted with amyl alcohol.

Tin(II) chloride reduces not only the reactive phosphomolybdic acid but also its relatively unreactive (e.g. ammonium or potassium) salts to 'molybdenum blue'. The composition of this compound is usually given as Mo_2O_5; the real molecular unit however contains 154 molybdenum atoms with attendant ligands in a cyclic arrangement[1]. However, antimony(III) salts do not reduce ammonium phosphomolybdate: Sn^{2+} may thus be detected in the presence of Sb^{3+}.

Impregnate filter paper with a solution of phosphomolybdic acid and then hold it for a short time over ammonia gas to form the yellow, sparingly soluble ammonium salt; dry. Place a drop of the test solution on this paper: a blue spot appears if tin(II) is present.

Sensitivity 0.03 µg Sn.

Concentration limit 1 in 650 000.

Atomic spectrometry

9. *Atomic spectrometric tests* tests for antimony by atomic spectrometry are not sensitive.

(a) *Flame atomic emission spectrometry* with an air/acetylene flame about 100 ppm antimony can be identified at 607.90 nm.

(b) *Flame atomic absorption spectrometry* using an antimony hollow-cathode lamp and an air/acetylene flame, 50 ppm can be identified at 206.83 nm.

(c) *Antimony standard, 1000 ppm* dissolve 1.0000 g analytical grade antimony metal in a solution made up from 10 ml concentrated hydrochloric acid, 10 g analytical grade tartaric acid and a few drops of concentrated nitric acid. After complete dissolution dilute with water to 1 litre.

3.16 Antimony, Sb (A_r: 121.75) – antimony(V)

The physical and chemical properties of antimony have been described in Section 3.15.

Sb(V) reactions

Reactions of antimony(V) ions Antimony(V) ions are derived from the amphoteric oxide Sb_2O_5. In **acids** this oxide dissolves under the formation of the antimony(V) cation Sb^{5+}:

$$Sb_2O_5 + 10H^+ \rightleftarrows 2Sb^{5+} + 5H_2O$$

In **alkalis**, on the other hand, the antimonate SbO_4^{3-} ion is formed:

$$Sb_2O_5 + 3OH^- \rightleftarrows 2SbO_4^{3-} + 3H^+$$

SbO_4^{3-} is a simplified expression of the composition of the antimonate ion; in fact it exists in the hydrated form, which may be termed hexahydroxoantimonate(V). Its formation from Sb_2O_5 with alkalis may be described by the reaction:

$$Sb_2O_5 + 2OH^- + 5H_2O \rightleftarrows 2[Sb(OH)_6]^-$$

For the study of these reactions, an acidified 0.2M solution of potassium hexahydroxoantimonate $K[Sb(OH)_6]$ can be used. Alternatively, antimony pentoxide Sb_2O_5 may be dissolved in concentrated hydrochloric acid.

1. A. Müller: *Angew. Chem.* 1996, **35**, 1206

H₂S
↓
orange Sb₂S₅

1. *Hydrogen sulphide (gas or saturated aqueous solution)* orange-red precipitate of antimony pentasulphide, Sb_2S_5, in moderately acid solutions. The precipitate is soluble in ammonium sulphide solution (yielding a thioantimonate), is soluble in alkali hydroxide solutions, and is also dissolved by concentrated hydrochloric acid with the formation of antimony trichloride and the separation of sulphur. The thio salt is decomposed by acids, the pentasulphide being precipitated.

$$2Sb^{5+} + 5H_2S \rightarrow Sb_2S_5\downarrow + 10H^+$$

$$Sb_2S_5\downarrow + 3S^{2-} \rightarrow 2SbS_4^{3-}$$

$$Sb_2S_5\downarrow + 6OH^- \rightarrow SbO_3S^{3-} + SbS_4^{3-} + 3H_2O$$

$$Sb_2S_5\downarrow + 6H^+ \rightarrow 2Sb^{3+} + 2S\downarrow + 3H_2S\uparrow$$

$$2SbS_4^{3-} + 6H^+ \rightarrow Sb_2S_5\downarrow + 3H_2S\uparrow$$

$$SbO_3S^{3-} + SbS_4^{3-} + 6H^+ \rightarrow Sb_2S_5\downarrow + 3H_2O$$

H₂O
↓
white basic salts

2. *Water* white precipitate of basic salts with various compositions; ultimately antimonic acid is formed:

$$Sb^{5+} + 4H_2O \rightleftharpoons H_3SbO_4\downarrow + 5H^+$$

The precipitate dissolves both in acids and alkalis (but not in alkali carbonates).

KI
↓
brown I₂

3. *Potassium iodide solution* in acidic solution iodine is formed:

$$Sb^{5+} + 2I^- \rightleftharpoons Sb^{3+} + I_2\downarrow$$

Dissolves in excess

If the Sb^{5+} ions are in excess, iodine crystals precipitate out and float on the surface of the solution. When heated the characteristic violet vapour of iodine appears. If the reagent is added in excess, brown tri-iodide ions are formed which screen the yellow colour of hexaiodoantimonate(III) ions:

$$Sb^{5+} + 9I^- \rightarrow [SbI_6]^{3-} + I_3^-$$

Zn or Sn
↓
black Sb

4. *Zinc or tin* black precipitate of antimony in the presence of hydrochloric acid:

$$2Sb^{5+} + 5Zn\downarrow \rightarrow 2Sb\downarrow + 5Zn^{2+}$$

$$2Sb^{5+} + 5Sn\downarrow \rightarrow 2Sb\downarrow + 5Sn^{2+}$$

Some stibine (SbH_3) is also generated by zinc.

5. *Rhodamine-B reagent* see antimony(III) compounds (Section 3.15, reaction 7). There is no need for preliminary oxidation in this case.

6. *Atomic spectrometric tests* these have been described in Section 3.15, reaction 9.

3.17 Special tests for small amounts of antimony

1. *Marsh's test* this test is carried out exactly as described for arsenic. The stibine, SbH_3 (mixed with hydrogen), which is evolved, burns with a faintly bluish-green flame and produces a dull black spot upon a cold porcelain dish held in the flame; this deposit is insoluble in sodium hypochlorite or bleaching powder solution, but is dissolved by a solution of tartaric acid (difference from arsenic).

The gas is also decomposed by passage though a tube heated to dull redness. A lustrous mirror of antimony is formed in a similar manner to the arsenic mirror, but it is deposited on both sides of the heated portion of the tube because of the greater instability of the stibine. This mirror may be converted into orange-red antimony trisulphide by dissolving it in a little boiling hydrochloric acid and passing hydrogen sulphide into the solution.

When the stibine-hydrogen mixture is passed into a solution of silver nitrate (Hofmann's test) a black precipitate of silver antimonide Ag_3Sb is obtained; this is decomposed by the excess of silver nitrate into silver and antimony(III) oxide:

$$SbH_3 + 3Ag^+ \rightarrow Ag_3Sb\downarrow + 3H^+$$

$$2Ag_3Sb\downarrow + 6Ag^+ + 3H_2O \rightarrow 12Ag\downarrow + Sb_2O_3\downarrow + 6H^+$$

It is best to dissolve the precipitate in a solution of tartaric acid and to test for antimony with hydrogen sulphide in the usual manner.

2. *Gutzeit's test* a brown stain is produced which is soluble in 80% alcohol, provided the concentration of antimony is not too high.

3. *Fleitmann's test* negative results are obtained when this is applied to antimony compounds (distinction from arsenic).

4. *Dry test (blowpipe test)* when antimony compounds are heated with sodium carbonate upon charcoal, a brittle metallic bead, surrounded by a white incrustation, is obtained.

3.18 Tin, Sn (A_r: 118.69) – tin(II)

Tin is a silver-white metal which is malleable and ductile at ordinary temperatures, but at low temperatures it becomes brittle due to transformation into a different allotropic modification. It melts at 231.8°C. The metal dissolves slowly in dilute hydrochloric and sulphuric acid with the formation of tin(II) salts:

$$Sn + 2H^+ \rightarrow Sn^{2+} + H_2\uparrow$$

Dilute nitric acid dissolves tin slowly without the evolution of any gas, tin(II) and ammonium ions are formed:

$$4Sn + 10H^+ + NO_3^- \rightarrow 4Sn^{2+} + NH_4^+ + 3H_2O$$

With concentrated nitric acid a vigorous reaction occurs. In this case a white solid, usually formulated as hydrated tin(IV) oxide $SnO_2 \cdot xH_2O$ and sometimes known as metastannic acid, is produced:

$$3Sn + 4HNO_3 + (x - 2)H_2O \rightarrow 4NO\uparrow + 3SnO_2 \cdot xH_2O\downarrow$$

In the presence of antimony and tartaric acid tin dissolves readily in nitric acid (induced dissolution) because of complex formation. If larger amounts of iron are present, the formation of metastannic acid is again prevented.

In hot, concentrated sulphuric acid tin(IV) ions are formed during dissolution:

$$Sn + 4H_2SO_4 \rightarrow Sn^{4+} + 2SO_4^{2-} + 2SO_2\uparrow + 4H_2O$$

Aqua regia dissolves tin readily, when again tin(IV) ions are formed:

$$3Sn + 4HNO_3 + 12HCl \rightarrow 3Sn^{4+} + 12Cl^- + 4NO\uparrow + 8H_2O$$

Tin can be bivalent and tetravalent in its compounds.

The tin(II) or stannous compounds are usually colourless. In acid solution the tin(II) ions Sn^{2+} are present, while in alkaline solutions tetrahydroxostannate(II) ions $[Sn(OH)_4]^{2-}$ are to be found. These two are readily transformed into each other:

$$Sn^{2+} + 4OH^- \rightleftharpoons [Sn(OH)_4]^{2-}$$

Tin(II) ions are strong reducing agents.

Tin(IV) or stannic compounds are more stable. In aqueous solution they can be present as the tin(IV) ions Sn^{4+} or as hexahydroxostannate(IV) ions $[Sn(OH)_6]^{2-}$. These again form an equilibrium system:

$$Sn^{4+} + 6OH^- \rightleftharpoons [Sn(OH)_6]^{2-}$$

In acid solutions the equilibrium is shifted towards the left, while in alkaline medium it is shifted towards the right.

Sn(II) reactions

Reactions of tin(II) ions   A 0.25M solution of tin(II) chloride, $SnCl_2 \cdot 2H_2O$, can be used for studying these reactions. The solution should contain free hydrochloric acid (100 ml concentrated HCl per litre).

H_2S
↓
brown SnS

1. *Hydrogen sulphide (gas or saturated aqueous solution)*   brown precipitate of tin(II) sulphide, SnS, from mildly acidic solutions (say in the presence of 0.25–0.3M hydrochloric acid or pH c. 0.6). The precipitate is soluble in concentrated hydrochloric acid (distinction from arsenic(III) sulphide and mercury(II) sulphide); it is also soluble in yellow $[(NH_4)_2S_x]$, but not in colourless $[(NH_4)_2S]$, ammonium sulphide solution to form a thiostannate. Treatment of the solution of ammonium thiostannate with an acid yields a yellow precipitate of tin(IV) sulphide, SnS_2.

$$Sn^{2+} + H_2S \rightarrow SnS\downarrow + 2H^+$$

$$SnS\downarrow + S_2^{2-} \rightarrow SnS_3^{2-}$$

$$SnS_3^{2-} + 2H^+ \rightarrow SnS_2\downarrow + H_2S\uparrow$$

Tin(II) sulphide is practically insoluble in solutions of caustic alkalis; hence, if potassium hydroxide solution is employed for separating Group IIA and Group IIB, the tin must be oxidized to the tetravalent state with hydrogen peroxide before precipitation with hydrogen sulphide.

NaOH
↓
white $Sn(OH)_2$
dissolves in excess

2. *Sodium hydroxide solution*   white precipitate of tin(II) hydroxide, soluble in excess alkali:

$$Sn^{2+} + 2OH^- \rightleftharpoons Sn(OH)_2\downarrow$$

$$Sn(OH)_2\downarrow + 2OH^- \rightleftharpoons [Sn(OH)_4]^{2-}$$

With ammonia solution, white tin(II) hydroxide is precipitated, which cannot be dissolved in excess ammonia.

$HgCl_2$
↓
white Hg_2Cl_2

3. *Mercury(II) chloride solution*   a white precipitate of mercury(I) chloride (calomel) is formed if a large amount of the reagent is added quickly:

$$Sn^{2+} + 2HgCl_2 \rightarrow Hg_2Cl_2\downarrow + Sn^{4+} + 2Cl^-$$

If however tin(II) ions are in excess, the precipitate turns grey, especially on warming, owing to further reduction to mercury metal:

$$Sn^{2+} + Hg_2Cl_2\downarrow \rightarrow 2Hg\downarrow + Sn^{4+} + 2Cl^-$$

All compounds of mercury are **POISONOUS!**

4. *Bismuth nitrate and sodium hydroxide solutions*   black precipitate of bismuth metal (cf. Section 3.9, reaction 5).

Zn
↓
spongy Sn

5. *Metallic zinc*   spongy tin is deposited which adheres to the zinc. If the zinc rests upon platinum foil, as described in Section 3.15, reaction 4, and the solution is weakly acid, the tin is partially deposited upon the zinc in a spongy form but does not stain the platinum. The precipitate should be dissolved in concentrated hydrochloric acid and the mercury(II) chloride test applied.

Cacotheline
↓
violet colour

6. *Cacotheline reagent (a nitro-derivative of brucine, $C_{21}H_{21}O_7N_3$)*   violet colouration with tin(II) salts. The test solution should be acidic (2M HCl), and if tin is in the tetravalent state, it should be reduced previously with aluminium or magnesium, and the solution filtered. **Both cacotheline and brucine are strongly toxic!**

The following interfere with the test: strong reducing agents (hydrogen sulphide, dithionites, sulphites and selenites); V, U, Te, Hg, Bi, Au, Pd, Se, Te, Sb, Mo, W, Co and Ni. The reaction is not selective, but is fairly sensitive: it can be used in the analysis of the Group IIB precipitate. Since iron(II) ions have no influence on the test, it may be applied to the tin solution which has been reduced with iron wire.

Impregnate some drop-reaction paper with the reagent and, before the paper is quite dry, add a drop of the test solution. A violet spot, surrounded by a less coloured zone, appears on the yellow paper.

Alternatively, treat a little of the test solution in a micro test-tube with a few drops of the reagent. A violet (purple) colouration is produced.

Sensitivity 0.2 µg Sn.
Concentration limit 1 in 250 000.

Diazine
↓
colour change
blue
↓
violet
↓
red

7. *Diazine green reagent (dyestuff formed by coupling diazotized safranine with dimethylaniline) (Janus green)* tin(II) chloride reduces the blue diazine green to the red safranine, hence the colour change is blue → violet → red. Titanium(III) chloride reacts similarly, but iron(II) salts and similar reducing agents have no effect. The reagent is therefore useful in testing for tin in the mixed sulphides of antimony and tin obtained in routine qualitative analysis. The solution of the sulphides in hydrochloric acid is reduced with iron wire, aluminium, or magnesium powder and a drop of the reduced solution employed for the test.

Mix 1 drop of the test solution on a spot plate with 1 ml of the reagent. The colour changes from blue to violet or red. It is advisable to carry out a blank test.

Sensitivity 2 µg Sn.
Concentration limit 1 in 25 000.

Atomic spectrometry

8. *Atomic spectrometric tests* the air/acetylene flame is not recommended for these, as several metals interfere in this relatively cold flame.

(a) *Flame atomic emission spectrometry* with a nitrous oxide/acetylene flame 1 ppm tin can be identified at 284.00 nm. At the same wavelength the air/acetylene flame sensitivity is increased twofold, but Cu, Pb, Ni and Zn interfere, together with larger amounts of other metals.

(b) *Flame atomic absorption spectrometry* using a tin hollow-cathode lamp and a nitrous oxide/acetylene flame, 1 ppm can be identified at 224.61 nm. With an air/acetylene flame at the same wavelength sensitivity is somewhat higher, but the danger of interference from the presence of other metals is considerable.

(c) *Tin standard, 1000 ppm* dissolve 1.0000 g analytical grade tin metal in a mixture of 200 ml concentrated hydrochloric acid and 5 ml of concentrated nitric acid. Dilute the solution with water to 1 litre.

3.19   Tin, Sn (A_r: 118.69) – tin(IV)

The properties of metallic tin were discussed at the beginning of Section 3.18.

Sn(IV) reactions

Reactions of tin(IV) ions To study these reactions use a 0.25M solution of ammonium hexachlorostannate(IV) by dissolving 92 g $(NH_4)_2[SnCl_6]$ in 250 ml concentrated hydrochloric acid and diluting the solution to 1 litre with water.

H_2S
↓
yellow SnS_2

1. *Hydrogen sulphide (gas or saturated aqueous solution)* yellow precipitate of tin(IV) sulphide SnS_2 from dilute acid solutions (0.3M). The precipitate is soluble in concentrated hydrochloric acid [distinction from arsenic(III) and mercury(II) sulphides], in solutions of alkali hydroxides, and also in ammonium sulphide and ammonium polysulphide. Yellow tin(IV) sulphide is precipitated upon acidification.

$$Sn^{4+} + 2H_2S \rightarrow SnS_2\downarrow + 4H^+$$

$$SnS_2\downarrow + S^{2-} \rightarrow SnS_3^{2-}$$

$$SnS_2\downarrow + 2S_2^{2-} \rightarrow SnS_3^{2-} + S_3^{2-}$$

$$SnS_3^{2-} + 2H^+ \rightarrow SnS_2\downarrow + H_2S\uparrow$$

No precipitation of tin(IV) sulphide occurs in the presence of oxalic acid, due to the formation of the stable complex ion of the type $[Sn(C_2O_4)_4(H_2O)_2]^{4-}$; this forms the basis of a method of separation of antimony and tin.

NaOH
↓
white Sn(OH)₄
dissolves in excess

2. *Sodium hydroxide solution* gelatinous white precipitate of tin(IV) hydroxide, $Sn(OH)_4$, soluble in excess of the precipitant forming hexa-hydroxostannate(IV)

$$Sn^{4+} + 4OH^- \rightarrow Sn(OH)_4\downarrow$$

$$Sn(OH)_4\downarrow + 2OH^- \rightleftharpoons [Sn(OH)_6]^{2-}$$

With ammonia and with sodium carbonate solutions, a similar precipitate is obtained which, however, is insoluble in excess reagent.

3. *Mercury(II) chloride solution* no precipitate (difference from tin(II)).

Fe
↓
redox reaction Sn(II)

4. *Metallic iron* reduces tin(IV) ions to tin(II):

$$Sn^{4+} + Fe \rightarrow Fe^{2+} + Sn^{2+}$$

If pieces of iron are added to a solution, and the mixture is filtered, tin(II) ions can be detected with mercury(II) chloride reagent. A similar result is obtained on boiling the solution with copper or antimony.

5. *Atomic spectrometric tests* these have been described in Section 3.18, reaction 8.

Dry tests

6. *Dry tests*

(a) *Blowpipe test* all tin compounds when heated with sodium carbon-ate, preferably in the presence of potassium cyanide, on charcoal give white, malleable and metallic globules of tin which do not mark paper. Part of the metal is oxidized to tin(IV) oxide, especially on strong heating, which forms a white incrustation on the charcoal.

(b) *Borax bead test* a borax bead, which has been coloured pale blue by a trace of copper salt, becomes a clear ruby red in the reducing flame if a minute quantity of tin is added.

3.20 **Third group of cations: iron(II) and (III), aluminium(III), chromium(III) and (VI), nickel(II), cobalt(II), manganese(II) and (VII), and zinc(II)**

Group reagent hydrogen sulphide (gas or saturated aqueous solution) in the presence of ammonia and ammonium chloride, or ammonium sulphide solution.

Group reaction   precipitates of various colours: iron(II) sulphide (black), aluminium hydroxide (white), chromium(III) hydroxide (green), nickel sulphide (black), cobalt sulphide (black), manganese(II) sulphide (pink), and zinc sulphide (white).

The metals of this group are not precipitated by the group reagents for Groups I and II, but are all precipitated, in the presence of ammonium chloride, by hydrogen sulphide from their solutions made alkaline with ammonia solution. The metals, with the exception of aluminium and chromium which are precipitated as the hydroxides owing to the complete hydrolysis of the sulphides in aqueous solution, are precipitated as the sulphides. Iron, aluminium, and chromium (often accompanied by a little manganese) are also precipitated as the hydroxides by ammonia solution in the presence of ammonium chloride, whilst the other metals of the group remain in solution and may be precipitated as sulphides by hydrogen sulphide. It is therefore usual to subdivide the group into the iron group (iron, aluminium, and chromium) or Group IIIA and the zinc group (nickel, cobalt, manganese, and zinc) or Group IIIB.

3.21   Iron, Fe (A_r: 55.85) – iron(II)

Chemically pure iron is a silver-white, tenacious, and ductile metal. It melts at 1535°C. The commercial metal is rarely pure and usually contains small quantities of carbide, silicide, phosphide, and sulphide of iron, and some graphite. These contaminants play an important role in the strength of iron structures. Iron can be magnetized. Dilute or concentrated hydrochloric acid and dilute sulphuric acid dissolve iron, when iron(II) salts and hydrogen gas are produced.

$$Fe + 2H^+ \rightarrow Fe^{2+} + H_2\uparrow$$

$$Fe + 2HCl \rightarrow Fe^{2+} + 2Cl^- + H_2\uparrow$$

Hot, concentrated sulphuric acid yields iron(III) ions and sulphur dioxide:

$$2Fe + 3H_2SO_4 + 6H^+ \rightarrow 2Fe^{3+} + 3SO_2\uparrow + 6H_2O$$

With cold dilute nitric acid, iron(II) and ammonium ions are formed:

$$4Fe + 10H^+ + NO_3^- \rightarrow 4Fe^{2+} + NH_4^+ + 3H_2O$$

Cold, concentrated nitric acid renders iron *passive*; in this state it does not react with dilute nitric acid nor does it displace copper from an aqueous solution of a copper salt. 8M or hot, concentrated nitric acid dissolves iron with the formation of nitrogen oxide gas and iron(III) ions:

$$Fe + HNO_3 + 3H^+ \rightarrow Fe^{3+} + NO\uparrow + 2H_2O$$

Iron forms two important series of salts.

The **iron(II)** (or ferrous) salts are derived from iron(II) oxide FeO. In solution they contain the cation Fe^{2+} and normally possess a slight-green colour. Intensively coloured ion-association and chelate complexes are also common. Iron(II) ions can be oxidized easily to iron(III); they are therefore strong reducing agents. The less acidic the solution the more pronounced this effect; in neutral or alkaline media even atmospheric oxygen will oxidize iron(II) ions. Solutions of iron(II) must be slightly acidic if they are to be kept for a longer time. An acidified solution of Mohr's salt (cf. below) is stable for several months.

Iron(III) (or ferric) salts are derived from iron(III) oxide Fe_2O_3. They are more stable than the iron(II) salts. In their solution the pale-yellow Fe^{3+} cations are present; if the solution contains chloride, the colour becomes stronger. Reducing agents convert iron(III) ions to iron(II).

Fe(II) reactions

Reactions of iron(II) ions Use a freshly prepared 0.5M solution of iron(II) sulphate $FeSO_4 \cdot 7H_2O$ or iron(II) ammonium sulphate (Mohr's salt; $FeSO_4 \cdot (NH_4)_2SO_4 \cdot 6H_2O$), acidified with 50 ml M H_2SO_4 per litre, for the study of these reactions.

NaOH
↓
white Fe(OH)₂,
turns dark

1. *Sodium hydroxide solution* white precipitate of iron(II) hydroxide, $Fe(OH)_2$, in the complete absence of air, insoluble in excess, but soluble in acids. Upon exposure to air, iron(II) hydroxide is rapidly oxidized, yielding ultimately reddish-brown iron(III) hydroxide. Under ordinary conditions it appears as a dirty-green precipitate; the addition of hydrogen peroxide immediately oxidizes it to iron(III) hydroxide.

$$Fe^{2+} + 2OH^- \rightarrow Fe(OH)_2\downarrow$$

$$4Fe(OH)_2 + 2H_2O + O_2 \rightarrow 4Fe(OH)_3\downarrow$$

$$2Fe(OH)_2 + H_2O_2 \rightarrow 2Fe(OH)_3\downarrow$$

NH₃
↓
white Fe(OH)₂,
turns dark

2. *Ammonia solution* precipitation of iron(II) hydroxide occurs (cf. reaction 1). If, however, larger amounts of ammonium ions are present, the dissociation of ammonium hydroxide is suppressed, and the concentration of hydroxyl ions is lowered to such an extent that the solubility product of iron(II) hydroxide, $Fe(OH)_2$, is not attained and precipitation does not occur. Similar remarks apply to the other divalent elements of Group III, nickel, cobalt, zinc and manganese and also to magnesium.

3. *Hydrogen sulphide* no precipitation takes place in acid solution since the sulphide ion concentration, $[S^{2-}]$, is insufficient to exceed the solubility product of iron(II) sulphide. If the hydrogen-ion concentration is reduced, and the sulphide-ion concentration correspondingly increased, by the addition of sodium acetate solution, partial precipitation of black iron(II) sulphide, FeS, occurs.

(NH₄)₂S
↓
black FeS

4. *Ammonium sulphide solution* black precipitate of iron(II) sulphide, readily soluble in acids with evolution of hydrogen sulphide. The moist precipitate becomes brown upon exposure to air, due to its oxidation to basic iron(III) sulphate $Fe_2O(SO_4)_2$.

$$Fe^{2+} + S^{2-} \rightarrow FeS\downarrow$$

$$FeS\downarrow + 2H^+ \rightarrow Fe^{2+} + H_2S\uparrow$$

$$4FeS\downarrow + 9O_2 \rightarrow 2Fe_2O(SO_4)_2\downarrow$$

KCN
↓
brown Fe(CN)$_2$,
dissolves in excess

5. *Potassium cyanide solution* (**POISON**)   yellowish-brown precipitate of iron(II) cyanide, soluble in excess reagent when a pale-yellow solution of hexacyanoferrate(II) (ferrocyanide) $[Fe(CN)_6]^{4-}$ ions is obtained:

$$Fe^{2+} + 2CN^- \rightarrow Fe(CN)_2\downarrow$$

$$Fe(CN)_2\downarrow + 4CN^- \rightarrow [Fe(CN)_6]^{4-}$$

The hexacyanoferrate(II) ion, being a complex ion, does not undergo the typical reactions of iron(II). The iron present in such solutions may be detected by decomposing the complex ion by boiling the solution with concentrated sulphuric acid *in a fume cupboard with good ventilation*, when carbon monoxide gas is formed (together with hydrogen cyanide, if potassium cyanide is present in excess):

$$[Fe(CN)_6]^{4-} + 6H_2SO_4 + 6H_2O \rightarrow Fe^{2+} + 6CO\uparrow + 6NH_4^+ + 6SO_4^{2-}$$

A dry sample which contains alkali hexacyanoferrate(II) decomposes on ignition to iron carbide, alkali cyanide, and nitrogen. By dissolving the residue in acid, iron can be detected in the solution. Because of the toxic nature of the carbon monoxide gas, **this operation must again be carried out in a fume cupboard.**

K$_4$[Fe(CN)$_6$]
↓
white K$_2$Fe[Fe(CN)$_6$]

6. *Potassium hexacyanoferrate(II) solution*   in the complete absence of air a white precipitate of potassium iron(II) hexacyanoferrate(II) is formed:

$$Fe^{2+} + 2K^+ + [Fe(CN)_6]^{4-} \rightarrow K_2Fe[Fe(CN)_6]\downarrow$$

Under ordinary atmospheric conditions a pale-blue precipitate is obtained (cf. reaction 7).

K$_3$[Fe(CN)$_6$]
↓
Turnbull's blue

7. *Potassium hexacyanoferrate(III) solution*   a dark-blue precipitate is obtained. First hexacyanoferrate(III) ions oxidize iron(II) to iron(III), when hexacyanoferrate(II) is formed:

$$Fe^{2+} + [Fe(CN)_6]^{3-} \rightarrow Fe^{3+} + [Fe(CN)_6]^{4-}$$

and these ions combine to a precipitate called Turnbull's blue:

$$4Fe^{3+} + 3[Fe(CN)_6]^{4-} \rightarrow Fe_4[Fe(CN)_6]_3$$

Note that the composition of this precipitate is identical to that of Prussian blue (cf. Section 3.22, reaction 6). Earlier it was suggested that its composition was iron(II) hexacyanoferrate(III), $Fe_3[Fe(CN)_6]_2$, hence the different name. The identical composition and structure of Turnbull's blue and Prussian blue has recently been proved by Mössbauer spectroscopy. The precipitate is decomposed by sodium or potassium hydroxide solution, iron(III) hydroxide being precipitated.

8. *Ammonium thiocyanate solution*   no colouration is obtained with pure iron(II) salts (distinction from iron(III) ions).

9. *2,2'-Bipyridyl reagent*

111

deep-red complex bivalent cation $[Fe(C_5H_4N)_2]^{2+}$ with iron(II) salts in mineral acid solution. Iron(III) ion does not react. Other metallic ions react with the reagent in acid solution, but the intensities of the resulting colours are so feeble that they do not interfere with the test for iron provided excess reagent is employed. Large amounts of halides and sulphates reduce the solubility of the iron(II) bipyridyl complex and a red precipitate may be formed.

2,2'-Bipyridyl
↓
red colour

Specific for Fe(II)

Treat a drop of the faintly acidified test solution with 1 drop of the reagent on a spot plate: a red colouration is obtained. Alternatively, treat drop-reaction paper (Whatman No. 3 M.M., 1st quality), which has been impregnated with the reagent and dried, with a drop of the test solution: a red or pink spot is produced.

Sensitivity 0.3 µg Fe^{2+}.

Concentration limit 1 in 1 600 000.

If appreciable quantities of iron(III) salts are present and traces of iron(II) salts are sought, it is best to carry out the reaction in a micro crucible lined with paraffin wax and to mask iron(III) ions as $[FeF_6]^{3-}$ by the addition of a few drops of potassium fluoride solution.

Dimethylglyoxime
↓
red colour

10. *Dimethylglyoxime reagent* soluble red iron(II) dimethylglyoxime in ammoniacal solution. Iron(III) salts give no colouration, but nickel, cobalt, and large quantities of copper salts interfere and must be absent. The test may be carried out in the presence of potassium cyanide solution in which nickel dimethylglyoxime (cf. Section 3.27, reaction 8) dissolves.

Mix a drop of the test solution with a small crystal of tartaric acid; introduce a drop of the reagent, followed by 2 drops ammonia solution. A red colouration appears.

Sensitivity 0.04 µg Fe^{2+}.

Concentration limit 1 in 125 000.

The colouration fades on standing owing to the oxidation of the iron(II) complex.

If it is desired to detect iron(III) ions by this test, they must first be reduced by a little hydroxylamine hydrochloride.

1,10-Phenanthroline
↓
red colour

11. *1,10-Phenanthroline reagent*

red colouration due to the complex cation $[Fe(C_{12}H_8N_2)_3]^{2+}$ in faintly acid solution. Iron(III) has no effect and must first be reduced to the bivalent state with hydroxylamine hydrochloride if the reagent is to be used in testing for iron.

Place a drop of the faintly acid test solution on a spot plate and add 1 drop of the reagent. A red colour is obtained.

Concentration limit 1 in 1 500 000.

Atomic spectrometry

12. *Atomic spectrometric tests* iron emits a large number of spectrum lines in the ultraviolet region. The lines recommended below are the most often used ones for quantitative work.

(a) *Flame atomic emission spectrometry*   with an air/acetylene flame 1 ppm iron can be identified at 371.99 nm. The nitrous oxide/acetylene flame allows a more sensitive test.

(b) *Flame atomic absorption spectrometry*   using an iron hollow-cathode lamp and an air/acetylene flame, 1 ppm can be identified at 248.33 nm.

(c) *Iron standard, 1000 ppm*   dissolve 1.0000 g analytical grade iron (e.g. iron wire) in a mixture of 20 ml concentrated hydrochloric acid and 5 ml concentrated nitric acid. After complete dissolution, heat the solution on a water bath until all chlorine and nitrous gases disappear. Cool, and dilute the solution to 1 litre.

3.22   Iron, Fe (A_r: 55.85) – iron(III)

The most important characteristics of the metal have been described in Section 3.21.

Fe(III) reactions

Reactions of iron(III) ions   use a 0.5M solution of iron(III) chloride $FeCl_3 \cdot 6H_2O$. The solution should be clear yellow. If it turns brown, due to hydrolysis, a few drops of hydrochloric acid should be added.

NH_3
↓
brown $Fe(OH)_3$

1. *Ammonia solution*   reddish-brown, gelatinous precipitate of iron(III) hydroxide $Fe(OH)_3$, insoluble in excess of the reagent, but soluble in acids:

$$Fe^{3+} + 3NH_3 + 3H_2O \rightarrow Fe(OH)_3\downarrow + 3NH_4^+$$

The solubility product of iron(III) hydroxide is so small (3.8×10^{-38}) that complete precipitation takes place even in the presence of ammonium salts (distinction from iron(II), nickel, cobalt, manganese, zinc, and magnesium). Precipitation does not occur in the presence of certain organic acids. Iron(III) hydroxide is converted during strong heating into iron(III) oxide; the ignited oxide is soluble with difficulty in dilute acids, but dissolves on vigorous boiling with concentrated hydrochloric acid.

$$2Fe(OH)_3\downarrow \rightarrow Fe_2O_3 + 3H_2O$$

$$Fe_2O_3 + 6H^+ \rightarrow 2Fe^{3+} + 3H_2O$$

NaOH
↓
brown $Fe(OH)_3$

2. *Sodium hydroxide solution*   reddish-brown precipitate of iron(III) hydroxide, insoluble in excess of the reagent (distinction from aluminium and chromium):

$$Fe^{3+} + 3OH^- \rightarrow Fe(OH)_3\downarrow$$

H_2S
↓
reduction to
Fe(II) + white S

3. *Hydrogen sulphide gas*   in acidic solution reduces iron(III) ions to iron(II) and sulphur is formed as a milky-white precipitate:

$$2Fe^{3+} + H_2S \rightarrow 2Fe^{2+} + 2H^+ + S\downarrow$$

Test for H_2S solutions

If a neutral solution of iron(III) chloride is added to a freshly prepared, saturated solution of hydrogen sulphide, a bluish colouration appears first, followed by the precipitation of sulphur. The blue colour is due to a colloidal solution of sulphur of extremely small particle size. This reaction can be used to test the freshness of hydrogen sulphide solutions.

The finely distributed sulphur cannot be readily filtered with ordinary filter papers. By boiling the solution with a few torn pieces of filter paper the precipitate coagulates and can be filtered.

(NH$_4$)$_2$S
↓
reduction +
precipitation of black
FeS + S

4. *Ammonium sulphide solution*   a black precipitate, consisting of iron(II) sulphide and sulphur is formed:

$$2Fe^{3+} + 3S^{2-} \rightarrow 2FeS\downarrow + S\downarrow$$

In hydrochloric acid the black iron(II) sulphide precipitate dissolves and the white colour of sulphur becomes visible:

$$FeS\downarrow + 2H^+ \rightarrow H_2S\uparrow + Fe^{2+}$$

From alkaline solutions black iron(III) sulphide is obtained:

$$2Fe^{3+} + 3S^{2-} \rightarrow Fe_2S_3\downarrow$$

On acidification with hydrochloric acid iron(III) ions are reduced to iron(II) and sulphur is formed:

$$Fe_2S_3\downarrow + 4H^+ \rightarrow 2Fe^{2+} + 2H_2S\uparrow + S\downarrow$$

The damp iron(II) sulphide precipitate, when exposed to air, is slowly oxidized to brown iron(III) hydroxide:

$$4FeS\downarrow + 6H_2O + 3O_2 \rightarrow 4Fe(OH)_3\downarrow + 4S\downarrow$$

The reaction is exothermal. Under certain conditions so much heat may be produced that the precipitate dries out, and the filter paper, with the finely distributed sulphur on it, catches fire. **Sulphide precipitates** therefore **should never be placed in a waste bin**, but should rather be washed away under running water; only the filter paper should be thrown away.

KCN
↓
brown Fe(CN)$_3$

5. *Potassium cyanide* (**POISON**)   when added slowly, produces a reddish-brown precipitate of iron(III) cyanide:

$$Fe^{3+} + 3CN^- \rightarrow Fe(CN)_3\downarrow$$

Dissolves in excess

In excess reagent the precipitate dissolves giving a yellow solution, when hexacyanoferrate(III) ions are formed:

$$Fe(CN)_3\downarrow + 3CN^- \rightarrow [Fe(CN)_6]^{3-}$$

These reactions should be carried out in a fume cupboard, as the free acid present in the iron(III) chloride solution forms hydrogen cyanide gas with the reagent:

$$H^+ + CN^- \rightarrow HCN\uparrow$$

Iron(III) ions cannot be detected in a solution of hexacyanoferrate(III) with the usual reactions. The complex has to be decomposed first by evaporating with concentrated sulphuric acid or by igniting a solid sample, as described with hexacyanoferrate(II) (cf. Section 3.21, reaction 5).

K$_4$[Fe(CN)$_6$]
↓
Prussian blue

6. *Potassium hexacyanoferrate(II) solution* intense blue precipitate of iron(III) hexacyanoferrate (Prussian blue):

$$4Fe^{3+} + 3[Fe(CN)_6]^{4-} \rightarrow Fe_4[Fe(CN)_6]_3$$

(cf. Section 3.21, reaction 7).

The precipitate is insoluble in dilute acids, but decomposes in concentrated hydrochloric acid. A large excess of the reagent dissolves it partly or entirely, when an intense blue solution is obtained. Sodium hydroxide turns the precipitate red as iron(III) oxide and hexacyanoferrate(II) ions are formed:

$$Fe_4[Fe(CN)_6]_3\downarrow + 12OH^- \rightarrow 4Fe(OH)_3\downarrow + 3[Fe(CN)_6]^{4-}$$

Oxalic acid also dissolves Prussian blue forming a blue solution; this process was once used to manufacture blue writing inks.

If iron(III) chloride is added to an excess of potassium hexacyanoferrate(II), a product with the composition KFe[Fe(CN)$_6$] is formed. This tends to form colloidal solutions ('Soluble Prussian Blue') and cannot be filtered.

K$_3$[Fe(CN)$_6$]
↓
brown colour

7. *Potassium hexacyanoferrate(III)* a brown colouration is produced, due to the formation of an undissociated complex, iron(III) hexacyanoferrate(III):

$$Fe^{3+} + [Fe(CN)_6]^{3-} \rightarrow Fe[Fe(CN)_6]$$

Upon adding hydrogen peroxide or some tin(II) chloride solution, the hexacyanoferrate(III) part of the compound is reduced and Prussian blue is precipitated.

Na$_2$HPO$_4$
↓
yellow FePO$_4$

8. *Disodium hydrogen phosphate solution* a yellowish-white precipitate of iron(III) phosphate is formed:

$$Fe^{3+} + HPO_4^{2-} \rightarrow FePO_4\downarrow + H^+$$

The reaction is reversible, because a strong acid is formed which dissolves the precipitate. It is advisable to add small amounts of sodium acetate, which acts as a buffer.

NaCH$_3$COO
↓
red colour or precipitate

9. *Sodium acetate solution* a reddish-brown colouration is obtained, attributed to the formation of a complex ion with the composition [Fe$_3$(OH)$_2$(CH$_3$COO)$_6$]$^+$. The reaction

$$3Fe^{3+} + 6CH_3COO^- + 2H_2O \rightleftharpoons [Fe_3(OH)_2(CH_3COO)_6]^+ + 2H^+$$

becomes complete only if the strong acid, which is formed, is removed by the addition of an excess of the reagent, which acts as a buffer.

If the solution is diluted and boiled, a reddish-brown precipitate of basic iron(III) acetate is formed:

$$[Fe_3(OH)_2(CH_3COO)_6]^+ + 4H_2O \rightarrow$$
$$\rightarrow 3Fe(OH)_2CH_3COO\downarrow + 3CH_3COOH + H^+$$

The excess of acetate ions acts again as a buffer and the reaction goes to completion.

Cupferron
↓
brown ppt

10. *Cupferron reagent, the ammonium salt of nitrosophenylhydroxylamine,* $C_6H_5N(NO)ONH_4$ reddish-brown precipitate is formed in the presence of hydrochloric acid:

$$Fe^{3+} + 3C_6H_5N(NO)ONH_4 \rightarrow Fe[C_6H_5N(NO)O]_3\downarrow + 3NH_4^+$$

The precipitate is soluble in ether. It is insoluble in acids, but can be decomposed by ammonia or alkali hydroxides, when iron(III) hydroxide precipitate is formed.

NH₄SCN
↓
red colour

11. *Ammonium thiocyanate solution* in slightly acidic solution a deep-red colouration is produced (difference from iron(II) ions), due to the formation of a non-dissociated iron(III) thiocyanate complex:

$$Fe^{3+} + 3SCN^- \rightarrow Fe(SCN)_3$$

Specific for Fe(III)

This neutral molecule can be extracted by ether or amyl alcohol. Fluorides and mercury(II) ions bleach the colour because of the formation of the more stable hexafluoroferrate(III) $[FeF_6]^{3-}$ complex and the non-dissociated mercury(II) thiocyanate species:

$$Fe(SCN)_3 + 6F^- \rightarrow [FeF_6]^{3-} + 3SCN^-$$

$$2Fe(SCN)_3 + 3Hg^{2+} \rightarrow 2Fe^{3+} + 3Hg(SCN)_2$$

The presence of nitrites should be avoided because in acidic solution they form nitrosyl thiocyanate NOSCN which yields a red colour, disappearing on heating.

The reaction can be adapted as a spot test and may be carried out as follows. Place a drop of the test solution on a spot plate and add 1 drop of 0.1M ammonium thiocyanate solution. A deep-red colouration appears.

Sensitivity 0.25 µg Fe^{3+}.

Concentration limit 1 in 200 000.

Coloured salts, e.g. those of copper, chromium, cobalt, and nickel, reduce the sensitivity of the test.

Ferron
↓
greenish-blue colour

12. *Ferron (8-hydroxy-7-iodoquinoline-5-sulphonic acid) reagent*

green or greenish-blue colouration with iron(III) salts in faintly acid solution (*p*H 2.5–3.0). Iron(II) does not react: only copper interferes.

Place a few drops of the slightly acid test solution in a micro test-tube and add 1 drop of the reagent. A green colouration appears.

Sensitivity 0.5 µg Fe^{3+}.

Concentration limit 1 in 1 000 000.

13. *Reduction of iron(III) to iron(II) ions* in acid solution this may be accomplished by various agents. Zinc or cadmium metal, or their amalgams (i.e. alloys with mercury) may be used:

$$2Fe^{3+} + Zn \rightarrow Zn^{2+} + 2Fe^{2+}$$

$$2Fe^{3+} + Cd \rightarrow Cd^{2+} + 2Fe^{2+}$$

The solution will contain zinc or cadmium ions, respectively, after the reduction. In acid solutions these metals will dissolve further with liberation of hydrogen; they should therefore be removed from the solution once the reduction has been accomplished.

Tin(II) chloride, potassium iodide, hydroxylamine hydrochloride, hydrazine sulphate, or ascorbic acid can also be used:

$$2Fe^{3+} + Sn^{2+} \rightarrow 2Fe^{2+} + Sn^{4+}$$

$$2Fe^{3+} + 3I^- \rightarrow 2Fe^{2+} + I_3^-$$

$$4Fe^{3+} + 2NH_2OH \rightarrow 4Fe^{2+} + N_2O + H_2O + 4H^+$$

$$4Fe^{3+} + N_2H_4 \rightarrow 4Fe^{2+} + N_2 + 4H^+$$

$$2Fe^{3+} + C_6H_8O_6 \rightarrow 2Fe^{2+} + C_6H_6O_6 + 2H^+$$

the product of the reduction with ascorbic acid being dehydroascorbic acid.

Hydrogen sulphide (cf. reaction 3) and sulphur dioxide gas reduce iron(III) ions also:

$$2Fe^{3+} + H_2S \rightarrow 2Fe^{2+} + S\downarrow + 2H^+$$

$$2Fe^{3+} + SO_2 + 2H_2O \rightarrow 2Fe^{2+} + SO_4^{2-} + 4H^+$$

14. *Oxidation of iron(II) ions to iron(III)* oxidation occurs slowly upon exposure to air. Rapid oxidation is effected by concentrated nitric acid, hydrogen peroxide, concentrated hydrochloric acid with potassium chlorate, aqua regia, potassium permanganate, potassium dichromate, and cerium(IV) sulphate in acid solution.

$$4Fe^{2+} + O_2 + 4H^+ \rightarrow 4Fe^{3+} + 2H_2O$$

$$3Fe^{2+} + HNO_3 + 3H^+ \rightarrow NO\uparrow + 3Fe^{3+} + 2H_2O$$

$$2Fe^{2+} + H_2O_2 + 2H^+ \rightarrow 2Fe^{3+} + 2H_2O$$

$$6Fe^{2+} + ClO_3^- + 6H^+ \rightarrow 6Fe^{3+} + Cl^- + 3H_2O$$

$$2Fe^{2+} + HNO_3 + 3HCl \rightarrow 2Fe^{3+} + NOCl\uparrow + 2Cl^- + 2H_2O$$

$$5Fe^{2+} + MnO_4^- + 8H^+ \rightarrow 5Fe^{3+} + Mn^{2+} + 4H_2O$$

$$6Fe^{2+} + Cr_2O_7^{2-} + 14H^+ \rightarrow 6Fe^{3+} + 2Cr^{3+} + 7H_2O$$

$$Fe^{2+} + Ce^{4+} \rightarrow Fe^{3+} + Ce^{3+}$$

Distinction of Fe(II) and Fe(III)

15. *Distinctive tests for iron(II) and iron(III) ions* **iron(II)** can be detected most reliably with 2,2'-bipyridyl (cf. reaction 9, Section 3.21); the test is conclusive also in the presence of iron(III). In turn, **iron(III) ions** can be detected with ammonium thiocyanate solution (cf. reaction 11). It must be remembered that even freshly prepared solutions of pure iron(II) salts contain some iron(III) and the thiocyanate test will be positive with these. If however iron(III) is reduced by one of the ways described in reaction 13, the thiocyanate test will be negative.

16. *Atomic spectrometric tests* these have been described in Section 3.21, reaction 12.

Dry tests

17. *Dry tests*

(a) *Blowpipe test* when iron compounds are heated on charcoal with sodium carbonate, grey metallic particles of iron are produced; these are ordinarily difficult to see, but can be separated from the charcoal by means of a magnet.

(b) *Borax bead test* with a small quantity of iron, the bead is yellowish-brown while hot (and yellow when cold) in an oxidizing flame, and pale green in a reducing flame; with large quantities of iron, the bead is reddish-brown in an oxidizing flame.

3.23 Aluminium, Al (A_r: 26.98)

Aluminium is a white, ductile and malleable metal; the powder is grey. It melts at 659°C. Exposed to air, aluminium objects are oxidized on the surface, but the oxide layer protects the object from further oxidation. Dilute hydrochloric acid dissolves the metal readily, dissolution is slower in dilute sulphuric or nitric acid:

$$2Al + 6H^+ \rightarrow 2Al^{3+} + 3H_2\uparrow$$

The dissolution process can be speeded up by the addition of some mercury(II) chloride to the mixture. Concentrated hydrochloric acid also dissolves aluminium:

$$2Al + 6HCl \rightarrow 2Al^{3+} + 3H_2\uparrow + 6Cl^-$$

Concentrated sulphuric acid dissolves aluminium with the liberation of sulphur dioxide:

$$2Al + 6H_2SO_4 \rightarrow 2Al^{3+} + 3SO_4^{2-} + 3SO_2\uparrow + 6H_2O$$

Concentrated nitric acid renders the metal passive. With alkali hydroxides a solution of tetrahydroxoaluminate is formed:

$$2Al + 2OH^- + 6H_2O \rightarrow 2[Al(OH)_4]^- + 3H_2\uparrow$$

Aluminium is tervalent in its compounds. Aluminium ions (Al^{3+}) form colourless salts with colourless anions. Its halides, nitrate, and sulphate are water-soluble; these solutions display acidic reactions owing to hydrolysis. Aluminium sulphide can be prepared in the dry state only, in aqueous solutions it hydrolyses and aluminium hydroxide, $Al(OH)_3$, is formed. Aluminium sulphate forms double salts with sulphates of monovalent cations, with attractive crystal shapes; these are called alums.

Al(III) reactions

Reactions of aluminium(III) ions   Use a 0.33M solution of aluminium chloride, $AlCl_3$, or a 0.166M solution of aluminium sulphate, $Al_2(SO_4)_3 \cdot 16H_2O$, or potash alum, $K_2SO_4 \cdot Al_2(SO_4)_3 \cdot 24H_2O$, to study these reactions.

NH_3
↓
white $Al(OH)_3$

1. *Ammonia solution*   white gelatinous precipitate of aluminium hydroxide, $Al(OH)_3$, slightly soluble in excess of the reagent. The solubility is decreased in the presence of ammonium salts. A small proportion of the precipitate passes into the solution as colloidal aluminium hydroxide (aluminium hydroxide sol): the sol is coagulated on boiling the solution or upon the addition of soluble salts (e.g. ammonium chloride) yielding a precipitate of aluminium hydroxide, known as aluminium hydroxide gel. To ensure complete precipitation the ammonia solution is added in slight excess and the mixture boiled until the liquid has a slight odour of ammonia. When freshly precipitated, it dissolves readily in strong acids and bases, but after boiling it becomes sparingly soluble:

$$Al^{3+} + 3NH_3 + 3H_2O \rightarrow Al(OH)_3\downarrow + 3NH_4^+$$

NaOH
↓
white $Al(OH)_3$

Dissolves in excess

2. *Sodium hydroxide solution*   white precipitate of aluminium hydroxide:

$$Al^{3+} + 3OH^- \rightarrow Al(OH)_3\downarrow$$

The precipitate dissolves in excess reagent, forming tetrahydroxoaluminate ions:

$$Al(OH)_3 + OH^- \rightarrow [Al(OH)_4]^-$$

The reaction is a reversible one, and any reagent which will reduce the hydroxyl-ion concentration sufficiently should cause the reaction to proceed from right to left with the consequent precipitation of aluminium hydroxide. This may be effected with a solution of ammonium chloride (the hydroxyl-ion concentration is reduced owing to the formation of the weak base ammonia, which can be readily removed as ammonia gas by heating) or by the addition of acid; in the latter case, a large excess of acid causes the precipitated hydroxide to redissolve.

$$[Al(OH)_4]^- + NH_4^+ \rightarrow Al(OH)_3\downarrow + NH_3\uparrow + H_2O$$

$$[Al(OH)_4]^- + H^+ \rightleftarrows Al(OH)_3\downarrow + H_2O$$

$$Al(OH)_3 + 3H^+\downarrow \rightleftarrows Al^{3+} + 3H_2O$$

The precipitation of aluminium hydroxide by solutions of sodium hydroxide and ammonia does not take place in the presence of tartaric acid, citric acid, sulphosalicylic acid, malic acid, sugars, and other organic hydroxy compounds, because of the formation of soluble complex salts. These organic substances must therefore be decomposed by gentle ignition or by evaporating with concentrated sulphuric or nitric acid before aluminium can be precipitated in the ordinary course of qualitative analysis.

(NH₄)₂S
↓
hydrolysis
white Al(OH)₃

3. *Ammonium sulphide solution*   a white precipitate of aluminium hydroxide:

$$2Al^{3+} + 3S^{2-} + 6H_2O \rightarrow 2Al(OH)_3\downarrow + 3H_2S\uparrow$$

The characteristics of the precipitate are the same as mentioned under reaction 2.

CH₃COONa
↓
white basic salt

4. *Sodium acetate solution*   no precipitate is obtained in cold, neutral solutions, but on boiling with excess reagent, a voluminous precipitate of basic aluminium acetate $Al(OH)_2CH_3COO$ is formed:

$$Al^{3+} + 3CH_3COO^- + 2H_2O \rightarrow Al(OH)_2CH_3COO\downarrow + 2CH_3COOH$$

Na₂HPO₄
↓
white AlPO₄

5. *Disodium hydrogen phosphate solution*   white gelatinous precipitate of aluminium phosphate:

$$Al^{3+} + HPO_4^{2-} \rightleftarrows AlPO_4\downarrow + H^+$$

The reaction is reversible; strong acids dissolve the precipitate. However, the precipitate is insoluble in acetic acid (difference from phosphates of alkaline earths, which are soluble). The precipitate can also be dissolved in sodium hydroxide.

Aluminon
↓
red lake on Al(OH)₃

6. *'Aluminon' reagent (tri-ammonium aurine-tricarboxylate)*

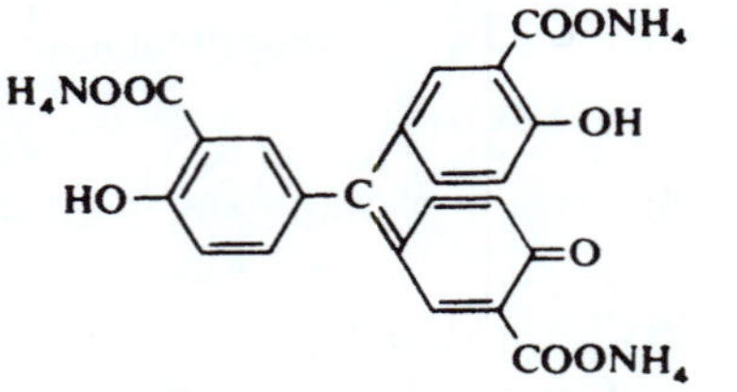

This dye is adsorbed by aluminium hydroxide giving a bright-red adsorption complex or 'lake'. The test is applied to the precipitate of aluminium hydroxide obtained in the usual course of analysis, since certain other elements interfere. Dissolve the aluminium hydroxide precipitate in 2 ml 2M hydrochloric acid, add 1 ml 6M ammonium acetate solution and 2 ml 0.1% aqueous solution of the reagent. Shake, allow to stand for 5 minutes, and add excess of ammoniacal ammonium carbonate solution to decolourize excess dyestuff and lakes due to traces of chromium(III) hydroxide and silica. A bright-red precipitate (or colouration), persisting in the alkaline solution, is obtained.

Iron interferes with the test and should be absent.

Alizarin
↓
red lake on Al(OH)₃

7. *Alizarin reagent*

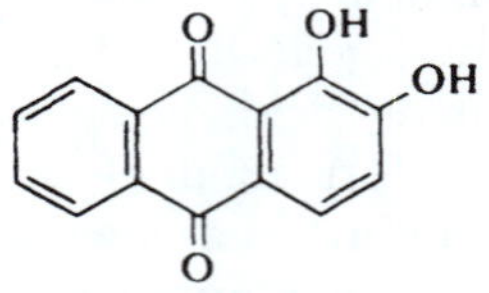

red lake with aluminium hydroxide.

Soak some quantitative filter (or drop-reaction) paper in a saturated alcoholic solution of alizarin and dry it. Place a drop of the acid test solution on the paper and hold it over ammonia fumes until a violet colour (due to ammonium alizarinate) appears. In the presence of large amounts of aluminium, the colour is visible almost immediately. It is best to dry the paper at 100°C when the violet colour due to ammonium alizarinate disappears owing to its conversion into ammonia and alizarin: the red colour of the alizarin lake is then clearly visible.

Sensitivity 0.15 µg Al.

Concentration limit 1 in 333 000.

Iron, chromium, uranium, thorium, titanium, and manganese interfere, but this may be obviated by using paper previously treated with potassium hexacyanoferrate(II). The interfering ions are thus 'fixed' on the paper as insoluble hexacyanoferrate(II)s, and the aluminium solution diffuses beyond as a damp ring. Upon adding a drop of a saturated alcoholic solution of alizarin, exposing to ammonia vapour and drying, a red ring of the alizarin-aluminium lake forms round the precipitate. Uranium hexacyanoferrate(II), owing to its slimy nature, has a tendency to spread outwards from the spot and thus obscure the aluminium lake; this difficulty is surmounted by dipping the paper after the alizarin treatment in ammonium carbonate solution which dissolves the uranium hexacyanoferrate(II).

Alizarin S
↓
red lake on Al(OH)$_3$

8. *Alizarin red S (or sodium alizarin sulphonate) reagent (0.1%)*

red precipitate or lake in ammoniacal solution, which is fairly stable to dilute acetic acid.

Place a drop of the test solution (which has been treated with just sufficient amount of M sodium hydroxide solution to give the tetrahydroxoaluminate ion $[Al(OH)_4]^-$) on a spot plate, add 1 drop of the reagent, then drops of acetic acid until the violet colour just disappears. Add 1 drop in excess. A red precipitate or colouration appears.

Sensitivity 0.7 µg Al.

Concentration limit 1 in 80 000.

A blank test should be made on the sodium hydroxide solution. Salts of Cu, Bi, Fe, Be, Zr, Ti, Co, Ce, rare earths, Zn, Th, U, Ca, Sr, and Ba interfere.

Quinalizarin
↓
red colour or ppt

9. *Quinalizarin (or 1,2,5,8-tetrahydroxyanthraquinone) reagent (0.05% in pyridine)*

red precipitate or colouration under the conditions given below.

Place a drop of the test solution upon the reagent paper; hold it for a short time over a bottle containing concentrated ammonia solution and then over glacial acetic acid until the blue colour (ammonium quinalizarinate) first formed disappears and the unmoistened paper regains the brown colour of free quinalizarin. A red-violet or red spot is formed.

Sensitivity 0.005 µg Al (drop of 0.1 ml).

Concentration limit 1 in 2 000 000.

Atomic spectrometry

10. *Atomic spectrometric tests*   tests for aluminium are less sensitive and require the nitrous oxide/acetylene flame. The dry test (see below) is perhaps more reliable and should be carried out to verify the spectrometric result.

(a) *Flame atomic emission spectrometry*   with a nitrous oxide/acetylene flame 1 ppm aluminium can be identified at 396.15 nm. The sensitivity with the air/acetylene flame is much reduced and some metals (e.g. iron) interfere.

(b) *Flame atomic absorption spectrometry*   using an aluminium hollow-cathode lamp and a nitrous oxide/acetylene flame, 1 ppm can be identified through the doublet at 309.28/29 nm.

(c) *Aluminium standard, 1000 ppm*   dissolve 17.583 g analytical grade potassium aluminium sulphate dodecahydrate $KAl(SO_4)_2 \cdot 12H_2O$ in water and dilute to 1 litre.

Dry tests

11. *Dry tests*   aluminium compounds when heated with sodium carbonate upon charcoal in the blowpipe flame give a white infusible solid, which glows when hot. If the residue is moistened with one or two drops of cobalt nitrate solution and again heated, a blue infusible mass (Thenard's blue, or cobalt meta-aluminate) is obtained. It is important not to use excess cobalt nitrate solution since this yields black cobalt oxide Co_3O_4 upon ignition, which masks the colour of the Thenard's blue.

$$2Al_2O_3 + 2Co^{2+} + 4NO_3^- \rightarrow 2CoAl_2O_4 + 4NO_2\uparrow + O_2\uparrow$$

An alternative method for carrying out this test is to soak some ashless filter paper in aluminium salt solution, add a drop or two of cobalt nitrate solution and then to ignite the filter paper in a crucible; the residue is coloured blue.

3.24   Chromium, Cr (A_r: 51.996) – chromium(III)

Chromium is a white, crystalline metal and is not appreciably ductile or malleable. It melts at 1765°C. The metal is soluble in dilute or concentrated hydrochloric acid. If air is excluded, chromium(II) ions are formed:

$$Cr + 2H^+ \rightarrow Cr^{2+} + H_2\uparrow$$

$$Cr + 2HCl \rightarrow Cr^{2+} + 2Cl^- + H_2\uparrow$$

In the presence of atmospheric oxygen, chromium gets partly or wholly oxidized to the tervalent state:

$$4Cr^{2+} + O_2 + 4H^+ \rightarrow 4Cr^{3+} + 2H_2O$$

Dilute sulphuric acid attacks chromium slowly, with the formation of hydrogen. In hot, concentrated sulphuric acid chromium dissolves readily, to form chromium(III) ions and sulphur dioxide:

$$2Cr + 6H_2SO_4 \rightarrow 2Cr^{3+} + 3SO_4^{2-} + 3SO_2\uparrow + 6H_2O$$

Both dilute and concentrated nitric acid render chromium passive, as does cold, concentrated sulphuric acid and aqua regia.

In aqueous solutions chromium forms three types of ions: the chromium(II) and chromium(III) cations and the chromate (and dichromate) anion, in which chromium has an oxidation state of +6.

The chromium(II) (or chromous) ion (Cr^{2+}) is derived from chromium(II) oxide CrO. These ions form blue coloured solutions. Chromium(II) ions are rather unstable, as they are strong reducing agents – they decompose even water slowly with the formation of hydrogen. Atmospheric oxygen oxidizes them readily to chromium(III) ions. As they are only rarely encountered in inorganic qualitative analysis, they will not be dealt with here.

Chromium(III) (or chromic) ions (Cr^{3+}) are stable and are derived from dichromium trioxide (or chromium trioxide) Cr_2O_3. In solution they are either green or violet. In the green solutions the pentaquomonochlorochromate(III) $[Cr(H_2O)_5Cl]^{2+}$ or the tetraquodichlorochromate(III) $[Cr(H_2O)_4Cl_2]^+$ complex is present (chloride may be replaced by another monovalent anion), while in the violet solutions the hexaquochromate(III) ion $[Cr(H_2O)_6]^{3+}$ is present. Chromium(III) sulphide, like aluminium sulphide, can be prepared only in the dry state; it hydrolyses readily with water to form chromium(III) hydroxide and hydrogen sulphide.

In the chromate, CrO_4^{2-}, or dichromate, $Cr_2O_7^{2-}$, anions, chromium is hexavalent with an oxidation state of $+6$. These ions are derived from chromium trioxide CrO_3. Chromate ions are yellow while dichromates have an orange colour. Chromates are readily transformed into dichromates on addition of acid:

$$2CrO_4^{2-} + 2H^+ \rightleftharpoons Cr_2O_7^{2-} + H_2O$$

The reaction is reversible. In neutral (or alkaline) solutions the chromate ion is stable, while, if acidified, dichromate ions will be predominant. Chromate and dichromate ions are strong oxidizing agents. Their reactions will be dealt with among the reactions of anions (Section 4.33).

Cr(III) reactions

Reactions of chromium(III) ions To study these reactions use a 0.33M solution of chromium(III) chloride $CrCl_3 \cdot 6H_2O$ or a 0.166M solution of chromium(III) sulphate $Cr_2(SO_4)_3 \cdot 15H_2O$.

NH_3
↓
grey-green $Cr(OH)_3$

Dissolves in excess

1. *Ammonia solution* grey-green to grey-blue gelatinous precipitate of chromium(III) hydroxide, $Cr(OH)_3$, slightly soluble in excess of the precipitant in the cold forming a violet or pink solution containing complex hexamminechromate(III) ion; upon boiling the solution, chromium hydroxide is precipitated. Hence for complete precipitation of chromium as the hydroxide, it is essential that the solution be boiling and excess aqueous

ammonia solution be avoided.

$$Cr^{3+} + 3NH_3 + 3H_2O \rightarrow Cr(OH)_3\downarrow + 3NH_4^+$$

$$Cr(OH)_3\downarrow + 6NH_3 \rightarrow [Cr(NH_3)_6]^{3+} + 3OH^-$$

In the presence of acetate ions and the absence of other trivalent metal ions, chromium(III) hydroxide is not precipitated. The precipitation of chromium(III) hydroxide is also prevented by tartrates and by citrates.

NaOH
↓
grey-green Cr(OH)₃

2. *Sodium hydroxide solution*   precipitate of chromium(III) hydroxide

$$Cr^{3+} + 3OH^- \rightarrow Cr(OH)_3\downarrow$$

Dissolves in excess

The reaction is reversible; on the addition of acids the precipitate dissolves. In excess reagent, the precipitate dissolves readily, tetrahydroxochromate(III) ions (or chromite ions) being formed:

$$Cr(OH)_3 + OH^- \rightleftharpoons [Cr(OH)_4]^-$$

The solution is green. The reaction is reversible; on (slight) acidification and also on boiling chromium(III) hydroxide precipitates again.

On adding hydrogen peroxide to the alkaline solution of tetrahydroxochromate(III), a yellow solution is obtained, owing to the oxidation of chromium(III) to chromate:

$$2[Cr(OH)_4]^- + 3H_2O_2 + 2OH^- \rightarrow 2CrO_4^{2-} + 8H_2O$$

After decomposing the excess of hydrogen peroxide by boiling, chromate ions may be identified in the solution by one of its reactions (cf. Section 4.33).

Na₂CO₃
↓
hydrolysis to Cr(OH)₃

3. *Sodium carbonate solution*   precipitate of chromium(III) hydroxide:

$$2Cr^{3+} + 3CO_3^{2-} + 3H_2O \rightarrow 2Cr(OH)_3\downarrow + 3CO_2\uparrow$$

(NH₄)₂S
↓
hydrolysis to Cr(OH)₃

4. *Ammonium sulphide solution*   precipitate of chromium(III) hydroxide:

$$2Cr^{3+} + 3S^{2-} + 6H_2O \rightarrow 2Cr(OH)_3\downarrow + 3H_2S\uparrow$$

5. *Chromate test*   chromium(III) ions can be oxidized to chromate in several ways.

Oxidation to Cr(VI)

(a) Adding an excess of sodium hydroxide to a chromium(III) salt followed by a few ml of 10% hydrogen peroxide (cf. reaction 2). The excess of hydrogen peroxide can be removed by boiling the mixture for a few minutes.

(b) Hydrogen peroxide can be replaced by a little solid sodium perborate, $NaBO_3 \cdot 4H_2O$, in experiment (a). Perborate ions, when hydrolysed, form hydrogen peroxide. In alkaline medium this reaction can be written as:

$$BO_3^- + 2OH^- \rightarrow BO_3^{3-} + H_2O_2$$

(c) Oxidation can be carried out with bromine water in alkaline solution (i.e. by hypobromite):

$$2Cr^{3+} + 3OBr^- + 10OH^- \rightarrow 2CrO_4^{2-} + 3Br^- + 5H_2O$$

The excess of bromine can be removed by the addition of phenol, which gives tribromophenol.

(d) In acid solution chromium(III) ions can be oxidized by potassium (or ammonium) peroxodisulphate:

$$2Cr^{3+} + 3S_2O_8^{2-} + 8H_2O \rightarrow 2CrO_4^{2-} + 16H^+ + 6SO_4^{2-}$$

One drop of dilute silver nitrate must be added to speed up the reaction. Silver ions act as catalysts. Halides must be absent; they can be removed easily by evaporating the solution with concentrated sulphuric acid until fumes of sulphur trioxide appear. After cooling, the solution can be diluted and the test carried out. The excess of peroxodisulphate can be decomposed by boiling:

$$2S_2O_8^{2-} + 2H_2O \rightarrow 4SO_4^{2-} + 4H^+ + O_2\uparrow$$

Identification as CrO_4^{2-}

6. *Identification of chromium after oxidation to chromate*   having carried out the oxidation with one of the methods described under reaction 5, chromate ions may be identified with one of the following tests:

(a) *Barium chloride test*   after acidifying the solution with acetic acid and adding barium chloride solution, a yellow precipitate of barium chromate is formed:

$$Ba^{2+} + CrO_4^{2-} \rightarrow BaCrO_4\downarrow$$

(b) *Chromium pentoxide (chromium peroxide, peroxochromic acid) test*   on acidifying the solution with dilute sulphuric acid, adding 2–3 ml of ether or amyl alcohol to the mixture and finally adding some hydrogen peroxide, a blue colouration is formed, which can be extracted into the organic phase by gently shaking. During the reaction chromium pentoxide is formed:

$$CrO_4^{2-} + 2H^+ + 2H_2O_2 \rightarrow CrO_5 + 3H_2O$$

Chromium pentoxide has the following structure:

$$\begin{array}{ccc} O & & O \\ | & \diagdown Cr \diagup & | \\ O & \diagup \, \| \, \diagdown & O \\ & O & \end{array}$$

Because of the two peroxide groups the compound is often called chromium peroxide. The name peroxochromic acid is less appropriate, because the compound does not contain hydrogen at all. In aqueous solution the blue colour fades rapidly, because chromium pentoxide decomposes to chromium(III) and oxygen:

$$4CrO_5 + 12H^+ \rightarrow 4Cr^{3+} + 7O_2\uparrow + 6H_2O$$

(c) *1,5-Diphenylcarbazide test*   in dilute mineral acid solution diphenylcarbazide produces a soluble violet colour, which is a characteristic test for chromium. During the reaction chromate is reduced to chromium(III), and diphenylcarbazone is formed; these reaction products in turn produce a complex with the characteristic colour:

$$
\underset{\substack{\text{NH}-\text{NH}-\text{C}_6\text{H}_5 \\ \diagup \\ 3\text{C}=\text{O} \\ \diagdown \\ \text{NH}-\text{NH}-\text{C}_6\text{H}_5}}{} + 4\text{CrO}_4^{2-} + 20\text{H}^+ \rightarrow \underset{\substack{\text{N}=\text{N}-\text{C}_6\text{H}_5 \\ \diagup \\ 3\text{C}=\text{O} \\ \diagdown \\ \text{N}=\text{N}-\text{C}_6\text{H}_5}}{} + \text{Cr}^{3+} + 16\text{H}_2\text{O}
$$

diphenylcarbazide diphenylcarbazone

$$
\underset{\substack{\text{N}=\text{N}-\text{C}_6\text{H}_5 \\ \diagup \\ \text{C}=\text{O} \\ \diagdown \\ \text{N}=\text{N}-\text{C}_6\text{H}_5}}{} + \text{Cr}^{3+} \rightarrow \left[\underset{\substack{\text{N}=\text{N}-\text{C}_6\text{H}_5 \\ \diagup \\ \text{C}-\text{O}-\text{Cr} \\ \diagdown \\ \text{N}=\text{N}-\text{C}_6\text{H}_5}}{} \right]^{3+}
$$

diphenylcarbazone-chromium(III)
complex

The reaction can be applied as a drop test for chromium; in this case the preliminary oxidation to chromate can also be made on the spot plate.

Place a drop of the test solution in mineral acid on a spot plate, introduce a drop of saturated bromine water followed by 2–3 drops 2M potassium hydroxide (the solution must be alkaline to litmus). Mix thoroughly, add a crystal of phenol, then a drop of the 1,5-diphenylcarbazide reagent (1% in ethanol), and finally M sulphuric acid dropwise until the red colour (from the reaction between diphenylcarbazide and alkali) disappears. A blue-violet colouration is obtained.

Sensitivity 0.25 µg Cr.

Concentration limit 1 in 2 000 000.

Alternatively, mix a drop of the acidified test solution on a spot plate with 2 drops 0.1M potassium peroxodisulphate solution and 1 drop 0.1M silver nitrate solution, and allow to stand for 2–3 minutes. Add a drop of the reagent. A violet or red colour is formed.

Sensitivity 0.8 µg Cr.

Concentration limit 1 in 600 000.

(d) *Chromotropic acid test* the sodium salt of 1,8-dihydroxynaphthalene-3,6-disulphonic acid (or chromotropic acid)

gives a red colouration (by transmitted light) with an alkali chromate in the presence of nitric acid. Chromium(III) salts may be oxidized with hydrogen peroxide and alkali to chromate, and then acidified with nitric acid before applying the test. The nitric acid serves to eliminate the influence of Fe, U, and Ti, which otherwise interfere.

Place a drop of the test solution in a semimicro test-tube, add a drop of the reagent, a drop of dilute nitric acid (8M), and dilute to about 2 ml. Chromates give a red colouration: this is best observed with a white light behind the tube.

Concentration limit 1 in 5000.

7. *Atomic spectrometric tests*

(a) *Flame atomic emission spectrometry*   with an air/acetylene flame 10 ppm chromium can be identified at 425.44 nm. With the nitrous oxide/acetylene flame the sensitivity increases tenfold.

(b) *Flame atomic absorption spectrometry*   using a chromium hollow-cathode lamp and an air/acetylene flame, 1 ppm can be identified at 357.87 nm. The nitrous oxide/acetylene flame allows somewhat better sensitivity, and the signal is less prone to interference by transition metals and phosphate.

(c) *Chromium standard, 1000 ppm*   dissolve 2.8290 g analytical grade potassium dichromate $K_2Cr_2O_7$ in water and dilute to 1 litre.

8. *Dry tests*

(a) *Blowpipe test*   all chromium compounds when heated with sodium carbonate on charcoal yield a green infusible mass of chromium(III) oxide Cr_2O_3.

(b) *Borax bead test*   the bead is coloured green in both oxidizing and reducing flames, but is not very characteristic.

(c) *Fusion*   with sodium carbonate and potassium nitrate in a loop of platinum wire or upon platinum foil or upon the lid of a nickel crucible results in the formation of a yellow mass of alkali chromate.

$$Cr_2(SO_4)_3 + 5Na_2CO_3 + 3KNO_3$$

$$\rightarrow 2Na_2CrO_4 + 3KNO_2 + 3Na_2SO_4 + 5CO_2\uparrow$$

3.25   Oxoanions of Group III metals: chromate and permanganate

The oxoanions of certain Group III metals, like chromate CrO_4^{2-} (and dichromate, $Cr_2O_7^{2-}$) as well as permanganate MnO_4^-, are reduced by hydrogen sulphide in hydrochloric acid media to chromium(III) and manganese(II) ions respectively. In the course of the systematic analysis of an unknown sample (cf. Chapter 5), these anions will already be converted into the corresponding Group III cations when the separating process reaches this stage. If therefore chromium(III) and/or manganese(II) are found, the initial oxidation states of these metals should be tested in the original sample.

The reactions of these ions are described in Chapter 4; those of chromates and dichromates in Section 4.33, and the reactions of permanganates in Section 4.34.

3.26   Cobalt, Co (A_r: 58.93)

Cobalt is a steel-grey, slightly magnetic metal. It melts at 1490°C. The metal dissolves readily in dilute mineral acids:

$$Co + 2H^+ \rightarrow Co^{2+} + H_2\uparrow$$

The dissolution in nitric acid is accompanied by the formation of nitrogen oxide:

$$3Co + 2HNO_3 + 6H^+ \rightarrow 3Co^{2+} + 2NO\uparrow + 4H_2O$$

In aqueous solutions cobalt is normally present as the cobalt(II) ion Co^{2+}; sometimes, especially in complexes, the cobalt(III) ion Co^{3+} is encountered. These two ions are derived from the oxides CoO and Co_2O_3 respectively. The cobalt(II)–cobalt(III) oxide Co_3O_4 is also known.

In the aqueous solutions of cobalt(II) compounds the red Co^{2+} ions are present. Anhydrous or undissociated cobalt(II) compounds are blue. If the dissociation of cobalt compounds is suppressed, the colour of the solution turns gradually to blue.

Cobalt(III) ions Co^{3+} are unstable, but their complexes are stable both in solution and in dry form. Cobalt(II) complexes can easily be oxidized to cobalt(III) complexes.

Co(II) reactions

Reactions of cobalt(II) ions The reactions of cobalt(II) ions can be studied with a 0.5M solution of cobalt(II) chloride $CoCl_2 \cdot 6H_2O$ or cobalt(II) nitrate $Co(NO_3)_2 \cdot 6H_2O$.

NaOH
↓
pink Co(OH)₂

1. *Sodium hydroxide solution* in cold a blue basic salt is precipitated:

$$Co^{2+} + OH^- + NO_3^- \rightarrow Co(OH)NO_3\downarrow$$

Upon warming with excess alkali (or sometimes merely upon addition of excess reagent) the basic salt is converted into a pink precipitate of cobalt(II) hydroxide:

$$Co(OH)NO_3\downarrow + OH^- \rightarrow Co(OH)_2\downarrow + NO_3^-$$

Some of the precipitate however passes into solution.

The hydroxide is slowly transformed into the brownish-black cobalt(III) hydroxide on exposure to the air:

$$4Co(OH)_2\downarrow + O_2 + 2H_2O \rightarrow 4Co(OH)_3\downarrow$$

Cobalt(II) hydroxide precipitate is readily soluble in ammonia or concentrated solutions of ammonium salts, provided that the mother liquor is alkaline.

NH₃
↓
basic salt

2. *Ammonia solution* in the absence of ammonium salts small amounts of ammonia precipitate the basic salt as in reaction 1:

$$Co^{2+} + NH_3 + H_2O + NO_3^- \rightarrow Co(OH)NO_3\downarrow + NH_4^+$$

Dissolves in excess

The excess of the reagent dissolves the precipitate, when hexamminecobaltate(II) ions are formed:

$$Co(OH)NO_3\downarrow + 6NH_3 \rightarrow [Co(NH_3)_6]^{2+} + NO_3^- + OH^-$$

The precipitation of the basic salt does not take place at all if larger amounts of ammonium ions are present, but the complex is formed in one step.

(NH₄)₂S
↓
black CoS

3. *Ammonium sulphide solution* black precipitate of cobalt(II) sulphide from neutral or alkaline solution:

$$Co^{2+} + S^{2-} \rightarrow CoS\downarrow$$

The precipitate is insoluble in dilute hydrochloric or acetic acids (though precipitation does not take place from such media). Hot, concentrated nitric acid or aqua regia dissolve the precipitate, and white sulphur remains:

$$3CoS\downarrow + 2HNO_3 + 6H^+ \rightarrow 3Co^{2+} + 3S\downarrow + 2NO\uparrow + 4H_2O$$

$$CoS\downarrow + HNO_3 + 3HCl \rightarrow Co^{2+} + S\downarrow + NOCl\uparrow + 2Cl^- + 2H_2O$$

On longer heating the mixture becomes clear because sulphur is oxidized to sulphate.

KCN
↓
brown Co(CN)$_2$

4. *Potassium cyanide solution* (**POISON**) reddish-brown precipitate of cobalt(II) cyanide:

$$Co^{2+} + 2CN^- \rightarrow Co(CN)_2\downarrow$$

Dissolves in excess

The precipitate dissolves in excess reagent; a brown solution of hexacyanocobaltate(II) is formed:

$$Co(CN)_2\downarrow + 4CN^- \rightarrow [Co(CN)_6]^{4-}$$

On acidification *in the cold* with dilute hydrochloric acid the precipitate reappears:

$$[Co(CN)_6]^{4-} + 4H^+ \rightarrow Co(CN)_2\downarrow + 4HCN\uparrow$$

The experiment must be carried out in a fume cupboard with good ventilation.

If the brown solution is boiled for a longer time in air, or if some hydrogen peroxide is added and the solution heated, it turns yellow because hexacyanocobaltate(III) ions are formed:

$$4[Co(CN)_6]^{4-} + O_2 + 2H_2O \rightarrow 4[Co(CN)_6]^{3-} + 4OH^-$$

$$2[Co(CN)_6]^{4-} + H_2O_2 \rightarrow 2[Co(CN)_6]^{3-} + 2OH^-$$

KNO$_2$
↓
yellow K$_3$[Co(NO$_2$)$_6$]

5. *Potassium nitrite solution* yellow precipitate of potassium hexanitritocobaltate(III), $K_3[Co(NO_2)_6]\cdot 3H_2O$:

$$Co^{2+} + 7NO_2^- + 2H^+ + 3K^+ \rightarrow K_3[Co(NO_2)_6]\downarrow + NO\uparrow + H_2O$$

Specific for Co

The test can be carried out most conveniently as follows: to a neutral solution of cobalt(II) add acetic acid, then a freshly prepared saturated solution of potassium nitrite. If the concentration of cobalt(II) in the test solution is high enough, the precipitate appears immediately. Otherwise the mixture should either be slightly heated, or the wall of the vessel should be rubbed with a glass rod.

The reaction can also be used for testing for potassium and for nitrite ions. Nickel ions do not react if acetic acid is present.

NH$_4$SCN
↓
blue colour

6. *Ammonium thiocyanate test (Vogel reaction)* on adding a few crystals of ammonium thiocyanate to a neutral or acid solution of cobalt(II) a blue colour appears owing to the formation of tetrathiocyanatocobaltate(II) ions:

$$Co^{2+} + 4SCN^- \rightarrow [Co(SCN)_4]^{2-}$$

If amyl alcohol or diethyl ether is added, the free acid $H_2[Co(SCN)_4]$ is formed and dissolved by the organic solvent (distinction from nickel). The test is rendered more sensitive if the solution is acidified with concentrated hydrochloric acid, when the equilibrium

$$2H^+ + [Co(SCN)_4]^{2-} \rightleftarrows H_2[Co(SCN)_4]$$

shifts towards the formation of the free acid, which then can be extracted with amyl alcohol or ether.

The reaction may be employed as a spot test as follows. On a spot plate, mix 1 drop test solution with 5 drops saturated solution of ammonium thiocyanate in acetone. A green to blue colouration appears.

Sensitivity 0.5 μg Co.

Concentration limit 1 in 100 000.

If iron is present mix 1–2 drops of the slightly acid test solution with a few milligrams of ammonium or sodium fluoride on a spot plate. Add 5 drops saturated solution of ammonium thiocyanate in acetone. A blue colouration is produced.

Sensitivity 1 μg Co in the presence of 100 times that amount of Fe.

Concentration limit 1 in 50 000.

1-Nitroso-2-naphthol
↓
reddish-brown ppt

7. *1-Nitroso-2-naphthol reagent*

Selective for Co

reddish-brown precipitate of slightly impure cobalt(III)-nitroso-2-naphthol $Co(C_{10}H_6O_2N)_3$ (a chelate complex) in solutions acidified with dilute hydrochloric acid or dilute acetic acid: the precipitate may be extracted by chloroform to give a claret-coloured solution.

The technique for using the reaction as a spot test is as follows. Place a drop of the faintly acid test solution on drop-reaction paper, and add a drop of the reagent. A brown stain is produced.

Sensitivity 0.05 μg Co.

Concentration limit 1 in 1 000 000.

Nitroso-R-salt
↓
red colour

8. *Sodium 1-nitroso-2-hydroxynaphthalene-3:6-disulphonate (nitroso-R-salt) reagent*

deep-red colouration. The test is applicable in the presence of nickel; tin and iron interfere and should be removed; the colouration produced by iron is prevented by the addition of an alkali fluoride.

Place a drop of the neutral test solution (buffered with sodium acetate) on a spot plate, and add 2–3 drops of the reagent. A red colouration is obtained.

Concentration limit 1 in 500 000.

Atomic spectrometry

9. *Atomic spectrometric tests*

(a) *Flame atomic emission spectrometry* reliable results can only be obtained with a high-temperature nitrous oxide/acetylene flame, when 5 ppm cobalt can be identified at 345.35 nm. With an air/acetylene flame only about 100 ppm cobalt produces a distinct analytical signal.

(b) *Flame atomic absorption spectrometry* using a cobalt hollow-cathode lamp and an air/acetylene flame, 1 ppm can be identified at 240.73 nm.

(c) *Cobalt standard, 1000 ppm* dissolve 4.9383 g analytical grade cobalt(II) nitrate hexahydrate $Co(NO_3)_2 \cdot 6H_2O$ in water and dilute to 1 litre.

Dry tests

10. *Dry tests*

(a) *Blowpipe test* all cobalt compounds when ignited with sodium carbonate on charcoal give grey, slightly metallic beads of cobalt. If these are removed, placed upon filter paper and dissolved by the addition of a few drops of dilute nitric acid, a few drops of concentrated hydrochloric acid added and the filter paper dried, the latter is coloured blue by the cobalt chloride produced.

(b) *Borax bead test* this gives a blue bead in both the oxidizing and reducing flames. Cobalt metaborate $Co(BO_2)_2$, or the complex salt $Na_2Co(BO_2)_4$, is formed. The presence of a large proportion of nickel does not interfere.

3.27 Nickel, Ni (A_r: 58.71)

Nickel is a hard, silver-white metal; it is ductile and malleable. It melts at 1455°C. It is slightly magnetic.

Hydrochloric acid (both dilute and concentrated) and dilute sulphuric acid dissolve nickel with the formation of hydrogen:

$$Ni + 2H^+ \rightarrow Ni^{2+} + H_2\uparrow$$

$$Ni + 2HCl \rightarrow Ni^{2+} + 2Cl^- + H_2\uparrow$$

These reactions accelerate if the solution is heated. Concentrated, hot sulphuric acid dissolves nickel with the formation of sulphur dioxide:

$$Ni + H_2SO_4 + 2H^+ \rightarrow Ni^{2+} + SO_2\uparrow + 2H_2O$$

Dilute and concentrated nitric acid dissolve nickel readily in cold:

$$3Ni + 2HNO_3 + 6H^+ \rightarrow 3Ni^{2+} + 2NO\uparrow + 4H_2O$$

The stable nickel(II) salts are derived from nickel(II) oxide, NiO, which is a green substance. The dissolved nickel(II) salts are green, owing to the colour of the hexaquonickelate(II) complex $[Ni(H_2O)_6]^{2+}$; in short, however, this will be regarded as the simple nickel(II) ion Ni^{2+}. A brownish-black nickel(III) oxide Ni_2O_3 also exists, but this dissolves in acids forming nickel(II) ions. With dilute hydrochloric acid this reaction yields chlorine gas:

$$Ni_2O_3 + 6H^+ + 2Cl^- \rightarrow 2Ni^{2+} + Cl_2\uparrow + 3H_2O$$

Ni(II) reactions

Reactions of nickel(II) ions For the study of these reactions use a 0.5M solution of nickel sulphate $NiSO_4 \cdot 7H_2O$ or nickel chloride $NiCl_2 \cdot 6H_2O$.

NaOH
↓
green $Ni(OH)_2$

1. *Sodium hydroxide solution* green precipitation of nickel(II) hydroxide:

$$Ni^{2+} + 2OH^- \rightarrow Ni(OH)_2\downarrow$$

The precipitate is insoluble in excess reagent. No precipitation occurs in the presence of tartrate or citrate, owing to complex formation. Ammonia dissolves the precipitate; in the presence of excess alkali hydroxide, ammonium salts will also dissolve the precipitate:

$$Ni(OH)_2\downarrow + 6NH_3 \rightarrow [Ni(NH_3)_6]^{2+} + 2OH^-$$

$$Ni(OH)_2\downarrow + 6NH_4^+ + 4OH^- \rightarrow [Ni(NH_3)_6]^{2+} + 6H_2O$$

The solution of hexamminenickelate(II) ions is deep blue; it can be easily mistaken for copper(II) ions, which form the blue tetramminecuprate(II) ions in an analogous reaction (cf. Section 3.10). The solution does not oxidize on boiling with free exposure to air or upon the addition of hydrogen peroxide (difference from cobalt).

The green nickel(II) hydroxide precipitate can be oxidized to black nickel(III) hydroxide with sodium hypochlorite solution:

$$2Ni(OH)_2 + ClO^- + H_2O \rightarrow 2Ni(OH)_3\downarrow + Cl^-$$

Hydrogen peroxide solution, however, does not oxidize nickel(II) hydroxide, but the precipitate catalyses the decomposition of hydrogen peroxide to oxygen and water

$$2H_2O_2 \xrightarrow{Ni(OH)_2\downarrow} 2H_2O + O_2\uparrow$$

without any other visible change.

NH_3
↓
green $Ni(OH)_2$

2. *Ammonia solution* green precipitate of nickel(II) hydroxide:

$$Ni^{2+} + 2NH_3 + 2H_2O \rightarrow Ni(OH)_2\downarrow + 2NH_4^+$$

Dissolves in excess

which dissolves in excess reagent:

$$Ni(OH)_2\downarrow + 6NH_3 \rightarrow [Ni(NH_3)_6]^{2+} + 2OH^-$$

the solution turns deep blue (cf. reaction 1). If ammonium salts are present, no precipitation occurs, but the complex is formed immediately.

$(NH_4)_2S$
↓
black NiS

3. *Ammonium sulphide solution* black precipitate of nickel sulphide from neutral or slightly alkaline solutions:

$$Ni^{2+} + S^{2-} \rightarrow NiS\downarrow$$

If the reagent is added in excess, a dark-brown colloidal solution is formed which runs through the filter paper. If the colloidal solution is boiled or if it is rendered slightly acid with acetic acid and boiled, the colloidal solution (hydrosol) is coagulated and can then be filtered. The presence of large quantities of ammonium chloride usually prevents the formation of the sol. Nickel sulphide is practically insoluble in cold dilute hydrochloric acid (distinction from the sulphides of manganese and zinc) and in acetic acid, but dissolves in hot concentrated nitric acid and in aqua regia with the separation of sulphur:

$$3NiS\downarrow + 2HNO_3 + 6H^+ \rightarrow 3Ni^{2+} + 2NO\uparrow + 3S\downarrow + 4H_2O$$

$$NiS\downarrow + HNO_3 + 3HCl \rightarrow Ni^{2+} + S\downarrow + NOCl\uparrow + 2Cl^- + 2H_2O$$

On longer heating, sulphur dissolves and the solution becomes clear:

$$S\downarrow + 2HNO_3 \rightarrow SO_4^{2-} + 2H^+ + 2NO\uparrow$$

$$S\downarrow + 3HNO_3 + 9HCl \rightarrow SO_4^{2-} + 6Cl^- + 3NOCl\uparrow + 8H^+ + 2H_2O$$

4. *Hydrogen sulphide (gas or saturated aqueous solution)*   only part of the nickel is precipitated as nickel sulphide from neutral solutions and the reaction is slow; no precipitation occurs from solutions containing mineral acid or much acetic acid. Complete precipitation occurs, however, from solutions made alkaline with ammonia solution, or from solutions containing excess of alkali acetate made slightly acid with acetic acid.

KCN
↓
green Ni(CN)₂

5. *Potassium cyanide solution* (**POISON**)   green precipitate of nickel(II) cyanide

$$Ni^{2+} + 2CN^- \rightarrow Ni(CN)_2\downarrow$$

Dissolves in excess

The precipitate is readily soluble in excess reagent, when a yellow solution appears owing to the formation of tetracyanonickelate(II) complex ions:

$$Ni(CN)_2\downarrow + 2CN^- \rightarrow [Ni(CN)_4]^{2-}$$

Dilute hydrochloric acid decomposes the complex, and the precipitate appears again. A fume cupboard with good ventilation must be used for this test:

$$[Ni(CN)_4]^{2-} + 2H^+ \rightarrow Ni(CN)_2\downarrow + 2HCN\uparrow$$

If the solution of tetracyanonickelate(II) is heated with sodium hypobromite solution (prepared *in situ* by adding bromine water to sodium hydroxide solution), the complex decomposes and a black nickel(III) hydroxide precipitate is formed (difference from cobalt ions):

$$2[Ni(CN)_4]^{2-} + OBr^- + 4OH^- + H_2O \rightarrow 2Ni(OH)_3\downarrow + 8CN^- + Br^-$$

Excess potassium cyanide and/or excess bromine water should be avoided, because these react with the formation of cyanogen bromide, **which is poisonous and causes watering of the eyes:**

$$CN^- + Br_2 \rightarrow BrCN\uparrow + Br^-$$

6. *Potassium nitrite solution*   no precipitate is produced in the presence of acetic acid (difference from cobalt).

1-Nitroso-2-naphthol
↓
brown ppt

7. *1-Nitroso-2-naphthol reagent (cf. Section 3.26, reaction 7)*   brown precipitate of composition $Ni(C_{10}H_6O_2N)_2$, which is soluble in hydrochloric acid (different from cobalt, which produces a reddish-brown precipitate, insoluble in dilute hydrochloric acid).

Dimethylglyoxime
↓
red ppt

Specific for Ni

8. *Dimethylglyoxime reagent* ($C_4H_8O_2N_2$) red precipitate of nickel dimethylglyoxime from solutions just made alkaline with ammonia or acid solutions buffered with sodium acetate:

$$Ni^{2+} + 2 \begin{array}{c} CH_3-C=N-OH \\ | \\ CH_3-C=N-OH \end{array} \longrightarrow [\text{nickel dimethylglyoxime complex}] + 2H^+$$

Iron(II) (red colouration), bismuth (yellow precipitate), and larger amounts of cobalt (brown colouration) interfere in ammoniacal solution. The influence of interfering elements (Fe^{2+} must be oxidized to Fe^{3+}, say, by hydrogen peroxide) can be eliminated by the addition of a tartrate. When large quantities of cobalt salts are present, they react with the dimethylglyoxime and a special procedure must be adopted (see below). Oxidizing agents must be absent. Palladium, platinum, and gold give precipitates in acid solution.

The spot test technique is as follows. Place a drop of the test solution on drop-reaction paper, add a drop of the above reagent and hold the paper over ammonia vapour. Alternatively, place a drop of the test solution and a drop of the reagent on a spot plate, and add a drop of dilute ammonia solution. A red spot or precipitate (or colouration) is produced.

Sensitivity 0.16 µg Ni.

Concentration limit 1 in 300 000.

Detection of traces of nickel in cobalt salts the solution containing the cobalt and nickel is treated with excess concentrated potassium cyanide (**POISON**) solution, followed by 30% hydrogen peroxide whereby the complex cyanides $[Co(CN)_6]^{3-}$ and $[Ni(CN)_4]^{4-}$ respectively are formed. Upon adding 40% formaldehyde solution the hexacyanocobaltate(III) is unaffected (and hence remains inactive to dimethylglyoxime) whereas the tetracyanato-nickelate(II) decomposes with the formation of nickel cyanide, which reacts immediately with the dimethylglyoxime.

$$[Ni(CN)_4]^{2-} + 2HCHO \rightarrow Ni(CN)_2\downarrow + 2CH_2(CN)O^-$$

$$Ni(CN)_2\downarrow + 2C_4H_8O_2N_2 \rightarrow Ni(C_4H_7O_2N_2)_2\downarrow + 2HCN\uparrow$$

Atomic spectrometry

9. *Atomic spectrometric tests*

(a) *Flame atomic emission spectrometry* with an air/acetylene flame 1 ppm nickel can be identified at 341.48 nm.

(b) *Flame atomic absorption spectrometry* using a nickel hollow-cathode lamp and an air/acetylene flame, 0.1 ppm can be identified at 341.48 nm.

(c) *Nickel standard, 1000 ppm* dissolve 4.9533 g analytical grade nickel(II) nitrate hexahydrate $Ni(NO_3)_2 \cdot 6H_2O$ in water and dilute to 1 litre.

10. *Dry tests*

(a) *Blowpipe test* all nickel compounds, when heated with sodium carbonate on charcoal, yield grey, slightly magnetic scales of metallic nickel. These can be placed on a strip of filter paper and dissolved by a few drops of 8M nitric acid. After adding a few drops of concentrated hydrochloric acid, the paper can be dried by moving it back and forth in a flame, or by placing it on the outside of a test-tube containing hot water. The paper acquires a green colour owing to the formation of nickel(II) chloride. On moistening the filter paper with ammonia solution and adding a few drops of dimethylglyoxime reagent, a red colour is produced.

(b) *Borax bead test* this is coloured brown in the oxidizing flame, due to the formation of nickel metaborate, $Ni(BO_2)_2$, or of the complex metaborate $Na_2[Ni(BO_2)_4]$, and grey, due to metallic nickel, in the reducing flame.

3.28 Manganese, Mn (A_r: 54.938) – manganese(II)

Manganese is a greyish-white metal, similar in appearance to cast iron. It melts at about 1250°C. It reacts with warm water forming manganese(II) hydroxide and hydrogen:

$$Mn + 2H_2O \rightarrow Mn(OH)_2\downarrow + H_2\uparrow$$

Dilute mineral acids and also acetic acid dissolve it with the production of manganese(II) salts and hydrogen:

$$Mn + 2H^+ \rightarrow Mn^{2+} + H_2\uparrow$$

When it is attacked by hot, concentrated sulphuric acid, sulphur dioxide is evolved:

$$Mn + 2H_2SO_4 \rightarrow Mn^{2+} + SO_4^{2-} + SO_2\uparrow + 2H_2O$$

Six oxides of manganese are known: MnO, Mn_2O_3, MnO_2, MnO_3, Mn_2O_7, and Mn_3O_4. The first five of these correspond to oxidation states $+2$, $+3$, $+4$, $+6$ and $+7$ respectively, while the last, Mn_3O_4, is a manganese(II)–manganese(III) oxide ($MnO \cdot Mn_2O_3$).

The manganese(II) cations are derived from manganese(II) oxide. They form colourless salts, although if the compound contains water of crystallization, and in solutions they are slightly pink; this is due to the presence of the hexaquomanganate(II) ion $[Mn(H_2O)_6]^{2+}$.

Manganese(III) ions are unstable; some complexes, containing manganese in the $+3$ oxidation state, are however known. They are easily reduced to manganese(II) ions. Although they can be derived from the manganese(III) oxide, Mn_2O_3, the latter, when treated with mineral acids, produces manganese(II) ions. If hydrochloric acid is used, chlorine is the by-product:

$$Mn_2O_3\downarrow + 6HCl \rightarrow 2Mn^{2+} + Cl_2\uparrow + 4Cl^- + 3H_2O$$

With sulphuric acid, oxygen is formed:

$$2Mn_2O_3 + 4H_2SO_4 \rightarrow 4Mn^{2+} + O_2\uparrow + 4SO_4^{2-} + 4H_2O$$

Manganese(IV) compounds, with the exception of the manganese(IV) oxide (or manganese dioxide), MnO_2, are unstable, as both the manganese(IV) ion, Mn^{4+}, and the manganate(IV) (or manganite) ion, MnO_3^{2-}, are easily reduced to manganese(II). When dissolved in concentrated hydrochloric or sulphuric acid, manganese(IV) oxide produces manganese(II) ions as well as chlorine and oxygen gas respectively:

$$MnO_2\downarrow + 4HCl \rightarrow Mn^{2+} + Cl_2\uparrow + 2Cl^- + 2H_2O$$

$$2MnO_2\downarrow + 2H_2SO_4 \rightarrow 2Mn^{2+} + O_2\uparrow + 2SO_4^{2-} + 2H_2O$$

Manganese(VI) compounds contain the manganate(VI) MnO_4^{2-} anion. This is stable in alkaline solutions, and possesses a green colour. Upon neutralization a disproportionation reaction takes place; manganese dioxide precipitate and manganate(VII) (permanganate) ions are formed:

$$3MnO_4^{2-} + 2H_2O \rightarrow MnO_2\downarrow + 2MnO_4^- + 4OH^-$$

If manganese(VI) oxide is treated with acids, manganese(II) ions are produced. With hot, concentrated sulphuric acid, the reaction

$$2MnO_3 + 2H_2SO_4 \rightarrow 2Mn^{2+} + 2O_2\uparrow + 2SO_4^{2-} + 2H_2O$$

takes place.

Manganese(VII) compounds contain the manganate(VII) or permanganate ion, MnO_4^-. Alkali permanganates are stable compounds, producing violet-coloured solutions. They are all strong oxidizing agents.

In this section, the reactions of manganese(II) ions will be dealt with, while reactions of permanganates will be described among the reactions of anions (cf. Section 4.34).

Mn(II) reactions

Reactions of manganese(II) ions For the study of these reactions a 0.25M solution of manganese(II) chloride $MnCl_2 \cdot 4H_2O$ or manganese(II) sulphate $MnSO_4 \cdot 4H_2O$ can be used.

NaOH
↓
white $Mn(OH)_2$

1. *Sodium hydroxide solution* an initially white precipitate of manganese(II) hydroxide:

$$Mn^{2+} + 2OH^- \rightarrow Mn(OH)_2\downarrow$$

Turns dark

The precipitate is insoluble in excess reagent. It rapidly oxidizes on exposure to air, becoming brown, when hydrated manganese dioxide, $MnO(OH)_2$, is formed:

$$2Mn(OH)_2\downarrow + O_2 \rightarrow 2MnO(OH)_2\downarrow$$

Hydrogen peroxide converts manganese(II) hydroxide rapidly into hydrated manganese dioxide:

$$Mn(OH)_2\downarrow + H_2O_2 \rightarrow MnO(OH)_2\downarrow + H_2O$$

NH₃
↓
white $Mn(OH)_2$

2. *Ammonia solution* partial precipitation of (initially) white manganese(II) hydroxide:

$$Mn^{2+} + 2NH_3 + 2H_2O \rightleftharpoons Mn(OH)_2\downarrow + 2NH_4^+$$

The precipitate is soluble in ammonium salts, when the reaction proceeds towards the left.

(NH$_4$)$_2$S
↓
pink MnS

3. *Ammonium sulphide solution*   pink precipitate of manganese(II) sulphide:

$$Mn^{2+} + S^{2-} \rightarrow MnS\downarrow$$

The precipitate also contains loosely-bound water. It is readily soluble in mineral acids (different from nickel and cobalt) and even in acetic acid (distinction from nickel, cobalt, and zinc):

$$MnS\downarrow + 2H^+ \rightarrow Mn^{2+} + H_2S\uparrow$$

$$MnS\downarrow + 2CH_3COOH \rightarrow Mn^{2+} + H_2S\uparrow + 2CH_3COO^-$$

Na$_2$HPO$_4$
↓
pink ppt

4. *Disodium hydrogen phosphate solution*   pink precipitate of manganese ammonium phosphate Mn(NH$_4$)PO$_4 \cdot$7H$_2$O, in the presence of ammonia (or ammonium ions):

$$Mn^{2+} + NH_3 + HPO_4^{2-} \rightarrow Mn(NH_4)PO_4\downarrow$$

If ammonium salts are absent, manganese(II) phosphate is formed:

$$3Mn^{2+} + 2HPO_4^{2-} \rightarrow Mn_3(PO_4)_2\downarrow + 2H^+$$

Both precipitates are soluble in acids.

PbO$_2$
↓
oxidation to
permanganate

5. *Lead dioxide and concentrated nitric acid*   on boiling a dilute solution of manganese(II) ions, free from hydrochloric acid and chlorides, with lead dioxide (or red lead, which yields the dioxide in the presence of nitric acid) and a little concentrated nitric acid, diluting somewhat and allowing the suspended solid containing unattacked lead dioxide to settle, the supernatant liquid acquires a violet-red (or purple) colour due to permanganic acid. The latter is decomposed by hydrochloric acid and hence chlorides should be absent.

$$5PbO_2 + 2Mn^{2+} + 4H^+ \rightarrow 2MnO_4^- + 5Pb^{2+} + 2H_2O$$

(NH$_4$)$_2$S$_2$O$_8$
↓
oxidation to
permanganate

6. *Ammonium or potassium peroxodisulphate*   solid (NH$_4$)$_2$S$_2$O$_8$ or K$_2$S$_2$O$_8$ is added to a dilute solution of manganese(II) ions, free of chloride. The solution is acidified with dilute sulphuric acid, and a few drops of dilute silver nitrate (which acts as a catalyst) are added; on boiling, a reddish-violet solution is formed, owing to the presence of permanganate:

$$2Mn^{2+} + 5S_2O_8^{2-} + 8H_2O \rightarrow 2MnO_4^- + 10SO_4^{2-} + 16H^+$$

The solution must be dilute (max 0.02M), otherwise manganese dioxide precipitate is formed.

Spot test   place a drop of the test solution in a micro crucible, add 1 drop 0.1M silver nitrate solution, and stir. Introduce a few mg solid ammonium peroxodisulphate and heat gently. The characteristic colour of permanganate appears.

Sensitivity 0.1 μg Mn.

Concentration limit 1 in 500 000.

NaBiO₃
↓
oxidation to
permanganate

7. *Sodium bismuthate (NaBiO₃)* when this solid is added to a cold solution of manganese(II) ions in dilute nitric acid or in dilute sulphuric acid, the mixture stirred and excess reagent filtered off (preferably through glass wool or a sintered glass funnel), a solution of permanganate is produced.

$$2Mn^{2+} + 5NaBiO_3 + 14H^+ \rightarrow 2MnO_4^- + 5Bi^{3+} + 5Na^+ + 7H_2O$$

The spot-test technique is as follows. Place a drop of the test solution on a spot plate, add a drop of concentrated nitric acid and then a little sodium bismuthate. The purple colour of permanganic acid appears. If the solution is so dark that the colour cannot be detected, dilute the mixture with water until the colour appears.

Sensitivity 25 μg Mn (in 5 ml).

Concentration limit 1 in 200 000.

Atomic spectrometry

8. *Atomic spectrometric tests*

(a) *Flame atomic emission spectrometry*   with an air/acetylene flame 10 ppm manganese can be identified at 403.08 nm. Using nitrous oxide/acetylene flame the sensitivity increases tenfold.

(b) *Flame atomic absorption spectrometry*   using a manganese hollow-cathode lamp and an air/acetylene flame, 0.1 ppm can be identified at the 279.48 nm resonance line.

(c) *Manganese standard, 1000 ppm*   dissolve 3.6021 g analytical grade manganese(II) chloride tetrahydrate $MnCl_2 \cdot 4H_2O$ in a mixture of 50 ml concentrated hydrochloric acid and 50 ml water. After complete dissolution, dilute with water to 1 litre.

Dry tests

9. *Dry tests*

(a) *Borax bead test*   The bead produced in the oxidizing flame by small amounts of manganese salts is violet whilst hot and amethyst-red when cold; with larger amounts of manganese the bead is almost brown and may be mistaken for that of nickel. In the reducing flame the manganese bead is colourless whilst that due to nickel is grey.

Fusion
↓
oxidation to manganate

(b) *Fusion test*   Fusion of any manganese compound with sodium carbonate and an oxidizing agent (potassium chlorate or potassium nitrate) gives a green mass of alkali manganate. The test may be carried out either by heating upon a piece of platinum foil with potassium nitrate and sodium carbonate (if platinum foil is not available, a piece of broken porcelain may be employed), or by fusing a bead of sodium carbonate with a small quantity of the manganese compound in the oxidizing flame and dipping the fused mass whilst hot into a little powdered potassium chlorate or nitrate and reheating.

$$MnSO_4 + 2KNO_3 + 2Na_2CO_3 = Na_2MnO_4 + 2KNO_2$$
$$+ Na_2SO_4 + 2CO_2\uparrow$$
$$3MnSO_4 + 2KClO_3 + 6Na_2CO_3 = 3Na_2MnO_4 + 2KCl$$
$$+ 3Na_2SO_4 + 6CO_2\uparrow$$

3.29   Zinc, Zn (A_r: 65.38)

Zinc is a bluish-white metal; it is fairly malleable and ductile at 110–150°C. It melts at 410°C and boils at 906°C.

The pure metal dissolves very slowly in acids and in alkalis; the presence of impurities or contact with platinum or copper, produced by the addition of a few drops of the solutions of the salts of these metals, accelerates the reaction. This explains the solubility of commercial zinc. The latter dissolves readily in dilute hydrochloric acid and in dilute sulphuric acid with the evolution of hydrogen:

$$Zn + 2H^+ \rightarrow Zn^{2+} + H_2\uparrow$$

Dissolution occurs in very dilute nitric acid, but no gas is evolved:

$$4Zn + 10H^+ + NO_3^- \rightarrow 4Zn^{2+} + NH_4^+ + 3H_2O$$

On increasing the concentration of nitric acid, dinitrogen oxide (N_2O) and nitrogen oxide (NO) are formed:

$$4Zn + 10H^+ + 2NO_3^- \rightarrow 4Zn^{2+} + N_2O\uparrow + 5H_2O$$

$$3Zn + 8HNO_3 \rightarrow 3Zn^{2+} + 2NO\uparrow + 6NO_3^- + 4H_2O$$

Concentrated nitric acid has little effect on zinc because of the low solubility of zinc nitrate in such a medium. With hot, concentrated sulphuric acid, sulphur dioxide is evolved:

$$Zn + 2H_2SO_4 \rightarrow Zn^{2+} + SO_2\uparrow + SO_4^{2-} + 2H_2O$$

Zinc also dissolves in alkali hydroxides, when tetrahydroxozincate(II) is formed:

$$Zn + 2OH^- + 2H_2O \rightarrow [Zn(OH)_4]^{2-} + H_2\uparrow$$

Zinc forms one series of salts only; these contain the zinc(II) cation, derived from zinc oxide, ZnO.

Zn(II) reactions

Reactions of zinc(II) ions   A 0.25M solution of zinc sulphate $ZnSO_4 \cdot 7H_2O$ can be used to study these reactions.

NaOH
↓
white $Zn(OH)_2$

1. *Sodium hydroxide solution*   white gelatinous precipitate of zinc hydroxide:

$$Zn^{2+} + 2OH^- \rightleftarrows Zn(OH)_2\downarrow$$

The precipitate is soluble in acids:

$$Zn(OH)_2\downarrow + 2H^+ \rightleftarrows Zn^{2+} + 2H_2O$$

Dissolves in excess

and also in the excess of the reagent:

$$Zn(OH)_2\downarrow + 2OH^- \rightleftarrows [Zn(OH)_4]^{2-}$$

Zinc hydroxide is thus an amphoteric compound.

NH_3
↓
white $Zn(OH)_2$
dissolves in excess

2. *Ammonia solution*   white precipitate of zinc hydroxide, readily soluble in excess reagent and in solutions of ammonium salts owing to the production of tetramminezincate(II). The non-precipitation of zinc hydroxide by ammonia solution in the presence of ammonium chloride is due to the

lowering of the hydroxyl-ion concentration to such a value that the solubility product of $Zn(OH)_2$ is not attained.

$$Zn^{2+} + 2NH_3 + 2H_2O \rightleftharpoons Zn(OH)_2\downarrow + 2NH_4^+$$

$$Zn(OH)_2\downarrow + 4NH_3 \rightleftharpoons [Zn(NH_3)_4]^{2+} + 2OH^-$$

$(NH_4)_2S$
↓
white ZnS

3. *Ammonium sulphide solution* white precipitate of zinc sulphide, ZnS, from neutral or alkaline solutions; it is insoluble in excess reagent, in acetic acid, and in solutions of caustic alkalis, but dissolves in dilute mineral acids. The precipitate thus obtained is partially colloidal; it is difficult to wash and tends to run through the filter paper, particularly on washing. To obtain the zinc sulphide in a form which can be readily filtered, the precipitation is conveniently carried out in boiling solution in the presence of excess ammonium chloride. The precipitate is washed with dilute ammonium chloride solution containing a little ammonium sulphide.

$$Zn^{2+} + S^{2-} \rightarrow ZnS\downarrow$$

4. *Hydrogen sulphide gas* partial precipitation of zinc sulphide in neutral solutions; when the concentration of acid produced is about 0.3M (pH about 0.6), the sulphide-ion concentration derived from the hydrogen sulphide is depressed so much by the hydrogen-ion concentration from the acid that it is too low to exceed the solubility product of ZnS, and consequently precipitation ceases.

$$Zn^{2+} + H_2S \rightleftharpoons ZnS\downarrow + 2H^+$$

When alkali acetate is added to the solution, the hydrogen-ion concentration is reduced because of the formation of the feebly dissociated acetic acid, the sulphide-ion concentration is correspondingly increased, and precipitation is almost complete.

$$Zn^{2+} + H_2S + 2CH_3COO^- \rightarrow ZnS\downarrow + 2CH_3COOH$$

Zinc sulphide is also precipitated from alkaline solutions of tetrahydroxozincate:

$$[Zn(OH)_4]^{2-} + H_2S \rightarrow ZnS\downarrow + 2OH^- + 2H_2O$$

Na_2HPO_4
↓
white $Zn_3(PO_4)_2$

5. *Disodium hydrogen phosphate solution* white precipitate of zinc phosphate:

$$3Zn^{2+} + 2HPO_4^{2-} \rightleftharpoons Zn_3(PO_4)_2\downarrow + 2H^+$$

In the presence of ammonium ions zinc ammonium phosphate is formed:

$$Zn^{2+} + NH_4^+ + HPO_4^{2-} \rightleftharpoons Zn(NH_4)PO_4\downarrow + H^+$$

Both precipitates are soluble in dilute acids, when the reactions are reversed. Also, both precipitates are soluble in ammonia:

$$Zn_3(PO_4)_2 + 12NH_3 \rightarrow 3[Zn(NH_3)_4]^{2+} + 2PO_4^{3-}$$

$$Zn(NH_4)PO_4 + 3NH_3 \rightarrow [Zn(NH_3)_4]^{2+} + HPO_4^{2-}$$

K₄[Fe(CN)₆]
↓
white ppt

6. Potassium hexacyanoferrate(II) solution white precipitate of variable composition; if the reagent is added in some excess, the composition of the precipitate is $K_2Zn_3[Fe(CN)_6]_2$:

$$3Zn^{2+} + 2K^+ + 2[Fe(CN)_6]^{4-} \rightarrow K_2Zn_3[Fe(CN)_6]_2$$

The precipitate is insoluble in dilute acids, but dissolves readily in sodium hydroxide:

$$K_2Zn_3[Fe(CN)_6]_2 + 12OH^- \rightarrow 2[Fe(CN)_6]^{4-} + 3[Zn(OH)_4]^{2-} + 2K^+$$

This reaction can be used to distinguish zinc from aluminium.

Quinaldic acid
↓
white ppt

7. Quinaldic acid reagent (quinoline-1-carboxylic acid, $C_9H_6N \cdot CO_2H$) upon the addition of a few drops of the reagent to a solution of a zinc salt which is faintly acid with acetic acid, a white precipitate of the zinc complex $Zn(C_{10}H_6NO_2)_2 \cdot H_2O$ is obtained. The precipitate is soluble in ammonia solution and in mineral acids, but is reprecipitated on neutralization. Copper, cadmium, uranium, iron, and chromium ions give precipitates with the reagent and should be absent. Cobalt, nickel, and manganese ions have no effect. This is an extremely sensitive test for zinc ions, and is a useful confirmatory test for zinc isolated from the Group IIIB separation. The reagent is, however, expensive. The reaction is best carried out on the semi-micro scale or as a spot test.

Test with
CuSO₄ + (NH₄)₂[Hg(SCN)₄]
↓
violet ppt

8. Ammonium tetrathiocyanatomercurate(II)–copper sulphate test Acidify the solution faintly with a few drops of M sulphuric or 2M acetic acid. Add 0.1 ml of 0.25M copper(II) sulphate solution, followed by 2 ml of ammonium tetrathiocyanatomercurate(II) reagent. A violet precipitate is obtained. The test is rendered still more sensitive by boiling the mixture for 1 minute, cooling and shaking with a little amyl alcohol; the violet precipitate collects at the interface. Iron salts produce a red colouration; this disappears when a little alkali fluoride is added.

Copper salts alone do not form a precipitate with the ammonium tetra-thiocyanatomercurate(II) reagent, whilst zinc ions, if present alone, form a white precipitate:

$$Zn^{2+} + [Hg(SCN)_4]^{2-} = Zn[Hg(SCN)_4]$$

In the presence of copper ions, the copper complex coprecipitates with that of zinc, and the violet (or blackish-purple) precipitate consists of mixed crystals of $Zn[Hg(SCN)_4] + Cu[Hg(SCN)_4]$.

Place a drop or two of the test solution, which is slightly acid (preferably with sulphuric acid), on a spot plate, add 1 drop 0.25M copper sulphate solution and 1 drop ammonium tetrathiocyanatomercurate(II) reagent. A violet (or blackish-purple) precipitate appears.

The reaction may be conducted in a semimicro test-tube; here 0.3–0.5 ml amyl alcohol is added. The violet precipitate collects at the interface.

Concentration limit 1 in 10 000.

Rinmann's green test

Specific for Zn

9. Rinmann's green test the test depends upon the production of Rinmann's green (largely cobalt zincate $CoZnO_2$) by heating the salts or oxides of zinc and cobalt. An excess of cobalt(II) oxide must be avoided for it leads to a red-brown colouration: oxidation to cobalt(III) oxide produces a darkening of colour. Optimum experimental conditions are obtained by converting the zinc into the hexacyanocobaltate(III).

$$K^+ + [Co(CN)_6]^{3-} + Zn^{2+} = KZn[Co(CN)_6]$$

Ignition of the latter leads to zinc oxide and cobalt(II) oxide in the correct proportion for the formation of cobalt zincate, whilst carbon from the filter paper (see below) prevents the formation of cobalt(III) oxide. All other metals must be removed.

Place a few drops of the neutral (litmus) test solution upon potassium hexacyanocobaltate(III) (or cobalticyanide, Rinmann's green) test paper. Dry the paper over a flame and ignite in a small crucible. Observe the colour of the ash against a white background: part of it will be green.

Sensitivity 0.6 μg Zn.

Concentration limit 1 in 3 000.

The potassium hexacyanocobaltate(III) (or cobalticyanide) test paper is prepared by soaking drop-reaction paper or quantitative filter paper in a solution containing 4 g potassium hexacyanocobaltate(III) and 1 g potassium chlorate in 100 ml water, and drying at room temperature or at 100°C. The paper is yellow and keeps well.

Dithizone
↓
red colour extractable

10. Diphenylthiocarbazone (dithizone) test dithizone forms complexes with a number of metal ions, which can be extracted with chloroform. The zinc complex, formed in neutral, alkaline or acetic acid solutions, is red:

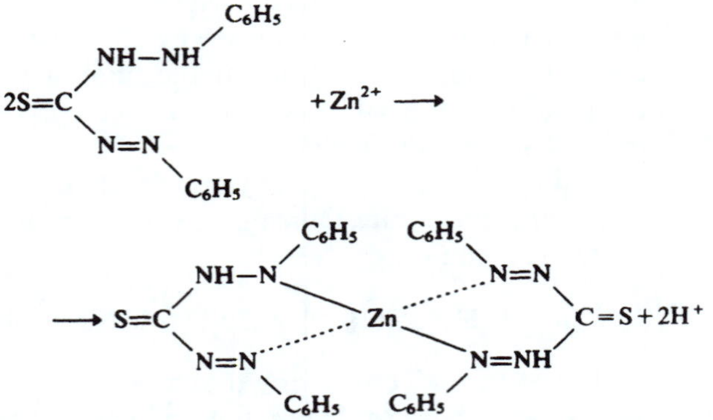

Acidify the test solution with acetic acid, and add a few drops of the reagent. The organic phase turns red in the presence of zinc. Cu^{2+}, Hg^{2+}, Hg_2^{2+}, Ag^+, Au^{3+}, and Pd^{2+} ions interfere. Upon adding sodium hydroxide solution to the mixture and shaking, the aqueous phase turns red also. This reaction is characteristic for zinc only. (In alkaline solutions the sensitivity of the test is less.)

Limit of detection in neutral medium 0.025 μg Zn^{2+}; in acetic acid or sodium hydroxide medium 0.9 μg Zn^{2+}.

The test can be applied in the presence of 2000 times those amounts of Ni and Al.

Atomic spectrometry

11. *Atomic spectrometric tests*

(a) *Flame atomic emission spectrometry* with an air/acetylene flame 10 ppm zinc can be identified at 213.86 nm. The nitrous oxide/acetylene flame provides a tenfold increase in sensitivity.

(b) *Flame atomic absorption spectrometry* using a zinc or brass hollow-cathode lamp and an air/acetylene flame, 0.1 ppm can be identified at 213.86 nm.

(c) *Zinc standard, 1000 ppm* dissolve 4.3980 g analytical grade zinc sulphate heptahydrate $ZnSO_4 \cdot 7H_2O$ in water and dilute to 1 litre.

Dry tests

12. *Dry tests (blowpipe tests)* compounds of zinc when heated upon charcoal with sodium carbonate yield an incrustation of the oxide, which is yellow when hot and white when cold. The metal cannot be isolated owing to its volatility and subsequent oxidation. If the incrustation is moistened with a drop of cobalt nitrate solution and again heated, a green mass (Rinmann's green), consisting largely of cobalt zincate $CoZnO_2$ is obtained.

An alternative method is to soak a piece of ashless filter paper in the zinc salt solution, add 1 drop cobalt nitrate solution and to ignite in a crucible or in a coil of platinum wire. The residue is coloured green.

3.30 Fourth group of cations: barium(II), strontium(II), and calcium(II)

Group reagent 1M solution of ammonium carbonate.

The reagent is colourless, and displays an alkaline reaction because of hydrolysis:

$$CO_3^{2-} + H_2O \rightleftharpoons HCO_3^- + OH^-$$

Fourth group of cations
Ba(II), Sr(II), Ca(II)

The reagent is decomposed by acids (even by acetic acid), when carbon dioxide gas is formed:

$$CO_3^{2-} + 2CH_3COOH \rightarrow CO_2\uparrow + H_2O + 2CH_3COO^-$$

The reagent has to be used in neutral or slightly alkaline media.

Commercial ammonium carbonate always contains ammonium hydrogen carbonate (NH_4HCO_3) and ammonium carbamate $NH_4O(NH_2)CO$. These compounds have to be removed before attempting the group reaction as thealkaline earth salts of both are soluble in water. This can be done by boiling the reagent solution for a while; both ammonium hydrogen carbonate and ammonium carbamate are converted to ammonium carbonate in this way:

$$2HCO_3^- \rightarrow CO_3^{2-} + CO_2\uparrow + H_2O$$

$$\begin{array}{c} {}^-O \\ \diagdown \\ \quad\quad C{=}O + H_2O \rightarrow NH_4^+ + CO_3^{2-} \\ \diagup \\ H_2N \end{array}$$

Group reaction cations of the fourth group react neither with hydrochloric acid, hydrogen sulphide nor ammonium sulphide, but ammonium carbonate (in the presence of moderate amounts of ammonia or ammonium ions) forms white precipitates. The test has to be carried out in neutral or alkaline solutions. In the absence of ammonia or ammonium ions, magnesium will also be precipitated. The white precipitates formed with the group reagent are: barium carbonate $BaCO_3$, strontium carbonate $SrCO_3$, and calcium carbonate $CaCO_3$.

The three alkaline earth metals decompose water at different rates, forming hydroxides and hydrogen gas. Their hydroxides are strong bases, although with different solubilities: barium hydroxide is the most soluble, while calcium hydroxide is the least soluble among them. Alkaline earth chlorides and nitrates are very soluble; the carbonates, sulphates, phosphates, and oxalates are insoluble. The sulphides can be prepared only in the dry state; they all hydrolyse in water, forming hydrogen sulphides and hydroxides, e.g.

$$2BaS + 2H_2O \rightarrow 2Ba^{2+} + 2SH^- + 2OH^-$$

Unless the anion is coloured, the salts form colourless solutions.

Because the alkaline earth ions behave very similarly to each other in aqueous solutions, it is very difficult to distinguish them and especially to separate them. There are, however, differences in the solubilities of some of their salts in non-aqueous media. Thus, 100 g anhydrous ethanol dissolves 12.5 g calcium chloride, 0.91 g strontium chloride, and only 0.012 g barium chloride (all anhydrous salts). One hundred grams of a $1+1$ mixture of diethyl ether and anhydrous ethanol dissolves more than 40 g anhydrous calcium nitrate; the solubilities of anhydrous strontium and barium nitrates in this solution are negligible. These differences can be utilized for separations.

3.31 Barium, Ba (A_r: 137.34)

Barium is a silver-white, malleable and ductile metal, which is stable in dry air. It reacts with the water in humid air forming oxide or hydroxide. It melts at 710°C. It reacts with water at room temperature forming barium hydroxide and hydrogen:

$$Ba + 2H_2O \rightarrow Ba^{2+} + H_2\uparrow + 2OH^-$$

Dilute acids dissolve barium readily with the evolution of hydrogen:

$$Ba + 2H^+ \rightarrow Ba^{2+} + H_2\uparrow$$

Barium is bivalent in its salts forming the barium(II) cation Ba^{2+}. Its chloride and nitrate are soluble, but on adding concentrated hydrochloric or nitric acids to barium solutions, barium chloride or nitrate may precipitate out.

Ba(II) reactions **Reactions of barium(II) ions** Use a 0.25M solution of barium chloride $BaCl_2 \cdot 2H_2O$ or barium nitrate $Ba(NO_3)_2$ for the study of these reactions.

1. *Ammonia solution*   no precipitate of barium hydroxide because of its relatively high solubility. If the alkaline solution is exposed to the atmosphere, some carbon dioxide is absorbed and a turbidity, due to barium carbonate, is produced.

A slight turbidity may occur when adding the reagent; this is due to small amounts of ammonium carbonate, often present in an aged reagent.

$(NH_4)_2CO_3$
$\downarrow$
white $BaCO_3$

2. *Ammonium carbonate solution*   white precipitate of barium carbonate, soluble in acetic acid and in dilute mineral acids.

$$Ba^{2+} + CO_3^{2-} \rightarrow BaCO_3\downarrow$$

The precipitate is slightly soluble in solutions of ammonium salts of strong acids.

$(COONH_4)_2$
$\downarrow$
white $Ba(COO)_2$

3. *Ammonium oxalate solution* white precipitate of barium oxalate $Ba(COO)_2$, slightly soluble in water (0.09 g per litre; $K_s = 1.7 \times 10^{-7}$), but readily dissolved by hot dilute acetic acid (distinction from calcium) and by mineral acids.

$$Ba^{2+} + (COO)_2^{2-} \rightleftharpoons Ba(COO)_2\downarrow$$

H_2SO_4
$\downarrow$
white $BaSO_4$

4. *Dilute sulphuric acid*   heavy, white, finely divided precipitate of barium sulphate $BaSO_4$, practically insoluble in water (2.5 mg l^{-1}; $K_s = 9.2 \times 10^{-11}$), almost insoluble in dilute acids and in ammonium sulphate solution, but appreciably soluble in boiling concentrated sulphuric acid. By precipitation in boiling solution, or preferably in the presence of ammonium acetate, a more readily filterable form is obtained.

$$Ba^{2+} + SO_4^{2-} \rightarrow BaSO_4\downarrow$$

$$BaSO_4\downarrow + H_2SO_4(\text{conc.}) \rightarrow Ba^{2+} + 2HSO_4^-$$

If barium sulphate is boiled with a concentrated solution of sodium carbonate, partial transformation into the less soluble barium carbonate occurs in accordance with the equation:

$$BaSO_4\downarrow + CO_3^{2-} \rightleftharpoons BaCO_3\downarrow + SO_4^{2-}$$

Owing to the reversibility of the reaction, the transformation is incomplete. Barium sulphate precipitate may also be dissolved in a hot 5% solution of disodium ethylenediamine tetraacetate (Na_2EDTA) in the presence of ammonia.

Saturated $CaSO_4$
$\downarrow$
white $BaSO_4$

5. *Saturated calcium sulphate solution*   immediate white precipitate of barium sulphate. A similar phenomenon occurs if saturated strontium sulphate reagent is used.

The explanation of these reactions is as follows: of the three alkaline earth sulphates barium sulphate is the least soluble. In the solutions of saturated calcium or strontium sulphate the concentration of sulphate ions is high enough to cause precipitation with larger amounts of barium, because the product of ion concentrations exceeds the value of the solubility product:

$$SO_4^{2-} + Ba^{2+} \rightleftharpoons BaSO_4\downarrow$$

K_2CrO_4
↓
yellow $BaCrO_4$

6. *Potassium chromate solution*   a yellow precipitate of barium chromate, practically insoluble in water ($3.2\,mg\,l^{-1}$, $K_s = 1.6 \times 10^{-10}$):

$$Ba^{2+} + CrO_4^{2-} \rightarrow BaCrO_4\downarrow$$

The precipitate is insoluble in dilute acetic acid (distinction from strontium and calcium), but readily soluble in mineral acids.

The addition of acid to potassium chromate solution causes the yellow colour of the solution to change to reddish-orange, owing to the formation of dichromate:

$$2CrO_4^{2-} + 2H^+ \rightleftharpoons Cr_2O_7^{2-} + H_2O$$

The solubility products for $SrCrO_4$ and $CaCrO_4$ are much larger than for $BaCrO_4$ and hence they require a larger CrO_4^{2-} ion concentration to precipitate them. The addition of acetic acid to the K_2CrO_4 solution lowers the CrO_4^{2-} ion concentration sufficiently to prevent the precipitation of $SrCrO_4$ and $CaCrO_4$ but it is maintained high enough to precipitate $BaCrO_4$.

Na rhodizonate
↓
reddish ppt

7. *Sodium rhodizonate (rhodizonic acid, sodium salt) reagent*

CO—CO—C.ONa
| ‖
CO--CO--C.ONa

reddish-brown precipitate of the barium salt of rhodizonic acid in neutral solution. Calcium and magnesium salts do not interfere; strontium salts react like those of barium, but only the precipitate due to the former is completely soluble in dilute hydrochloric acid. Other elements, e.g. those precipitated by hydrogen sulphide and by ammonium sulphide, should be absent. The reagent should be confined to testing for elements in Group IV.

Place a drop of the neutral or faintly acidic test solution upon drop-reaction paper and add a drop of the reagent. A brown or reddish-brown spot is obtained.

Sensitivity 0.25 µg Ba.

Concentration limit 1 in 200 000.

In the presence of strontium the reddish-brown stain of barium rhodizonate is treated with 0.5M hydrochloric acid; the strontium rhodizonate dissolves, whilst the barium derivative is converted into the brilliant-red acid salt. The reaction is best carried out on drop-reaction paper as above.

Treat the reddish-brown spot with a drop of 0.5M hydrochloric acid when a bright-red stain is formed if barium is present. If barium is absent, the spot disappears.

Sensitivity 0.5 µg Ba in the presence of 50 times that amount of Sr.

Concentration limit 1 in 90 000.

Organic solvents

8. *Anhydrous ethanol and ether*   a $1 + 1$ mixture of these solvents does not dissolve anhydrous barium nitrate or barium chloride (distinct from strontium and calcium). The salts must be heated to 180°C before the test to remove all water of crystallization. The test can be used for the separation of barium from strontium and/or calcium.

Atomic spectrometry

9. *Atomic spectrometric tests*   in air/acetylene flames a number of ions (l___ phosphate, sulphate, aluminium) decrease the emission and absorption signals of barium (and all of the alkaline earth metals). Adding an equal volume of a 25% solution of lanthanum chloride heptahydrate $LaCl_3 \cdot 7H_2O$ eliminates these interferences. Addition of lanthanum is not required if the nitrous oxide/acetylene flame is used.

   (a) *Flame atomic emission spectrometry*   with an air/acetylene flame 1 ppm barium can be identified at 455.4 nm.

   (b) *Flame atomic absorption spectrometry*   using a barium hollow-cathode lamp and an air/acetylene flame, 0.1 ppm can be identified at 553.55 nm.

   (c) *Barium standard, 1000 ppm*   dissolve 1.4369 g analytical grade barium carbonate $BaCO_3$ in 15 ml 2M hydrochloric acid. After complete dissolution dilute with water to 1 litre.

Flame colour: green

10. *Dry test (flame colouration)*   barium salts, when heated in a non-luminous Bunsen flame, impart a yellowish-green colour to the flame. Since most barium salts, with the exception of the chloride, are non-volatile, the platinum wire is moistened with concentrated hydrochloric acid before being dipped into the substance. The sulphate is first reduced to the sulphide in the reducing flame, then moistened with concentrated hydrochloric acid, and reintroduced into the flame.

3.32   Strontium, Sr (A_r: 87.62)

Strontium is a silver-white, malleable and ductile metal. It melts at 771°C. Its properties are similar to those of barium.

Sr(II) reactions

Reactions of strontium(II) ions   For the study of these reactions a 0.25M solution of strontium chloride $SrCl_2 \cdot 6H_2O$ or strontium nitrate $Sr(NO_3)_2$ can be used.

1. *Ammonia solution*   no precipitate.

$(NH_4)_2CO_3$
↓
white $SrCO_3$

2. *Ammonium carbonate solution*   white precipitate of strontium carbonate:

$$Sr^{2+} + CO_3^{2-} \rightarrow SrCO_3\downarrow$$

Strontium carbonate is somewhat less soluble than barium carbonate; otherwise its characteristics (slight solubility in ammonium salts, decomposition with acids) are similar to those of the latter.

H_2SO_4
↓
white $SrSO_4$

3. *Dilute sulphuric acid*   white precipitate of strontium sulphate:

$$Sr^{2+} + SO_4^{2-} \rightarrow SrSO_4\downarrow$$

The solubility of the precipitate is low but not negligible (0.097 g l^{-1}, $K_s = 2.8 \times 10^{-7}$). The precipitate is insoluble in ammonium sulphate solution even on boiling (distinction from calcium) and slightly soluble in boiling hydrochloric acid. It is almost completely converted into the corresponding carbonate by boiling with a concentrated solution of sodium carbonate:

$$SrSO_4 + CO_3^{2-} \rightleftarrows SrCO_3\downarrow + SO_4^{2-}$$

Strontium carbonate is less soluble than strontium sulphate (solubility: 5.9 mg $SrCO_3$ l^{-1}, $K_s = 1.6 \times 10^{-9}$ at room temperature).

After filtering the solution the precipitate can be dissolved in hydrochloric acid, thus strontium ions can be transferred into the solution.

Saturated CaSO₄
↓
white SrSO₄

4. *Saturated calcium sulphate solution* white precipitate of strontium sulphate, formed slowly in the cold but more rapidly on boiling (distinction from barium).

(COONH₄)₂
↓
white Sr(COO)₂

5. *Ammonium oxalate solution* white precipitate of strontium oxalate

$$Sr^{2+} + (COO)_2^{2-} \rightarrow Sr(COO)_2\downarrow$$

The precipitate is sparingly soluble in water ($0.039\,g\,l^{-1}$, $K_s = 5 \times 10^{-8}$). Acetic acid does not attack it; mineral acids, however, dissolve the precipitate.

K₂CrO₄
↓
yellow SrCrO₄

6. *Potassium chromate solution* yellow precipitate of strontium chromate:

$$Sr^{2+} + CrO_4^{2-} \rightarrow SrCrO_4\downarrow$$

The precipitate is appreciably soluble in water ($1.2\,g\,l^{-1}$, $K_s = 3.5 \times 10^{-5}$), no precipitate occurs therefore in dilute solutions of strontium. The precipitate is soluble in acetic acid (distinction from barium) and in mineral acids, for the same reasons as described under barium (cf. Section 3.31, reaction 6).

Na rhodizonate
↓
brown ppt

7. *Sodium rhodizonate reagent* reddish-brown precipitate of strontium rhodizonate in neutral solution. The test is applied to the elements of Group IV. Barium reacts similarly and a method for the detection of barium in the presence of strontium has already been described (Section 3.31, reaction 7). To detect strontium in the presence of barium, the latter is converted into the insoluble barium chromate. Barium chromate does not react with sodium rhodizonate, but the more soluble strontium chromate reacts normally.

If barium is absent, place a drop of the neutral test solution on drop-reaction paper or on a spot plate, and add a drop of the reagent. A brownish-red colouration or precipitate is produced.

Sensitivity 4 μg Sr.

Concentration limit 1 in 13 000.

If barium is present, proceed as follows. Impregnate some quantitative filter paper or drop-reaction paper with a saturated solution of potassium chromate, and dry it. Place a drop of the test solution on this paper and, after a minute, place 1 drop of the reagent on the moistened spot. A brownish-red spot or ring is formed.

Sensitivity 4 μg Sr in the presence of 80 times that amount of Ba.

Concentration limit 1 in 13 000.

For further details of the reagent, see Section 3.31, reaction 7.

Organic solvents

8. *Anhydrous ethanol and ether* a 1 + 1 mixture of these solvents does *not* dissolve anhydrous strontium nitrate, but *does* dissolve anhydrous strontium chloride. The test can be utilized for the separation of calcium, strontium, and barium.

The test can be carried out as follows: precipitate strontium as the carbonate. Filter the precipitate, dissolve one part of it in hydrochloric acid, another in nitric acid. Evaporate the two solutions on separate watch glasses to dryness, heat the residue to 180°C for 30 minutes, and try to dissolve the residues in a few millilitres of the solvent.

Atomic spectrometry

9. *Atomic spectrometric tests* the atomic spectrometric behaviour of strontium is very similar to that of barium (cf. Section 3.31, reaction 9). The use of lanthanum to eliminate interferences is recommended in air/acetylene flames.

(a) *Flame atomic emission spectrometry* with an air/acetylene flame 1 ppm strontium can be identified at 460.73 nm.

(b) *Flame atomic absorption spectrometry* using a strontium hollow-cathode lamp and an air/acetylene flame, 0.1 ppm can be identified at 460.73 nm.

(c) *Strontium standard, 1000 ppm* dissolve 1.6849 g analytical grade strontium carbonate $SrCO_3$ in 15 ml 2M hydrochloric acid. After complete dissolution, dilute with water to 1 litre.

Flame colour: red

10. *Dry test (flame colouration)* volatile strontium compounds, especially the chloride, impart a characteristic carmine-red colour to the non-luminous Bunsen flame (see remarks under barium).

3.33 Calcium, Ca (A_r: 40.08)

Calcium is a silver-white, rather soft metal. It melts at 845°C. It is attacked by atmospheric oxygen and humidity, when calcium oxide and/or calcium hydroxide is formed. Calcium decomposes water forming calcium hydroxide and hydrogen.

Calcium forms the calcium(II) cation Ca^{2+} in aqueous solutions. Its salts are normally white powders and form colourless solutions, unless the anion is coloured. Solid calcium chloride is hygroscopic and is often used as a drying agent. Calcium chloride and calcium nitrate dissolve readily in ethanol or in a $1 + 1$ mixture of anhydrous ethanol and diethyl ether.

Ca(II) reactions

Reactions of calcium(II) ions To study these reactions a 0.5M solution of calcium chloride $CaCl_2 \cdot 6H_2O$ can be used.

1. *Ammonia solution* no precipitate, as calcium hydroxide is fairly soluble. With an aged precipitant a turbidity may occur owing to the formation of calcium carbonate (cf. Section 3.31, reaction 1).

$(NH_4)_2CO_3$
↓
white $CaCO_3$

2. *Ammonium carbonate solution* white amorphous precipitate of calcium carbonate:

$$Ca^{2+} + CO_3^{2-} \rightarrow CaCO_3\downarrow$$

On boiling, the precipitate becomes crystalline. The precipitate is soluble in water which contains excess carbonic acid (e.g. freshly prepared soda water) because of the formation of soluble calcium hydrogen carbonate:

$$CaCO_3\downarrow + H_2O + CO_2 \rightleftarrows Ca^{2+} + 2HCO_3^-$$

On boiling the precipitate appears again, because carbon dioxide is removed during the process and the reaction proceeds towards the left. Barium and strontium ions react in a similar way.

The precipitate is soluble in acids, even in acetic acid.

H_2SO_4
↓
white $CaSO_4$

3. *Dilute sulphuric acid*   white precipitate of calcium sulphate

$$Ca^{2+} + SO_4^{2-} \rightarrow CaSO_4\downarrow$$

The precipitate is appreciably soluble in water (0.61 g Ca^{2+}, 2.06 g $CaSO_4$ or 2.61 g $CaSO_4 \cdot 2H_2O\,l^{-1}$, $K_s = 2.3 \times 10^{-4}$); that is, more soluble than barium or strontium sulphate. In the presence of ethanol the solubility is much less.

The precipitate dissolves in hot, concentrated sulphuric acid:

$$CaSO_4 + H_2SO_4 \rightleftarrows 2H^+ + [Ca(SO_4)_2]^{2-}$$

The same complex is formed if a precipitate is heated with a 10% solution of ammonium sulphate, leading to partial dissolution.

4. *Saturated calcium sulphate*   no precipitate is formed (difference from strontium and barium).

$(COONH_4)_2$
↓
white $Ca(COO)_2$

5. *Ammonium oxalate solution*   white precipitate of calcium oxalate, immediately from concentrated and slowly from dilute solutions:

$$Ca^{2+} + (COO)_2^{2-} \rightarrow Ca(COO)_2\downarrow$$

Precipitation is facilitated by making the solution alkaline with ammonia. The precipitate is practically insoluble in water (6.53 mg $Ca(COO)_2\,l^{-1}$, $K_s = 2.6 \times 10^{-9}$), insoluble in acetic acid, but readily soluble in mineral acids.

6. *Potassium chromate solution*   no precipitate from dilute solutions, nor from concentrated solutions in the presence of acetic acid (cf. Section 3.31, reaction 6).

$K_4[Fe(CN)_6]$
↓
white $K_2Ca[Fe(CN)_6]$

7. *Potassium hexacyanoferrate(II) solution*   white precipitate of a mixed salt:

$$Ca^{2+} + 2K^+ + [Fe(CN)_6]^{4-} \rightarrow K_2Ca[Fe(CN)_6]\downarrow$$

In the presence of ammonium chloride the test is more sensitive. In this case potassium is replaced by ammonium ions in the precipitate. The tests can be used to distinguish calcium from strontium; barium and magnesium ions however interfere.

Na dihydroxytartrate
osazone
· ↓
yellow ppt

8. *Sodium dihydroxytartrate osazone [sodium salt of succinic acid-dioxo-2,3-bis(phenylhydrazone)] reagent*

yellow sparingly-soluble precipitate of the calcium salt. All other metals, with the exception of alkali and ammonium salts, must be absent. Magnesium does not interfere provided its concentration does not exceed 10 times that of the calcium.

Place a drop of the neutral test solution on a black spot plate or upon a black watch glass, and add a tiny fragment of the solid reagent. If calcium is absent, the reagent dissolves completely. The presence of calcium is revealed by the formation over the surface of the liquid of a white film which ultimately separates as a dense precipitate.

Sensitivity 0.01 μg Ca.

Concentration limit 1 in 5 000 000.

This reagent is useful *inter alia* for the rapid differentiation between tap and distilled water: a positive result is obtained with a mixture of 1 part of tap water and 30 parts of distilled water.

Picrolonic acid
↓
white ppt of
rectangular crystals

9. *Picrolonic acid (or 1-p-nitrophenyl-3-methyl-4-nitro-5-pyrazolone) reagent*

characteristic rectangular crystals of calcium picrolonate

$$Ca(C_{10}H_7O_5N_4)_2 \cdot 8H_2O\downarrow$$

in neutral or acetic acid solutions. Strontium and barium form precipitates but of different crystalline form. Numerous elements, including copper, lead, thorium, iron, aluminium, cobalt, nickel, and barium, interfere.

Place a drop of the test solution (either neutral or acidified with acetic acid) in the depression of a *warm* spot plate and add 1 drop of a saturated solution of picrolonic acid. Characteristic rectangular crystals are produced.

Sensitivity 100 μg (in 5 ml).

Concentration limit 1 in 50 000.

The sensitivity is 0.01 μg (in 0.01 ml) under the microscope.

Microscopic test

10. *Calcium sulphate dihydrate (microscope) test*   This is an excellent confirmatory test for calcium in Group IV; it involves the use of a microscope (magnification about 110×). The salts should preferably be present as nitrates.

Evaporate a few drops of the test solution on a watch glass to dryness on a water bath, dissolve the residue in a few drops of water, transfer to a microscope slide, and add a minute drop of dilute sulphuric acid. (It may be necessary to warm the slide gently on a water bath until crystallization just sets in at the edges.) Upon observation through a microscope, bundles of needles or elongated prisms will be visible if calcium is present.   ·

Concentration limit 1 in 6000.

Organic solvents

11. *Anhydrous ethanol*   or a 1 + 1 mixture of anhydrous ethanol and diethyl ether, dissolves both anhydrous calcium chloride and calcium nitrate. For practical details see Section 3.32, reaction 8.

Atomic spectrometry

12. *Atomic spectrometric tests* the atomic spectrometric behaviour of calcium is very similar to that of barium (cf. Section 3.31, reaction 9). The use of lanthanum to eliminate interferences is recommended in air/acetylene flames.

(a) *Flame atomic emission spectrometry* with an air/acetylene flame 0.1 ppm calcium can be identified at 422.67 nm.

(b) *Flame atomic absorption spectrometry* using a calcium hollow-cathode lamp and an air/acetylene flame, 0.1 ppm can be identified at 422.67 nm.

(c) *Calcium standard, 1000 ppm* dissolve 2.4972 g analytical grade calcium carbonate $CaCO_3$ in 15 ml 2M hydrochloric acid. After complete dissolution dilute with water to 1 litre.

Flame colour:
yellowish-red

13. *Dry test (flame colouration)* volatile calcium compounds impart a yellowish-red colour to the Bunsen flame (see remarks under barium).

3.34 Fifth group of cations: magnesium(II), sodium(I), potassium(I), and ammonium(I)

Fifth group of cations
Mg(II), Na(I), K(I),
NH_4(I)

Group reagent there is no common reagent for the cations of this group.
Group reaction cations of the fifth group do not react with hydrochloric acid, hydrogen sulphide, ammonium sulphide or (in the presence of ammonium salts) with ammonium carbonate. Special reactions or flame tests can be used for their identification.

Of the cations of this group, magnesium displays similar reactions to those of cations in the fourth group. However, magnesium carbonate, in the presence of ammonium salts, is soluble, and therefore during the course of systematic analysis (when considerable amounts of ammonium salts are building up in the solution) magnesium will not precipitate with the cations of the fourth group.

The reactions of ammonium ions are quite similar to those of the potassium ion, because the ionic radii of these two ions are almost identical.

3.35 Magnesium, Mg (A_r: 24.305)

Magnesium is a white, malleable and ductile metal. It melts at 650°C. It burns readily in air or oxygen with a brilliant white light, forming the oxide MgO and some nitride Mg_3N_2. The metal is slowly decomposed by water at ordinary temperature, but at the boiling point of water the reaction proceeds rapidly:

$$Mg + 2H_2O \rightarrow Mg(OH)_2\downarrow + H_2\uparrow$$

Magnesium hydroxide, if ammonium salts are absent, is practically insoluble. Magnesium dissolves readily in acids:

$$Mg + 2H^+ \rightarrow Mg^{2+} + H_2\uparrow$$

Magnesium forms the bivalent cation Mg^{2+}. Its oxide, hydroxide, carbonate, and phosphate are insoluble; the other salts are soluble. They taste bitter. Some of the salts are hygroscopic.

Reactions of magnesium(II) ions   To study these reactions a 0.5M solution of magnesium chloride $MgCl_6 \cdot 6H_2O$ or magnesium sulphate $MgSO_4 \cdot 7H_2O$ can be used.

1. *Ammonia solution*   partial precipitation of white, gelatinous magnesium hydroxide:

$$Mg^{2+} + 2NH_3 + 2H_2O \rightarrow Mg(OH)_2\downarrow + 2NH_4^+$$

The precipitate is very sparingly soluble in water ($12\,mg\,l^{-1}$, $K_s = 3.4 \times 10^{-11}$), but readily soluble in ammonium salts.

2. *Sodium hydroxide solution*   white precipitate of magnesium hydroxide, insoluble in excess reagent, but readily soluble in solutions of ammonium salts:

$$Mg^{2+} + 2OH^- \rightarrow Mg(OH)_2\downarrow$$

3. *Ammonium carbonate solution*   in the absence of ammonium salts a white precipitate of basic magnesium carbonate:

$$5Mg^{2+} + 6CO_3^{2-} + 7H_2O \rightarrow 4MgCO_3 \cdot Mg(OH)_2 \cdot 5H_2O\downarrow + 2HCO_3^-$$

In the presence of ammonium salts no precipitation occurs, because the equilibrium

$$NH_4^+ + CO_3^{2-} \rightleftarrows NH_3 + HCO_3^-$$

is shifted towards the formation of hydrogen carbonate ions. The solubility product of the precipitate being high (K_s of pure $MgCO_3$ is 1×10^{-5}), the concentration of carbonate ions necessary to produce a precipitate is not attained.

4. *Sodium carbonate solution*   white, voluminous precipitate of basic carbonate (cf. reaction 3), insoluble in solutions of bases, but readily soluble in acids and in solutions of ammonium salts.

5. *Disodium hydrogen phosphate solution*   white crystalline precipitate of magnesium ammonium phosphate $Mg(NH_4)PO_4 \cdot 6H_2O$ in the presence of ammonium chloride (to prevent precipitation of magnesium hydroxide) and ammonia solutions:

$$Mg^{2+} + NH_3 + HPO_4^{2-} \rightarrow Mg(NH_4)PO_4\downarrow$$

The precipitate is sparingly soluble in water, soluble in acetic acid and in mineral acids. The precipitate separates slowly from dilute solutions because of its tendency to form supersaturated solutions; this may usually be overcome by cooling and by rubbing the test-tube or beaker beneath the surface of the liquid with a glass rod.

A white flocculant precipitate of magnesium hydrogen phosphate, $MgHPO_4$, is produced in neutral solutions.

$$Mg^{2+} + HPO_4^{2-} \rightarrow MgHPO_4\downarrow$$

Diphenyl carbazide
↓
violet colour

6. *Diphenylcarbazide reagent* $(C_6H_5\cdot NH\cdot NH\cdot CO\cdot NH\cdot NH\cdot C_6H_5)$ *(0.2%)* The magnesium salt solution is treated with sodium hydroxide solution – a precipitate of magnesium hydroxide will be formed – then with a few drops of the diphenylcarbazide reagent and the solution filtered. On washing the precipitate with hot water, it will be seen to have acquired a violet-red colour due to the formation of a complex salt or an adsorption complex. Metals of Groups II and III interfere and should therefore be absent.

Oxine
↓
yellow ppt

7. *8-Hydroxyquinoline or 'oxine' reagent (2% in acetic acid)*

when a solution of a magnesium salt, containing a little ammonium chloride, is treated with 1–2 ml reagent which has been rendered strongly ammoniacal by the addition of 3–4 ml dilute ammonia solution and the mixture heated to the boiling point, a yellow precipitate of the complex salt $Mg(C_9H_6ON)_2\cdot 4H_2O$ is obtained. All other metals, except sodium and potassium, must be absent.

4-(4-Nitrophenylazo)-
resorcinol
↓
blue lake on $Mg(OH)_2$

8. *4-(4-Nitrophenylazo)-resorcinol (or magneson I) reagent (0.5%)*

this test depends upon the adsorption of the reagent, which is a dyestuff, upon $Mg(OH)_2$ in alkaline solution whereby a blue lake is produced. Two ml test solution, acidified slightly with hydrochloric acid, is treated with 1 drop of the reagent and sufficient 2M sodium hydroxide solution to render the solution strongly alkaline (say 2–3 ml). A blue precipitate appears.

The spot-test technique is as follows. Place a drop of the test solution on a spot plate and add 1–2 drops of the reagent. It is essential that the solution be strongly alkaline; the addition of 1 drop of 2M sodium hydroxide may be advisable. According to the concentration of magnesium a blue precipitate is formed or the reddish-violet reagent assumes a blue colour. A comparative test on distilled water should be carried out.

Sensitivity 0.5 µg Mg.

Concentration limit 1 in 100 000.

Filter or drop-reaction paper should not be used.

An alternative agent is 4-nitrobenzeneazo-1-naphthol or magneson II

$$O_2N-\left\langle\bigcirc\right\rangle-N{=}N-\overset{\textstyle\bigcirc\!\!\bigcirc}{}-OH$$

It yields the same colour changes as magneson I, but has the advantage that it is more sensitive (sensitivity: 0.2 µg Mg; concentration limit: 1 in 250 000) and the blank test is not so deeply coloured. Its mode of use is identical with that described above for magneson I.

Titan yellow
↓
red colour or ppt

9. *Titan yellow reagent*   titan yellow (also known as clayton yellow) is a water-soluble yellow dyestuff. It is adsorbed by magnesium hydroxide producing a deep-red colour or precipitate. Barium and calcium do not react but intensify the red colour. All elements of Groups I to III should be removed before applying the test.

Place a drop of the test solution on a spot plate, introduce a drop of reagent and a drop of 2M sodium hydroxide. A red colour or precipitate is produced.

An alternative technique is to treat 0.5 ml neutral or slightly acidic test solution with 0.2 ml 2M sodium hydroxide solution. A red precipitate or colouration is produced.

Sensitivity 1.5 µg Mg.

Concentration limit 1 in 33 000.

Quinalizarin
↓
blue colour or ppt

10. *Quinalizarin reagent (0.05% in NaOH)*   blue precipitate or cornflower-blue colouration with magnesium salts. The colouration can be readily distinguished from the blue-violet colour of the reagent. Upon the addition of a little bromine water, the colour disappears (difference from beryllium). The alkaline earth metals and aluminium do not interfere under the conditions of the test, but all elements of Groups I to III should be removed. Phosphates and large amounts of ammonium salts decrease the sensitivity of the reaction.

Place a drop of the test solution and a drop of distilled water in adjacent cavities of a spot plate and add 2 drops of the reagent to each. If the solution is acid, it will be coloured yellowish-red by the reagent. Add 2M sodium hydroxide until the colour changes to violet and a further excess to increase the volume by 25–50%. A blue precipitate or colouration appears. The blank test has a blue-violet colour.

Sensitivity 0.25 µg Mg.

Concentration limit 1 in 200 000.

Atomic spectrometry

11. *Atomic spectrometric tests*   the atomic spectrometric behaviour of magnesium is very similar to that of barium (cf. Section 3.31, reaction 9). The use of lanthanum to eliminate interferences is recommended in air/acetylene flames.

(a) *Flame atomic emission spectrometry*   with an air/acetylene flame 1 ppm magnesium can be identified at 285.21 nm.

(b) *Flame atomic absorption spectrometry* using a magnesium hollow-cathode lamp and an air/acetylene flame, 0.1 ppm can be identified at 285.21 nm.

(c) *Magnesium standard, 1000 ppm* dissolve 1.0000 g analytical grade magnesium metal in 50 ml 6M hydrochloric acid, and dilute with water to 1 litre.

Dry test

12. *Dry test (blowpipe test)* all magnesium compounds when ignited on charcoal in the presence of sodium carbonate are converted into white magnesium oxide, which glows brightly when hot. Upon moistening with a drop or two of cobalt nitrate solution and reheating strongly, a pale-pink mass is obtained.

3.36 Potassium, K (A_r: 39.098)

Potassium is a soft, silver-white metal. Potassium melts at 63.5°C. It remains unchanged in dry air, but is rapidly oxidized in moist air, becoming covered with a blue film. The metal decomposes water violently, evolving hydrogen and burning with a violet flame:

$$2K^+ + 2H_2O \rightarrow 2K^+ + 2OH^- + H_2\uparrow$$

Potassium is usually kept under liquid paraffin.

Potassium salts contain the monovalent cation K^+. These salts are usually soluble and form colourless solutions, unless the anion is coloured.

K(I) reactions

Reactions of potassium(I) ions A M solution of potassium chloride, KCl, can be used for these tests.

$Na_3[Co(NO_2)_6]$
↓
yellow $K_3[Co(NO_2)_6]$

1. *Sodium hexanitritocobaltate(III) solution $Na_3[Co(NO_2)_6]$* yellow precipitate of potassium hexanitritocobaltate(III):

$$3K^+ + [Co(NO_2)_6]^{3-} \rightarrow K_3[Co(NO_2)_6]\downarrow$$

The precipitate is insoluble in dilute acetic acid. If larger amounts of sodium are present (or if the reagent is added in excess) a mixed salt, $K_2Na[Co(NO_2)_6]$, is formed. The precipitate forms immediately in concentrated solutions and slowly in diluted solutions; precipitation may be accelerated by warming. Ammonium salts give a similar precipitate and must be absent. In alkaline solutions a brown or black precipitate of cobalt(III) hydroxide $Co(OH)_3$ is obtained. Iodides and other reducing agents interfere and should be removed before applying the test.

Tartaric acid
↓
white K-H-tartrate

2. *Tartaric acid solution (or sodium hydrogen tartrate solution)* white crystalline precipitate of potassium hydrogen tartrate:

$$K^+ + H_2 \cdot C_4H_4O_6 \rightleftharpoons KHC_4H_4O_6\downarrow + H^+ \tag{a}$$

and

$$K^+ + H \cdot C_4H_4O_6^- \rightleftharpoons KHC_4H_4O_6 \tag{b}$$

If tartaric acid is used, the solution should be buffered with sodium acetate, because the strong acid, formed in reaction (a), dissolves the precipitate. Strong alkalis also dissolve the precipitate.

The precipitate is slightly soluble in water ($3.26\,\mathrm{g\,l^{-1}}$, $K_s = 3 \times 10^{-4}$), but quite insoluble in 50% ethanol. Precipitation is accelerated by vigorous agitation of the solution, by scratching the sides of the vessel with a glass rod, and by adding alcohol. Ammonium salts yield a similar precipitate and must be absent.

$HClO_4$
↓
white $KClO_4$

3. *Perchloric acid solution (HClO₄)*   white crystalline precipitate of potassium perchlorate $KClO_4$ from moderately dilute solutions.

$$K^+ + ClO_4^- \rightarrow KClO_4\downarrow$$

The precipitate is slightly soluble in water ($3.2\,\mathrm{g\,l^{-1}}$ and $198\,\mathrm{g\,l^{-1}}$ at $0\,°C$ and $100\,°C$ respectively), and virtually insoluble in absolute alcohol. The alcoholic solution should not be heated as a dangerous explosion may result. This reaction is unaffected by the presence of ammonium salts.

$H_2[PtCl_6]$
↓
white $K_2[PtCl_6]$

4. *Hexachloroplatinic(IV) acid (H₂[PtCl₆]) reagent*   yellow precipitate of potassium hexachloroplatinate(IV):

$$2K^+ + [PtCl_6]^{2-} \rightarrow K_2[PtCl_6]\downarrow$$

Precipitation is instantaneous from concentrated solutions; in dilute solutions, precipitation takes place slowly on standing, but may be hastened by cooling and by rubbing the sides of the vessel with a glass rod. The precipitate is slightly soluble in water, but is almost insoluble in 75% alcohol. Ammonium salts give a similar precipitate and must be absent.

As it is expensive, only small quantities of the reagent should be employed and all precipitates placed in the platinum residues bottle.

$Na_3[Co(NO_2)_6] + KNO_3$
↓
yellow ppt

5. *Sodium hexanitritocobaltate(III)–silver nitrate test*   this is a modification of reaction 1 and is applicable to halogen-free solutions. Precipitation of potassium salts with sodium hexanitritocobaltate(III) and silver nitrate solution gives the compound $K_2Ag[Co(NH_2)_6]$, which is less soluble than the corresponding sodium compound $K_2Na[Co(NO_2)_6]$ and hence the test is more sensitive. Lithium, thallium, and ammonium salts must be absent because they give precipitates with sodium hexanitritocobaltate(III) solution.

Place a drop of the neutral or acetic acid test solution on a black spot plate, and add a drop of 0.1M silver nitrate solution and a small amount of finely powdered sodium hexanitritocobaltate(III). A yellow precipitate of turbidity appears.

Sensitivity 1 µg K.

Concentration limit 1 in 50 000.

If silver nitrate solution is not added, the sensitivity is 4 µg K.

p-Dipicrylamine
↓
orange ppt

6. *p-Dipicrylamine (or hexanitrodiphenylamine) reagent*

the hydrogen atom of the NH group is replaceable by metals, the sodium salt is soluble in water to yield a yellow solution. With solutions of potassium salts, the latter gives a crystalline orange-red precipitate of the potassium derivative. The test is applicable in the presence of 80 times as much sodium and 130 times as much lithium. Ammonium salts should be removed before applying the test. Magnesium does not interfere.

Place a drop of the neutral test solution upon drop-reaction paper and immediately add a drop of the slightly alkaline reagent. An orange-red spot is obtained, which is unaffected by treatment with 1–2 drops of 2M hydrochloric acid.

Sensitivity 3 µg K.

Concentration limit 1 in 10 000.

Na tetraphenylboron
↓
white ppt

7. *Sodium tetraphenylboron test* potassium forms a white precipitate in neutral solutions or in the presence of acetic acid:

$$K^+ + [B(C_6H_5)_4]^- \rightarrow K[B(C_6H_5)_4]\downarrow$$

The precipitate is almost insoluble in water ($0.053\,\mathrm{g\,l^{-1}}$, $K_s = 2.25 \times 10^{-8}$); potassium is precipitated quantitatively if a small excess of the reagent is applied (0.1–0.2%). The precipitate is soluble in strong acids and alkalis, and also in acetone. Rubidium, caesium, thallium(I), and ammonium ions interfere.

Atomic spectrometry

8. *Atomic spectrometric tests* to minimize ionization of potassium, low-temperature flames should be applied.

(a) *Flame atomic emission spectrometry* with an air/acetylene flame 0.1 ppm potassium can be identified at 766.49 nm.

(b) *Flame atomic absorption spectrometry* using a potassium hollow-cathode lamp (or an electrodeless potassium discharge tube) and an air/acetylene flame, 0.1 ppm can be identified at 766.49 nm.

(c) *Potassium standard, 1000 ppm* dissolve 1.9067 g analytical grade potassium chloride KCl in water and dilute to 1 litre.

Flame colour: violet

9. *Dry test (flame colouration)* potassium compounds, preferably the chloride, colour the non-luminous Bunsen flame (violet). The yellow flame produced by small quantities of sodium obscures the violet colour, but by viewing the flame through two thicknesses of cobalt blue glass, the yellow sodium rays are absorbed and the reddish-violet potassium flame becomes visible. A solution of chrome alum ($310\,\mathrm{g\,l^{-1}}$), 3 cm thick, also makes a good filter.

3.37   Sodium, Na (A_r: 22.99)

Sodium is a silver-white soft metal melting at 97.5°C. It oxidizes rapidly in moist air and is therefore kept under liquid paraffin. The metal reacts violently with water forming sodium hydroxide and hydrogen:

$$2Na + 2H_2O \rightarrow 2Na^+ + 2OH^- + H_2\uparrow$$

In its salts sodium is present as the monovalent cation Na^+. These salts form colourless solutions unless the anion is coloured; almost all sodium salts are soluble in water.

Na(I) reactions

Reactions of sodium(I) ions   To study these reactions a M solution of sodium chloride, NaCl, can be used.

Uranyl-Mg-acetate
↓
yellow ppt

1. *Uranyl magnesium acetate solution*   yellow, crystalline precipitate of sodium magnesium uranyl acetate $NaMg(UO_2)_3(CH_3COO)_9 \cdot 9H_2O$ from concentrated solutions. The addition of about one-third volume of alcohol helps the precipitation.

$$Na^+ + Mg^{2+} + 3UO_2^{2+} + 9CH_3COO^- \rightarrow NaMg(UO_2)_3(CH_3COO)_9\downarrow$$

2. *Chloroplatinic acid, tartaric acid or sodium hexanitritocobaltate(III) solution*   no precipitate with solutions of sodium salts.

Uranyl-Zn-acetate
↓
yellow ppt

3. *Uranyl zinc acetate reagent*   as a delicate test for sodium, the uranyl zinc acetate reagent is sometimes preferred to that employing uranyl magnesium acetate. The yellow crystalline sodium zinc uranyl acetate, $NaZn(UO_2)_3(CH_3COO)_9 \cdot 9H_2O$, is obtained. The reaction is fairly selective for sodium. The sensitivity of the reaction is affected by copper, mercury, cadmium, aluminium, cobalt, nickel, manganese, zinc, calcium, strontium, barium, and ammonium when present in concentrations exceeding $5\,g\,l^{-1}$; potassium and lithium salts are precipitated if their concentration in solution exceeds $5\,g\,l^{-1}$ and $1\,g\,l^{-1}$ respectively.

Place a drop of the neutral test solution on a black spot plate or upon a black watch glass, add 8 drops of the reagent, and stir with a glass rod. A yellow cloudiness or precipitate forms.

Sensitivity 12.5 µg Na.

Concentration limit 1 in 4000.

Atomic spectrometry

4. *Atomic spectrometric tests*   these tests, taken at the 589.00/589.59 nm doublet, may be too sensitive for qualitative analysis, as contaminations from reagents can cause significantly high analytical signals. The test should be repeated at the 330.23/330.30 nm doublet, where sensitivities are lower. Colder air/acetylene flames are sufficient.

(a) *Flame atomic emission spectrometry*   with an air/acetylene flame 0.01 ppm sodium can be identified at the 589.00/589.59 nm doublet.

(b) *Flame atomic absorption spectrometry*   using a sodium hollow-cathode lamp (or an electrodeless sodium discharge tube) and an air/acetylene flame, 0.01 ppm can be identified at the 589.00/589.59 nm doublet.

(c) *Sodium standard, 1000 ppm*   dissolve 2.5424 g analytical grade sodium chloride NaCl in water and dilute to 1 litre.

Flame colour: yellow

5. *Dry test (flame colouration)* a non-luminous **Bunsen** flame is coloured an intense yellow by vapours of sodium salts. The colour is not visible when viewed through two thicknesses of cobalt blue glass. Minute quantities of sodium salts produce this colour and it is only when the colour is intense and persistent that appreciable quantities of sodium are present.

3.38 Ammonium ion, NH_4^+ (M_r: 18.038)

Ammonium ions are derived from ammonia, NH_3, and the hydrogen ion H^+. The characteristics of these ions are similar to those of alkali metal ions. By electrolysis with a mercury cathode, ammonium amalgam can be prepared, which has similar properties to the amalgams of sodium or potassium.

Ammonium salts are generally water-soluble compounds, forming colourless solutions (unless the anion is coloured). On heating, all ammonium salts decompose to ammonia and the appropriate acid. Unless the acid is non-volatile, ammonium salts can be quantitatively removed from dry mixtures by heating.

The reactions of ammonium ions are in general similar to those of potassium, because the sizes of the two ions are almost identical.

$NH_4(I)$ reactions

Reactions of ammonium(I) ions To study these reactions a M solution of ammonium chloride NH_4Cl can be used.

NaOH
↓
NH_3 gas

1. *Sodium hydroxide solution* ammonia gas is evolved on warming.

$$NH_4^+ + OH^- \rightarrow NH_3\uparrow + H_2O$$

This may be identified (*a*) by its odour (cautiously smell the vapour after removing the test-tube or small beaker from the flame); (*b*) by the formation of white fumes of ammonium chloride when a glass rod moistened with concentrated hydrochloric acid is held in the vapour; (*c*) by its turning moistened red litmus paper blue or turmeric paper brown; (*d*) by its ability to turn filter paper moistened with mercury(I) nitrate solution black (this is a very dependable test);[1] and (*e*) filter paper moistened with a solution of manganese(II) chloride and hydrogen peroxide gives a brown colour, due to the oxidation of manganese by the alkaline solution thus formed.

In test 1(*d*) a mixture of mercury(II) amidonitrate (white precipitate) and mercury (black precipitate) is formed:

$$2NH_3 + Hg_2^{2+} + NO_3^- \rightarrow Hg(NH_2)NO_3\downarrow + Hg\downarrow + NH_4^+$$

In test 1(*e*) hydrated manganese(IV) oxide is formed:

$$2NH_3 + Mn^{2+} + H_2O_2 + H_2O \rightarrow MnO(OH)_2\downarrow + 2NH_4^+$$

Nessler test
↓
brown colour or ppt

2. *Nessler's reagent (alkaline solution of potassium tetraiodomercurate(II))* brown precipitate or brown or yellow colouration is produced

---

[1] Arsine, however, blackens mercury(I) nitrate paper, and must therefore be absent.

according to the amount of ammonia or ammonium ions present. The precipitate is a basic mercury(II) amido-iodine:

$$NH_4^+ + 2[HgI_4]^{2-} + 4OH^- \rightarrow HgO \cdot Hg(NH_2)I\downarrow + 7I^- + 3H_2O$$

The test is an extremely delicate one and will detect traces of ammonia present in drinking water. All metals, except sodium or potassium, must be absent.

The spot-test technique is as follows. Mix a drop of the test solution with a drop of concentrated sodium hydroxide solution on a watch glass. Transfer a micro drop of the resulting solution or suspension to drop-reaction paper and add a drop of Nessler's reagent. A yellow or orange-red stain or ring is produced.

Sensitivity 0.3 μg NH_3 (in 0.002 ml).

A better procedure is to employ the technique described under the manganese(II) nitrate–silver nitrate reagent in reaction 9 below. A drop of Nessler's solution is placed on the glass knob of the apparatus. After the reaction is complete, the drop of the reagent is touched with a piece of drop-reaction or quantitative filter paper when a yellow colouration will be apparent.

Sensitivity 0.25 μg NH_3.

$Na_3[Co(NO_2)_6]$
↓
yellow
$(NH_4)_3[Co(NO_2)_6]$

3. *Sodium hexanitritocobaltate(III), ($Na_3[Co(NO_2)_6]$)* yellow precipitate of ammonium hexanitritocobaltate(III), $(NH_4)_3[Co(NO_2)_6]$, similar to that produced by potassium ions:

$$3NH_4^+ + [Co(NO_2)_6]^{3-} \rightarrow (NH_4)_3[Co(NO_2)_6]\downarrow$$

$H_2[PtCl_6]$
↓
yellow $(NH_4)_2PtCl_6$

4. *Hexachloroplatinic(IV) acid ($H_2[PtCl_6]$)* yellow precipitate of ammonium hexachloroplatinate(IV)

$$2NH_4^+ + [PtCl_6]^{2-} \rightarrow (NH_4)_2[PtCl_6]\downarrow$$

The characteristics of the precipitate are similar to that of the corresponding potassium salt, but differ from it in being decomposed by warming with sodium hydroxide solution with the evolution of ammonia gas.

Na-H-tartrate
↓
white NH_4-H-tartrate

5. *Saturated sodium hydrogen tartrate solution ($NaH \cdot C_4H_4O_6$)* white precipitate of ammonium acid tartrate $NH_4 \cdot H \cdot C_4H_4O_6$, similar to but slightly more soluble than the corresponding potassium salt, from which it is distinguished by the evolution of ammonia gas on being heated with sodium hydroxide solution.

$$NH_4^+ + HC_4H_4O_6^- \rightarrow NH_4HC_4H_4O_6\downarrow$$

6. *Perchloric acid or sodium perchlorate solution* no precipitate (distinction from potassium).

Tannic acid + $AgNO_3$
↓
black Ag

7. *Tannic acid–silver nitrate test* the basis of this test is the reducing action of tannic acid (a glucoside of digallic acid) upon the silver ammine complex $[Ag(NH_3)_2]^+$ to yield black silver: it therefore precipitates silver in the presence of ammonia but not from a slightly acid silver nitrate solution.

Mix 2 drops 5% tannic acid (tannin) solution with 2 drops 20% silver nitrate solution, and place the mixture upon drop-reaction paper or upon a little cotton wool. Hold the paper in the vapour produced by heating an ammonium salt with sodium hydroxide solution. A black stain is formed on the paper or upon the cotton wool. The test is a sensitive one.

4-Nitrobenzene-
diazonium
chloride
↓
red colour

8. *4-Nitrobenzene-diazonium chloride reagent* the reagent (1) yields a red colouration (due to 2) with an ammonium salt in the presence of sodium hydroxide solution.

$$O_2N-\langle\bigcirc\rangle-N{=}N{-}Cl + NH_4^+ + 2OH^- \longrightarrow$$

(1)

$$\longrightarrow O_2N-\langle\bigcirc\rangle-N{=}NONH_4 + Cl^- + H_2O$$

(2)

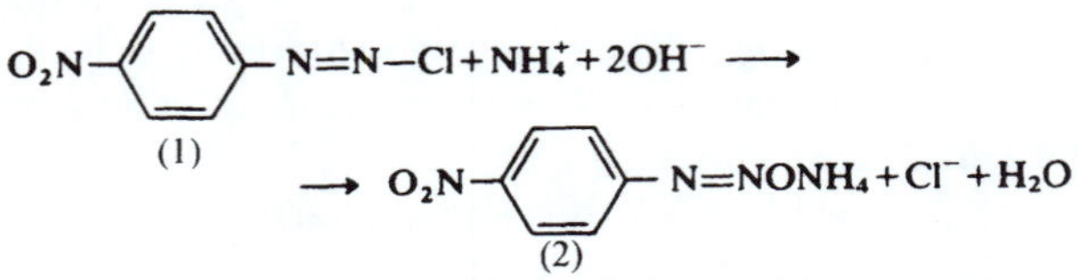

Place a drop of the neutral or slightly acid test solution on a spot plate, followed by a drop of the reagent and a small granule of calcium oxide between the two drops. A red zone forms around the calcium oxide. A blank test should be carried out on a drop of water.

Sensitivity 0.7 μg NH_3.
Concentration limit 1 in 75 000.

Test by liberating NH_3 gas

9. *Ammonia-formation test* this is a more sensitive version of reaction 1. The apparatus is shown in Fig. 2.52 and consists of a small glass tube of 1 ml capacity, which can be closed with a small ground-glass stopper carrying a small glass hook at the lower end.

Place a drop of the test solution or a little of the solid in the micro test-tube, and add a drop of 2M sodium hydroxide solution. Fix a small piece of red litmus paper on the glass hook and insert the stopper into position. Warm to 40°C for 5 minutes. The paper assumes a blue colour.

Sensitivity 0.01 μg NH_3.
Concentration limit 1 in 5 000 000.

Cyanides should be absent, for they give ammonia with alkalis:

$$CN^- + 2H_2O \rightarrow HCOO^- + NH_3\uparrow$$

If, however, a little mercury(II) oxide or a mercury(II) salt is added, the alkali-stable mercury(II) cyanide $Hg(CN)_2$ is formed and the interfering effect of cyanides is largely eliminated.

An alternative method for carrying out the test is to employ the manganese(II) nitrate–silver nitrate reagent. Upon treating a neutral solution of manganese(II) and silver salts with ammonia, a black precipitate is formed:

$$4NH_3 + Mn^{2+} + 2Ag^+ + 3H_2O \rightarrow MnO(OH)_2\downarrow + 2Ag\downarrow + 4NH_4^+$$

Sensitivity 0.005 μg NH_3.
Concentration limit 1 in 10 000 000.

Dry test

10. *Dry test* all ammonium salts are either volatilized or decomposed when heated to just below red heat. In some cases, where the acid is volatile, the vapours recombine on cooling to form a sublimate of the salt, e.g. ammonium chloride.

Chapter 4

Reactions of the anions

4.1 Scheme of classification

The methods available for the detection of anions are not as systematic as those which have been described in the previous chapter for cations. No really satisfactory scheme has yet been proposed which permits of the separation of the common anions into major groups, and the subsequent unequivocal separation of each group into its independent constituents.

The following scheme of classification has been found to work well in practice; it is not a rigid one since some of the anions belong to more than one of the subdivisions. Essentially the processes employed may be divided into Class A, those involving the identification by volatile products obtained on treatment with acids, and Class B, those dependent upon reactions in solution. Class A is subdivided into (1) gases evolved with dilute hydrochloric acid or dilute sulphuric acid, and (2) gases or vapours evolved with concentrated sulphuric acid. Class B is subdivided into (1) precipitation reactions, and (2) oxidation and reduction in solution.

Class A

Class A

(1) Gases evolved with dilute hydrochloric acid or dilute sulphuric acid	(2) Gases or acid vapours evolved with concentrated sulphuric acid: all anions in (1) plus	
Carbonate	Fluoride	Hexafluorosilicate[a]
Hydrogen carbonate	Chloride	Chlorate: **DANGER**
Sulphite	Bromide	Perchlorate
Thiosulphate	Iodide	Permanganate: **DANGER**
Sulphide	Nitrate	Bromate
Nitrite	Borate[a]	Hexacyanoferrate(II) and (III)
Hypochlorite	Formate	Thiocyanate
Cyanide	Acetate	Tartrate
Cyanate	Oxalate	Citrate

[a] These are often included in Class B1.

Class B

Class B

(1) Precipitation reactions		(2) Oxidation/reduction in solution
Sulphate	Peroxodisulphate[b]	Manganate
Phosphate	Phosphite	Permanganate
Succinate	Hypophosphite	Chromate
Arsenate	Arsenite	Dichromate
Chromate	Dichromate	
Silicate	Hexafluorosilicate	
Salicylate	Benzoate	

[b] Strictly speaking, peroxodisulphates should be grouped with Class B2, but are best studied together with sulphates.

The reactions of all these anions will be systematically studied in the following pages. For convenience the reactions of certain organic acids are grouped together; these include acetates, formates, oxalates, tartrates, citrates, salicylates, benzoates, and succinates. It may be pointed out that acetates, formates, salicylates, benzoates, and succinates themselves form another group; all give a characteristic colouration or precipitate upon the addition of iron(III) chloride solution to a practically neutral solution.

4.2 Carbonates, CO_3^{2-}

Solubility all normal carbonates, with the exception of those of the alkali metals and of ammonium, are insoluble in water. The hydrogen carbonates (or bicarbonates) of calcium, strontium, barium, magnesium, and possibly of iron exist in aqueous solution; they are formed by the action of excess carbonic acid upon the normal carbonates either in aqueous solution or suspension and are decomposed on boiling the solutions.

$$CaCO_3\downarrow + H_2O + CO_2 \rightarrow Ca^{2+} + 2HCO_3^-$$

The hydrogen carbonates of the alkali metals are soluble in water, but are less soluble than the corresponding normal carbonates.

CO_3^{2-} reactions

To study these reactions a 0.5M solution of sodium carbonate $Na_2CO_3 \cdot 10H_2O$ can be used.

HCl
↓
CO_2 gas

1. *Dilute hydrochloric acid* decomposition with effervescence, due to the evolution of carbon dioxide:

$$CO_3^{2-} + 2H^+ \rightarrow CO_2\uparrow + H_2O$$

The gas can be identified by its property of rendering lime water (or baryta water) turbid:

$$CO_2 + Ca^{2+} + 2OH^- \rightarrow CaCO_3\downarrow + H_2O$$

$$CO_2 + Ba^{2+} + 2OH^- \rightarrow BaCO_3\downarrow + H_2O$$

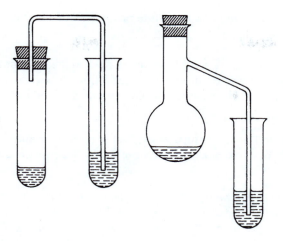

Fig 4.1

Some natural carbonates, such as magnesite, $MgCO_3$, siderite, $FeCO_3$, and dolomite, $(Ca,Mg)CO_3$, do not react appreciably in the cold; they must be finely powdered and the reaction mixture warmed.

The lime water or baryta water test is best carried out in the apparatus shown in Fig. 4.1. An all-glass apparatus can be constructed of Quickfit test-tube MF 24/1 or round bottom flask FR 50/1S, combined with joint cone with stem MF 15/1. The latter has to be bent by a glassblower. The solid substance is placed in the test-tube or small distilling flask (10–25 ml capacity), dilute hydrochloric acid added, and the cork immediately replaced. The gas which is evolved (warming may be necessary) is passed into lime water or baryta water contained in the test-tube; the production of a turbidity indicates the presence of a carbonate. It must be remembered that, with prolonged passage of carbon dioxide, the turbidity slowly disappears as a result of the formation of a soluble hydrogen carbonate:

$$CaCO_3\downarrow + CO_2 + H_2O \rightarrow Ca^{2+} + 2HCO_3^-$$

Any acid which is stronger than carbonic acid ($K_1 = 4.31 \times 10^{-7}$) will displace it, especially on warming. Thus, even acetic acid ($K = 1.76 \times 10^{-5}$) will decompose carbonates; the weak boric acid ($K_1 = 5.8 \times 10^{-10}$) and hydrocyanic acid ($K = 4.79 \times 10^{-10}$) will not.

2. *Barium chloride (or calcium chloride) solution* white precipitate of barium (or calcium) carbonate (soda water):

$BaCl_2$
$\downarrow$
white $BaCO_3$

$$CO_3^{2-} + Ba^{2+} \rightarrow BaCO_3\downarrow$$
$$CO_3^{2-} + Ca^{2+} \rightarrow CaCO_3\downarrow$$

Only normal carbonates react; hydrogen carbonates do not. The precipitate is soluble in mineral acids and carbonic acid (soda water):

$$BaCO_3 + 2H^+ \rightarrow Ba^{2+} + CO_2\uparrow + H_2O$$
$$BaCO_3 + CO_2 + H_2O \rightarrow Ba^{2+} + 2HCO_3^-$$

165

AgNO$_3$
↓
white Ag$_2$CO$_3$

3. *Silver nitrate solution*   white precipitate of silver carbonate:

$$CO_3^{2-} + 2Ag^+ \rightarrow Ag_2CO_3\downarrow$$

The precipitate is soluble in nitric acid and in ammonia. The precipitate

Turns dark

becomes yellow or brown upon addition of excess reagent owing to the formation of silver oxide; the same happens if the mixture is boiled:

$$Ag_2CO_3\downarrow \rightarrow Ag_2O\downarrow + CO_2\uparrow$$

The solution obtained after dissolving in ammonia should be discarded quickly to avoid a serious **explosion!** (Cf. Section 3.6, reaction 1.)

Phenolphthalein
↓
decolourized
by CO$_2$

4. *Sodium carbonate–phenolphthalein test*   this test depends upon the fact that phenolphthalein is turned pink by soluble carbonate and colourless by soluble hydrogen carbonates. Hence if the carbon dioxide liberated by dilute acids from carbonates is allowed to come into contact with phenolphthalein solution coloured pink by sodium carbonate solution, it may be identified by the decolourization which takes place

$$CO_2 + CO_3^{2-} + H_2O \rightarrow 2HCO_3^-$$

The concentration of the sodium carbonate solution must be such as not to be decolourized under the conditions of the experiment by the carbon dioxide in the atmosphere.

Place 1–2 drops test solution (or a small quantity of the test solid) in the apparatus shown in Fig. 2.50 and place 1 drop sodium carbonate-phenolphthalein reagent on the knob of the stopper. Add 3–4 drops M sulphuric acid and insert the stopper into position. The drop is decolourized either immediately or after a short time according to the quantity of carbon dioxide formed. Perform a blank test in a similar apparatus.

Sensitivity 4 μg CO$_2$ (in 2 drops of solution).

Concentration limit 1 in 12 500.

Sulphides, sulphites, thiosulphates, cyanides, cyanates, fluorides, nitrites, and acetates interfere. The sulphur-containing anions can be quantitatively oxidized to sulphates by hydrogen peroxide. The modified procedure in the presence of these anions is therefore to stir a drop of the test solution with 4 drops 3% hydrogen peroxide, then to add 2 drops M sulphuric acid, and to continue as above. Cyanides are rendered innocuous by treating the test solution with 4 drops of a saturated solution of mercury(II) chloride, followed by 2 drops sulphuric acid, etc.; the slightly dissociated mercury(II) cyanide is formed. Nitrites can be removed by treatment with aniline hydrochloride.

4.3   Hydrogen carbonates, HCO$_3^-$

Most of the reactions of hydrogen carbonates are similar to those of carbonates. The tests described here are suitable for distinguishing hydrogen carbonates from carbonates.

A freshly prepared 0.5M solution of sodium hydrogen carbonate, NaHCO$_3$, or potassium hydrogen carbonate, KHCO$_3$, can be used to study these reactions.

Boiling
↓
decomposes

1. *Boiling*   when boiling, hydrogen carbonates decompose:

$$2HCO_3^- \rightarrow CO_3^{2-} + H_2O + CO_2\uparrow$$

carbon dioxide, formed in this way, can be identified with lime water or baryta water (cf. Section 4.2, reaction 1).

MgSO$_4$ and HgCl$_2$
↓
differentiation of
carbonate and
hydrogen carbonate

2. *Magnesium sulphate*   when magnesium sulphate is added to a cold solution of hydrogen carbonate, no precipitation occurs, while a white precipitate of magnesium carbonate, MgCO$_3$, is formed with normal carbonates.

On heating the mixture, a white precipitate of magnesium carbonate is formed:

$$Mg^{2+} + 2HCO_3^- \rightarrow MgCO_3 + H_2O + CO_2\uparrow$$

The carbon dioxide gas, formed in the reaction, can be detected with lime water or baryta water (cf. Section 4.2, reaction 1).

3. *Mercury (II) chloride* (**POISON**)   no precipitate is formed with hydrogen carbonate ions, while in a solution of normal carbonates a reddish-brown precipitate of basic mercury(II) carbonate ($3HgO \cdot HgCO_3 = Hg_4O_3CO_3$) is formed:

$$CO_3^{2-} + 4Hg^{2+} + 3H_2O \rightarrow Hg_4O_3CO_3\downarrow + 6H^+$$

the excess of carbonate acts as a buffer, reacting with the hydrogen ions formed in the reaction:

$$CO_3^{2-} + 2H^+ \rightarrow CO_2\uparrow + H_2O$$

Dry test
↓
decomposition

4. *Dry test*   on heating some solid alkali hydrogen carbonate in a dry test-tube carbon dioxide is evolved:

$$2NaHCO_3 \rightarrow Na_2CO_3 + H_2O + CO_2\uparrow$$

The gas can be identified with lime water or baryta water (cf. Section 4.2, reaction 1). The residue evolves carbon dioxide if, after cooling, dilute hydrochloric acid is poured on it:

$$Na_2CO_3 + 2H^+ \rightarrow 2Na^+ + CO_2\uparrow + H_2O$$

Test for HCO$_3^-$ in
presence of CO$_3^{2-}$

5. *Test for hydrogen carbonate in the presence of normal carbonate*   adding an excess of calcium chloride to a mixture of carbonate and hydrogen carbonate the former is precipitated quantitatively:

$$CO_3^{2-} + Ca^{2+} \rightarrow CaCO_3\downarrow$$

On filtering the solution rapidly hydrogen carbonate ions pass into the filtrate. On adding ammonia to the filtrate, a white precipitate or cloudiness is obtained if hydrogen carbonates are present:

$$2HCO_3^- + 2Ca^{2+} + 2NH_3 \rightarrow 2CaCO_3\downarrow + 2NH_4^+$$

4.4 Sulphites, SO_3^{2-}

Solubility only the sulphites of the alkali metals and of ammonium are soluble in water; the sulphites of the other metals are either sparingly soluble or insoluble. The hydrogen sulphites of the alkali metals are soluble in water; the hydrogen sulphites of the alkaline earth metals are known only in solution.

SO_3^{2-} reactions

A freshly prepared 0.5M solution of sodium sulphite $Na_2SO_3 \cdot 7H_2O$ can be used to study these reactions.

Dilute acids
↓
decomposition

1. *Dilute hydrochloric acid (or dilute sulphuric acid)* decomposition, more rapidly on warming, with the evolution of sulphur dioxide:

$$SO_3^{2-} + 2H^+ \rightarrow SO_2\uparrow + H_2O$$

The gas may be identified (i) by its suffocating odour of burning sulphur, (ii) by the green colouration, due to the formation of chromium(III) ions, produced when a filter paper, moistened with acidified potassium dichromate solution, is held over the mouth of the test-tube.

$$3SO_2 + Cr_2O_7^{2-} + 2H^+ \rightarrow 2Cr^{3+} + 3SO_4^{2-} + H_2O$$

Another method of identifying the gas is (iii) to hold a filter paper, moistened with potassium iodate and starch solution, in the vapour, when a blue colour, owing to the formation of iodine, is observable:

$$5SO_2 + 2IO_3^- + 4H_2O \rightarrow I_2 + 5SO_4^{2-} + 8H^+$$

$BaCl_2$
↓
white $BaSO_3$

2. *Barium chloride or strontium chloride solution* white precipitate of barium (or strontium) sulphite:

$$SO_3^{2-} + Ba^{2+} \rightarrow BaSO_3\downarrow$$

Dissolves in HCl

the precipitate dissolves in dilute hydrochloric acid, when sulphur dioxide is evolved:

$$BaSO_3\downarrow + 2H^+ \rightarrow Ba^{2+} + SO_2\uparrow + H_2O$$

On standing, the precipitate is slowly oxidized to the sulphate and is then insoluble in dilute mineral acids; this change is rapidly effected by warming with bromine water or a little concentrated nitric acid or with hydrogen peroxide.

$$2BaSO_3\downarrow + O_2 \rightarrow 2BaSO_4\downarrow$$

$$BaSO_3\downarrow + Br_2 + H_2O \rightarrow BaSO_4\downarrow + 2Br^- + 2H^+$$

$$3BaSO_3\downarrow + 2HNO_3 \rightarrow 3BaSO_4\downarrow + 2NO\uparrow + H_2O$$

$$BaSO_3\downarrow + H_2O_2 \rightarrow BaSO_4\downarrow + H_2O$$

The solubilities at 18°C of the sulphites of calcium, strontium, and barium are respectively $1.25\,g\,l^{-1}$, $0.033\,g\,l^{-1}$, $0.022\,g\,l^{-1}$.

AgNO$_3$
↓
white Ag$_2$SO$_3$

3. *Silver nitrate solution* no visible change occurs initially because of the formation of sulphitoargentate ions:

$$SO_3^{2-} + Ag^+ \rightarrow [AgSO_3]^-$$

on the addition of more reagent, a white, crystalline precipitate of silver sulphite is formed:

$$[AgSO_3]^- + Ag^+ \rightarrow Ag_2SO_3\downarrow$$

The precipitate dissolves if sulphite ions are added in excess:

$$Ag_2SO_3\downarrow + SO_3^{2-} \rightarrow 2[AgSO_3]^-$$

On boiling the solution of the complex salt, or an aqueous suspension of the precipitate, grey metallic silver is precipitated:

$$2[AgSO_3]^- \rightarrow 2Ag\downarrow + SO_4^{2-} + SO_2\uparrow$$

$$Ag_2SO_3\downarrow + H_2O \rightarrow 2Ag\downarrow + SO_4^{2-} + 2H^+$$

The precipitate is soluble in dilute nitric acid, and sulphur dioxide gas is evolved:

$$Ag_2SO_3\downarrow + 2H^+ \rightarrow SO_2\uparrow + 2Ag^+ + H_2O$$

The precipitate also dissolves in ammonia. The solution obtained after dissolving in ammonia should be discarded quickly to avoid a serious **explosion!** (Cf. Section 3.6, reaction 1.)

KMnO$_4$ or K$_2$Cr$_2$O$_7$
↓
oxidizes to SO$_4^{2-}$

4. *Potassium permanganate solution, acidified with dilute sulphuric acid before the test* decolourization owing to reduction to manganese(II) ions:

$$5SO_3^{2-} + 2MnO_4^- + 6H^+ \rightarrow 2Mn^{2+} + 5SO_4^{2-} + 3H_2O$$

5. *Potassium dichromate solution, acidified with dilute sulphuric acid before the test* a green colouration, owing to the formation of chromium(III) ions:

$$3SO_3^{2-} + Cr_2O_7^{2-} + 8H^+ \rightarrow 2Cr^{3+} + 3SO_4^{2-} + 4H_2O$$

Pb(CH$_3$COO)$_2$
↓
white PbSO$_3$

Dissolves in HNO$_3$

6. *Lead acetate or lead nitrate solution* white precipitate of lead sulphite:

$$SO_3^{2-} + Pb^{2+} \rightarrow PbSO_3\downarrow$$

The precipitate dissolves in dilute nitric acid. On boiling, the precipitate is oxidized by atmospheric oxygen and lead sulphate is formed:

$$2PbSO_3\downarrow + O_2 \rightarrow 2PbSO_4\downarrow$$

This reaction can be used to distinguish sulphites and thiosulphates; the latter produce a black precipitate (cf. Section 4.5, reaction 5) on boiling.

Zn + H$_2$SO$_4$
↓
reduction to H$_2$S

7. *Zinc and sulphuric acid* hydrogen sulphide gas is evolved, which may be detected by holding lead acetate paper to the mouth of the test-tube (cf. Section 4.6, reaction 1):

$$SO_3^{2-} + 3Zn + 8H^+ \rightarrow H_2S\uparrow + 3Zn^{2+} + 3H_2O$$

Lime water
↓
white CaSO₃

8. *Lime water*   this test is carried out by adding dilute hydrochloric acid to the solid sulphite, and bubbling the evolved sulphur dioxide through lime water (Fig. 4.1); a white precipitate of calcium sulphite $CaSO_3$ is formed.

$$SO_3^{2-} + Ca^{2+} \rightarrow CaSO_3\downarrow$$

The precipitate dissolves on prolonged passage of the gas, due to the formation of hydrogen sulphite ions:

$$CaSO_3\downarrow + SO_2 + H_2O \rightarrow Ca^{2+} + 2HSO_3^-$$

A turbidity is also produced by carbonates; sulphur dioxide must therefore be first removed when testing for the latter. This may be effected by adding potassium dichromate solution to the test-tube before acidifying. The dichromate oxidizes and destroys the sulphur dioxide without affecting the carbon dioxide (cf. Section 4.2).

Fuchsin
↓
decolourization
through reduction

9. *Fuchsin test*   dilute solutions of triphenylmethane dyestuffs, such as fuchsin (for formula, see Section 4.15, reaction 9) and malachite green, are immediately decolourized by neutral sulphites. Sulphur dioxide also decolourizes fuchsin solution, but the reaction is not quite complete: nevertheless it is a very useful test for sulphur and acid sulphites; carbon dioxide does not interfere, but nitrogen dioxide does. If the test solution is acidic, it should preferably be just neutralized with sodium hydrogen carbonate. Thiosulphates do not interfere but sulphides, polysulphides, and free alkali do. Zinc, lead, and cadmium salts reduce the sensitivity of the test; hence the interference of sulphides cannot be obviated by the addition of these salts.

Place 1 drop of the fuchsin reagent on a spot plate and add 1 drop of the neutral test solution. The reagent is decolourized.

Sensitivity 1 µg SO_2.
Concentration limit 1 in 50 000.

Zn[Fe(CN)₅NO]
↓
red colour

10. *Sodium nitroprusside–zinc sulphate test*   sodium nitroprusside solution reacts with a solution of a zinc salt to yield a salmon-coloured precipitate of zinc nitroprusside $Zn[Fe(CN)_5NO]$. The latter reacts with moist sulphur dioxide to give a red compound of unknown composition; the test is rendered more sensitive when the reaction product is held over ammonia vapour, which decolourizes the unused zinc nitroprusside.

Place a drop of the test solution (or a grain of the solid test sample) in the tube of Fig. 2.50 and coat the knob of the glass stopper with a thin layer of the zinc nitroprusside paste. Add a drop of 2M hydrochloric or sulphuric acid and close the apparatus. After the sulphur dioxide has been evolved, hold the stopper for a short time in ammonia vapour. The paste is coloured more or less deep red.

Sensitivity 3.5 µg SO_2.
Concentration limit 1 in 14 000.

The zinc nitroprusside paste is prepared by precipitating sodium nitroprusside solution with an excess of zinc sulphate solution and boiling for a few minutes: the precipitate is filtered and washed, and kept in a dark glass bottle or tube.

The test is not applicable in the presence of sulphides and/or thiosulphates. These can be removed by the addition of mercury(II) chloride which reacts forming the acid-stable mercury(II) sulphide:

$$Hg^{2+} + S^{2-} \rightarrow HgS\downarrow$$

$$Hg^{2+} + S_2O_3^{2-} + H_2O \rightarrow HgS\downarrow + SO_4^{2-} + 2H^+$$

Place a drop of the test solution and 2 drops of saturated mercury(II) chloride solution in the same apparatus (Fig. 2.50) and, after a minute, acidify with 2M hydrochloric or M sulphuric acid, and proceed as above. 20 µg Na_2SO_3 can be detected in the presence of 900 µg $Na_2S_2O_3$ and 1500 µg Na_2S.

Distinction between SO_3^{2-} **and** HSO_3^-

11. *Distinction between sulphites and hydrogen sulphites* solutions of normal alkali sulphites show an alkaline reaction against litmus paper, because of hydrolysis:

$$SO_3^{2-} + H_2O \rightleftarrows HSO_3^- + OH^-$$

while the solution of alkali hydrogen sulphites is neutral. Adding a neutral solution of dilute hydrogen peroxide to the solution of normal sulphites, sulphate ions are formed and the solution becomes neutral:

$$SO_3^{2-} + H_2O_2 \rightarrow SO_4^{2-} + H_2O$$

With hydrogen sulphites, the same test yields hydrogen ions:

$$HSO_3^- + H_2O_2 \rightarrow SO_4^{2-} + H^+ + H_2O$$

and the solution shows a definite acid reaction. It must be emphasized that these tests alone are not specific for sulphites or hydrogen sulphites; their presence must be confirmed first by other reactions.

4.5 Thiosulphates, $S_2O_3^{2-}$

Solubility   most of the thiosulphates that have been prepared are soluble in water; those of lead, silver, and barium are very sparingly soluble. Many of them dissolve in excess sodium thiosulphate solution forming complex salts.

$S_2O_3^{2-}$ reactions

To study these reactions use a 0.5M solution of sodium thiosulphate $Na_2S_2O_3 \cdot 5H_2O$.

HCl
↓
slowly white ppt of sulphur

1. *Dilute hydrochloric acid*   no immediate change in the cold with a solution of a thiosulphate; the acidified liquid soon becomes turbid owing to the separation of sulphur, and sulphurous acid is present in solution. On warming the solution, sulphur dioxide is evolved which is recognized by its odour and its action upon filter paper moistened with acidified potassium dichromate solution. The sulphur first forms a colloidal solution, which is gradually coagulated by the free acid present. Side reactions also occur giving rise to thionic acids.

$$S_2O_3^{2-} + 2H^+ \rightarrow S\downarrow + SO_2\uparrow + H_2O$$

I_3^-
↓
reduced to I^-

2. *Potassium triiodide solution (solution of iodine in potassium iodide)* de-colourized when a colourless solution of tetrathionate ions is formed:

$$I_3^- + 2S_2O_3^{2-} \rightarrow 3I^- + S_4O_6^{2-}$$

This reaction has important practical applications in the iodometric and iodimetric methods of titrimetric analysis.[1]

$BaCl_2$
↓
white BaS_2O_3

3. *Barium chloride solution* white precipitate of barium thiosulphate, BaS_2O_3, from moderately concentrated solutions.

$$S_2O_3^{2-} + Ba^{2+} \rightarrow BaS_2O_3\downarrow$$

Precipitation is accelerated by agitation and by rubbing the sides of the vessel with a glass rod. The solubility is $0.5\,g\,l^{-1}$ at 18°C. No precipitate is obtained with calcium chloride solution since calcium thiosulphate is fairly soluble in water.

$AgNO_3$
↓
white $Ag_2S_2O_3$

4. *Silver nitrate solution* white precipitate of silver thiosulphate:

$$S_2O_3^{2-} + 2Ag^+ \rightarrow Ag_2S_2O_3\downarrow$$

At first no precipitation occurs because the soluble dithiosulphato-argentate(I) complex is formed:

$$2S_2O_3^{2-} + Ag^+ \rightarrow [Ag(S_2O_3)_2]^{3-}$$

The precipitate is unstable, turning dark on standing, when silver sulphide is formed:

$$Ag_2S_2O_3\downarrow + H_2O \rightarrow Ag_2S\downarrow + 2H^+ + SO_4^{2-}$$

This hydrolytic decomposition can be accelerated by warming.

$Pb(CH_3COO)_2$
↓
white PbS_2O_3

5. *Lead acetate or lead nitrate solution* at first there is no change, but on further addition of the reagent a white precipitate of lead thiosulphate is formed:

$$S_2O_3^{2-} + Pb^{2+} \rightarrow PbS_2O_3\downarrow$$

The precipitate is soluble in excess thiosulphate; for this reason no precipitation occurs initially. On boiling the suspension the precipitate darkens, forming finally a black precipitate of lead sulphide:

$$PbS_2O_3\downarrow + H_2O \rightarrow PbS\downarrow + 2H^+ + SO_4^{2-}$$

This reaction can be applied to distinguish sulphite and thiosulphate ions (cf. Section 4.4, reaction 6).

KCN
↓
formation of SCN^-

6. *Potassium cyanide solution* (**POISON**) after first alkalizing the test solution with sodium hydroxide and adding potassium cyanide, thiocyanate ions are formed on boiling:

$$S_2O_3^{2-} + CN^- \rightarrow SCN^- + SO_3^{2-}$$

---

[1] See *Vogel's Textbook of Quantitative Inorganic Analysis, including Elementary Instrumental Analysis* (4th edn) edited by J. Basset, R. C. Denney, G. H. Jeffery and J. Mendham, Longman 1978, p. 370 ff.

On acidifying the cold solution, in a well-ventilated fume cupboard, with hydrochloric acid and adding iron(III) chloride, the red colour of iron(III) thiocyanate can be observed:

$$3SCN^- + Fe^{3+} \rightarrow Fe(SCN)_3$$

$(NH_4)_2MoO_4$
↓
blue ring

7. *Blue ring test*   when a solution of thiosulphate mixed with ammonium molybdate solution is poured *slowly* down the side of a test tube which contains concentrated sulphuric acid, a blue ring is formed temporarily at the contact zone. The colour is caused by the reduction of molybdate to molybdenum blue (see Section 6.6, reaction 3).

$FeCl_3$
↓
violet complex

8. *Iron(III) chloride solution*   a dark-violet colouration appears, probably due to the formation of a dithiosulphatoiron(III) complex:

$$2S_2O_3^{2-} + Fe^{3+} \rightarrow [Fe(S_2O_3)_2]^-$$

On standing
reduction to Fe^{2+}

on standing the colour disappears rapidly, while tetrathionate and iron(II) ions are formed:

$$[Fe(S_2O_3)_2]^- + Fe^{3+} \rightarrow 2Fe^{2+} + S_4O_6^{2-}$$

The overall reaction can be written as the reduction of iron(III) by thiosulphate:

$$2S_2O_3^{2-} + 2Fe^{3+} \rightarrow S_4O_6^{2-} + 2Fe^{2+}$$

$[Ni(en)_3](NO_3)_2$
↓
violet ppt

9. *Nickel ethylenediamine nitrate reagent* $[Ni(NH_2 \cdot CH_2 \cdot CH_2 \cdot NH_2)_3]$-$(NO_3)_2$, abbreviated to $[Ni(en)_3](NO_3)_2$. When a neutral or slightly alkaline solution of a thiosulphate is treated with the reagent, a crystalline, violet precipitate of the complex thiosulphate is obtained:

$$[Ni(en)_3]^{2+} + S_2O_3^{2-} \rightarrow [Ni(en)_3]S_2O_3\downarrow$$

Sulphites, sulphates, tetrathionates, and thiocyanates do not interfere, but hydrogen sulphide and ammonium sulphide decompose the reagent with the precipitation of nickel sulphide.

Concentration limit 1 in 25 000.

By obvious modification the reaction may be used for the detection of nickel; it is applicable in the presence of copper, cobalt, iron, and chromium.

$NaN_3 + I_3^-$
↓
reaction catalysed

10. *Catalytic test*   solutions of sodium azide, NaN_3, and iodine (as I_3^-) do not react, but on addition of a trace of thiosulphate, which acts as a catalyst, there is an immediate vigorous evolution of nitrogen:

$$2N_3^- + I_3^- \rightarrow 3I^- + 3N_2\uparrow$$

Sulphides and thiocyanates act similarly and must therefore be absent.

Mix a drop of the test solution and a drop of the iodine–azide reagent on a watch glass. A vigorous evolution of bubbles (nitrogen) ensues.

Sensitivity 0.15 µg $Na_2S_2O_3$.

Concentration limit 1 in 330 000.

4.6 Sulphides, S^{2-}

Solubility the acid, normal, and polysulphides of alkali metals are soluble in water; their aqueous solutions exhibit an alkaline reaction because of hydrolysis.

$$S^{2-} + H_2O \rightleftarrows SH^- + OH^-$$

$$SH^- + H_2O \rightleftarrows H_2S + OH^-$$

The normal sulphides of most other metals are insoluble; those of the alkaline earths are sparingly soluble, but are gradually changed by contact with water into soluble hydrogen sulphides:

$$CaS + H_2O \rightarrow Ca^{2+} + SH^- + OH^-$$

The sulphides of aluminium, chromium, and magnesium can only be prepared under dry conditions, as they are completely hydrolysed by water:

$$Al_2S_3 + 6H_2O \rightarrow 2Al(OH)_3\downarrow + 3H_2S\uparrow$$

The characteristic colours and solubilities of many metallic sulphides have already been discussed in connection with the reactions of the cations in Chapter 3. The sulphides of iron, manganese, zinc, and the alkali metals are decomposed by dilute hydrochloric acid with the evolution of hydrogen sulphide; those of lead, cadmium, nickel, cobalt, antimony, and tin(IV) require concentrated hydrochloric acid for decomposition; others, such as mercury(II) sulphide, are insoluble in concentrated hydrochloric acid, but dissolve in aqua regia with the separation of sulphur. The presence of sulphide in insoluble sulphides may be detected by reduction with nascent hydrogen (derived from zinc or tin and hydrochloric acid) to the metal and hydrogen sulphide, the latter being identified with lead acetate paper (see reaction 1 below). An alternative method is to fuse the sulphide with anhydrous sodium carbonate, extract the mass with water, and to treat the filtered solution with freshly prepared sodium nitroprusside solution, when a purple colour will be obtained; the sodium carbonate solution may also be treated with lead nitrate solution when black lead sulphide is precipitated.

S^{2-} reactions

For the study of these reactions a 2M solution of sodium sulphide $Na_2S \cdot 9H_2O$ can be used.

HCl
↓
H_2S gas

1. *Dilute hydrochloric or sulphuric acid* hydrogen sulphide gas is evolved, which may be identified by its characteristic odour, and by the blackening of filter paper moistened with lead acetate solution:

$$S^{2-} + 2H^+ \rightarrow H_2S\uparrow$$

$$H_2S + Pb^{2+} \rightarrow PbS\downarrow + 2H^+$$

Alternatively, a filter paper moistened with cadmium acetate solution turns yellow:

$$H_2S + Cd^{2+} \rightarrow CdS\downarrow + 2H^+$$

A more sensitive test is attained by the use of sodium tetrahydroxo-plumbate(II) solution, prepared by adding sodium hydroxide to lead acetate until the initial precipitate of lead hydroxide has just dissolved:

$$Pb^{2+} + 2OH^- \rightarrow Pb(OH)_2\downarrow$$

$$Pb(OH)_2\downarrow + 2OH^- \rightarrow [Pb(OH)_4]^{2-}$$

$$[Pb(OH)_4]^{2-} + H_2S \rightarrow PbS\downarrow + 2OH^- + 2H_2O$$

Hydrogen sulphide is a good reducing agent. It reduces (1) acidified potassium permanganate, (2) acidified potassium dichromate, and (3) potassium triiodide (iodine) solution:

$$2MnO_4^- + 5H_2S + 6H^+ \rightarrow 2Mn^{2+} + 5S\downarrow + 8H_2O \qquad (1)$$

$$Cr_2O_7^{2-} + 3H_2S + 8H^+ \rightarrow 2Cr^{3+} + 3S\downarrow + 7H_2O \qquad (2)$$

$$I_3^- + H_2S \rightarrow 3I^- + 2H^+ + S\downarrow \qquad (3)$$

In each case sulphur is precipitated. Small quantities of chlorine may be produced in (1) and (2) if the hydrochloric acid is other than very dilute; this is avoided by using dilute sulphuric acid.

AgNO$_3$
↓
black Ag$_2$S

2. *Silver nitrate solution*   black precipitate of silver sulphide Ag$_2$S, insoluble in cold, but soluble in hot, dilute nitric acid:

$$S^{2-} + 2Ag^+ \rightarrow Ag_2S\downarrow$$

3. *Lead acetate solution*   black precipitate of lead sulphide PbS (Section 3.4, reaction 2).

4. *Barium chloride solution*   no precipitate.

Ag
↓
black Ag$_2$S stain

5. *Silver*   when a solution of a sulphide is brought into contact with a bright silver foil, a brown to black stain of silver sulphide is produced. The result is obtained more expeditiously by the addition of a few drops of dilute hydrochloric acid. The stain may be removed by rubbing the foil with moist lime.

Na$_2$[Fe(CN)$_5$NO]
↓
purple colour

6. *Sodium nitroprusside solution (Na₂[Fe(CN)₅NO])*   transient purple colour in the presence of solutions of alkalis. No reaction occurs with solutions of hydrogen sulphide or with the free gas: if, however, filter paper is moistened with a solution of the reagent made alkaline with sodium hydroxide or ammonia solution, a purple colouration is produced with free hydrogen sulphide.

$$S^{2-} + [Fe(CN)_5NO]^{2-} \rightarrow [Fe(CN)_5NOS]^{4-}$$

The spot-test technique is as follows. Mix on a spot plate a drop of the alkaline test solution with a drop of sodium nitroprusside. A violet colour appears. Alternatively, filter paper impregnated with an ammoniacal (2M) solution of sodium nitroprusside may be employed.

Sensitivity 1 µg Na$_2$S.

Concentration limit 1 in 50 000.

NN-Dimethyl-*p*-phenylenediamine
↓
blue colour

7. *Methylene blue test*   *NN*-Dimethyl-*p*-phenylenediamine is converted by iron(III) chloride and hydrogen sulphide in strongly acid solution into the water-soluble dyestuff, methylene blue:

This is a sensitive test for soluble sulphides and hydrogen sulphide.

Place a drop of the test solution on a spot plate, add a drop of concentrated hydrochloric acid, mix, then dissolve a few grains of *NN*-dimethyl-*p*-phenylenediamine in the mixture (or add 1 drop 1% solution of the chloride or sulphate) and add a drop of 0.5M iron(III) chloride solution. A clear blue colouration appears after a short time (2–3 minutes).

Sensitivity 1 µg H$_2$S.

Concentration limit 1 in 50 000.

NaN$_3$ + I$_3^-$
reaction catalysed

8. *Catalysis of iodine–azide reaction test*   solutions of sodium azide, NaN$_3$, and of iodine (as I$_3^-$) do not react, but on the addition of a trace of a sulphide, which acts as a catalyst, there is an immediate evolution of nitrogen:

$$2N_3^- + I_3^- \rightarrow 3I^- + 3N_2\uparrow$$

Thiosulphates and thiocyanates act similarly and must therefore be absent. The sulphide can, however, be separated by precipitation with zinc or cadmium carbonate. The precipitated sulphide may then be introduced, say, at the end of a platinum wire into a semimicro test-tube or centrifuge tube containing the iodine–azide reagent, when the evolution of nitrogen will be seen.

Mix a drop of the test solution and a drop of the reagent on a watch glass. An immediate evolution of gas in the form of fine bubbles occurs.

Sensitivity 0·3 µg Na$_2$S.

Concentration limit 1 in 166 000.

4.7   Nitrites, NO$_2^-$

Solubility   silver nitrite is sparingly soluble in water. All other nitrites are soluble in water.

NO$_2^-$ reactions

Use freshly prepared 0.1M solution of potassium nitrite, KNO$_2$, to study these reactions.

HCl
↓
blue colour
+
NO$_2$ gas

1. *Dilute hydrochloric acid*   cautious addition of the acid to a solid nitrite in the cold yields a transient, pale-blue liquid (due to the presence of free nitrous acid, HNO$_2$, or its anhydride, N$_2$O$_3$) and the evolution of brown fumes of nitrogen dioxide, the latter being largely produced by combination of nitric oxide with the oxygen of the air. Similar results are obtained with the aqueous solution.

$$NO_2^- + H^+ \rightarrow HNO_2$$

$$(2HNO_2 \rightarrow H_2O + N_2O_3)$$

$$3HNO_2 \rightarrow HNO_3 + 2NO\uparrow + H_2O$$

$$2NO\uparrow + O_2\uparrow \rightarrow 2NO_2\uparrow$$

FeSO$_4$ + acid
↓
brown ring

2. *Iron(II) sulphate solution*   when the nitrite solution is added carefully to a saturated solution of iron(II) sulphate acidified with dilute acetic acid or with dilute sulphuric acid, a brown ring, due to the compound [Fe, NO]SO$_4$, is formed at the junction of the two liquids. If the addition has not been made cautiously, a brown colouration results. This reaction is similar to the brown ring test for nitrates (see Section 4.18, reaction 3), for which a stronger acid (concentrated sulphuric acid) must be employed.

$$NO_2^- + CH_3COOH \rightarrow HNO_2 + CH_3COO^-$$

$$3HNO_2 \rightarrow H_2O + HNO_3 + 2NO\uparrow$$

$$Fe^{2+} + SO_4^{2-} + NO\uparrow \rightarrow [Fe, NO]SO_4$$

Iodides, bromides, coloured ions, and anions that give coloured compounds with iron(II) ions must be absent.

3. *Barium chloride solution*   no precipitate.

AgNO$_3$
↓
white AgNO$_2$

4. *Silver nitrate solution*   white crystalline precipitate of silver nitrite from concentrated solutions:

$$NO_2^- + Ag^+ \rightarrow AgNO_2\downarrow$$

KI
↓
brown I$_3^-$

5. *Potassium iodide solution*   the addition of a nitrite solution to a solution of potassium iodide, followed by acidification with acetic acid or with dilute sulphuric acid, results in the liberation of iodine, which may be identified by the blue colour produced with starch paste. A similar result is obtained by dipping potassium iodide–starch paper moistened with a little dilute acid into the solution. An alternative method is to extract the liberated iodine with chloroform (see Section 4.16, reaction 4).

$$2NO_2^- + 3I^- + 4CH_3COOH \rightarrow I_3^- + 2NO\uparrow + 4CH_3COO^- + 2H_2O$$

KMnO$_4$
↓
decolourized
by reduction

6. *Acidified potassium permanganate solution*   decolourized by a solution of a nitrite, but no gas is evolved.

$$5NO_2^- + 2MnO_4^- + 6H^+ \rightarrow 5NO_3^- + 2Mn^{2+} + 3H_2O$$

NH_4Cl
↓
N_2 gas

7. *Ammonium chloride* by boiling a solution of a nitrite with excess of the solid reagent, nitrogen is evolved and the nitrite is completely destroyed.

$$NO_2^- + NH_4^+ \rightarrow N_2\uparrow + 2H_2O$$

Urea
↓
$N_2 + CO_2$ gases

8. *Urea $CO(NH_2)_2$* when a solution of a nitrite is treated with solid urea and the mixture acidified with dilute hydrochloric acid, the nitrite is decomposed, and nitrogen and carbon dioxide are evolved.

$$CO(NH_2)_2 + 2HNO_2 \rightarrow 2N_2\uparrow + CO_2\uparrow + 3H_2O$$

Thiourea
↓
N_2 gas $+ SCN^-$

9. *Thiourea $CS(NH_2)_2$* when a dilute acetic acid solution of a nitrite is treated with a little solid thiourea, nitrogen is evolved and thiocyanic acid is produced. The latter may be identified by the red colour produced with dilute HCl and $FeCl_3$ solution.

$$CS(NH_2)_2 + HNO_2 \rightarrow N_2\uparrow + H^+ + SCN^- + 2H_2O$$

Thiocyanates and iodides interfere and, if present, must be removed either with excess of solid Ag_2SO_4 or with dilute $AgNO_3$ solution before adding the acetic acid and thiourea.

Sulphamic acid
↓
N_2 gas

10. *Sulphamic acid ($HO \cdot SO_2 \cdot NH_2$)* when a solution of a nitrite is treated with solid sulphamic acid, it is completely decomposed:

$$HO \cdot SO_2 \cdot NH_2 + HNO_2 \rightarrow N_2\uparrow + 2H^+ + SO_4^{2-} + H_2O$$

No nitrate is formed in this reaction, and it is therefore an excellent method for the complete removal of nitrite. Traces of nitrates are formed with ammonium chloride, urea, and thiourea (reactions 7, 8, and 9).

Griess–Ilosvay test
↓
red colour

11. *Sulphanilic acid – 1-naphthylamine reagent (Griess–Ilosvay test)* this test depends upon the diazotization of sulphanilic acid by nitrous acid, followed by coupling with 1-naphthylamine to form a red azo dye:

Specific for NO_2^-

Iron(III) ions must be masked by tartaric acid. The test solution must be very dilute; otherwise the reaction does not go beyond the diazotization stage: $0.2\,mg\ NO_2^-\ l^{-1}$ is the optimal concentration.

Place a drop of the neutral or acid test solution on a spot plate and mix it with a drop of the sulphanilic acid reagent, followed by a drop of the 1-naphthylamine reagent. A red colour is formed.

Sensitivity 0.01 µg HNO$_2$.
Concentration limit 1 in 5 000 000.

Indole
↓
red colour

12. *Indole reagent*

a red-coloured nitroso-indole is formed.

Place a drop of the test solution in a semimicro test-tube, add 10 drops of the reagent and 5 drops of 8M sulphuric acid. A purplish-red colouration appears.

Concentration limit 1 in 1 000 000.

4.8   Cyanides, CN$^-$

Solubility   only the cyanides of the alkali and alkaline earth metals are soluble in water; the solutions have an alkaline reaction owing to hydrolysis.

$$CN^- + H_2O \rightarrow HCN + OH^-$$

Mercury(II) cyanide, Hg(CN)$_2$, is also soluble in water, but is practically a non-electrolyte and therefore does not exhibit the ionic reactions of cyanides. Many of the metallic cyanides dissolve in solutions of potassium cyanide to yield complex salts.

CN$^-$ reactions

Use a freshly prepared solution of 0.1M potassium cyanide, KCN, to study these reactions.

All cyanides are highly poisonous. The free acid, HCN, is volatile and is particularly dangerous so that all experiments in which the gas is likely to be evolved, or those in which cyanides are heated, must be carried out in the fume cupboard.

HCl
↓
HCN gas

1. *Dilute hydrochloric acid*   hydrocyanic acid, HCN, with an odour reminiscent of bitter almonds, is evolved in the cold. It should be smelt with great caution. A more satisfactory method for identifying hydrocyanic acid is to convert it into ammonium thiocyanate by allowing the vapour to come into contact with a little ammonium polysulphide on filter paper. The paper may be conveniently placed over the test-tube or dish in which the substance is being treated with the dilute acid. Upon adding a drop of iron(III) chloride solution and a drop of dilute hydrochloric acid to the filter paper, the characteristic red colouration, due to the iron(III) thiocyanate complex, Fe(SCN)$_3$, is obtained (see reaction 6 below). Mercury(II) cyanide is not decomposed by dilute acids.

$$CN^- + H^+ \rightarrow HCN\uparrow$$

AgNO$_3$
↓
white AgCN

2. *Silver nitrate solution*   white precipitate of silver cyanide, AgCN, readily soluble in excess of the cyanide solution forming the complex ion, dicyano-argentate(I) [Ag(CN)$_2$]$^-$ (cf. Section 3.6, reaction 7).

H₂SO₄
↓
CO gas

3. *Concentrated sulphuric acid* heat a little of the solid salt with concentrated sulphuric acid; carbon monoxide is evolved which may be ignited and burns with a blue flame. All cyanides, complex and simple, are decomposed by this treatment.

$$2KCN + 2H_2SO_4 + 2H_2O \rightarrow 2CO\uparrow + K_2SO_4 + (NH_4)_2SO_4$$

Prussian blue test

4. *Prussian blue test* this is a delicate test and is carried out in the following manner. The solution of the cyanide is rendered strongly alkaline with sodium hydroxide solution, a few millilitres of a freshly prepared solution of iron(II) sulphate added (if only traces of cyanide are present, it is best to use a saturated solution of iron(II) sulphate) and the mixture boiled. Hexacyanoferrate(II) ions are thus formed. Upon acidifying with hydrochloric acid (in order to neutralize any free alkali which may be present), a clear solution is obtained, which gives a precipitate of Prussian blue upon the addition of a little iron(III) chloride solution. If only a little cyanide has been used, or is present, in the solution to be tested, a green solution is obtained at first; this deposits Prussian blue on standing.

$$6CN^- + Fe^{2+} \rightarrow [Fe(CN)_6]^{4-}$$

$$3[Fe(CN)_6]^{4-} + 4Fe^{3+} \rightarrow Fe_4[Fe(CN)_6]_3\downarrow$$

Hg₂(NO₃)₂
↓
Hg metal

5. *Mercury(I) nitrate solution* (**POISON**) grey precipitate of metallic mercury (difference from chloride, bromide, and iodide):

$$2CN^- + Hg_2^{2+} \rightarrow Hg\downarrow + Hg(CN)_2$$

Mercury(II) cyanide is only slightly ionized in solution. To detect cyanide in the presence of mercury, an excess of potassium iodide should be added to the sample, when cyanide ions are liberated:

$$Hg(CN)_2 + 4I^- \rightarrow [HgI_4]^{2-} + 2CN^-$$

Mercury(II) cyanide is decomposed by hydrogen sulphide, when mercury(II) sulphide is precipitated ($K_s = 4 \times 10^{-53}$). If the precipitate is filtered off, cyanide ions can be tested for in the solution:

$$Hg(CN)_2 + H_2S \rightarrow HgS\downarrow + 2HCN$$

(NH₄)₂Sₓ
↓
SCN⁻
red colour with Fe³⁺

6. *Iron(III) thiocyanate test* this is another excellent test for cyanides and depends upon the direct combination of alkali cyanides with sulphur (best derived from an alkali or ammonium polysulphide). A little ammonium polysulphide solution is added to the potassium cyanide solution contained in a porcelain dish, and the contents evaporated to dryness on the water bath in the fume cupboard. The residue contains alkali and ammonium thiocyanates together with any residual polysulphide. The latter is destroyed by the addition of a few drops of hydrochloric acid. One or two drops of iron(III) chloride solution are then added. A blood-red colouration, due to the iron(III) thiocyanate complex, is produced immediately:

$$CN^- + S_2^{2-} \rightarrow SCN^- + S^{2-}$$

$$3SCN^- + Fe^{3+} \rightarrow Fe(SCN)_3 \text{ (cf. Section 4.10, reaction 6)}$$

The spot-test technique is as follows. Stir a drop of the test solution with a drop of yellow ammonium sulphide on a watch glass and warm until a rim of sulphur is formed round the liquid (evaporation to dryness, other than on a water bath, should be avoided). Add 1–2 drops dilute hydrochloric acid, allow to cool, and add 1–2 drops 0.5M iron(III) chloride solution. A red colouration is obtained.

Sensitivity 1 µg CN⁻.

Concentration limit 1 in 50 000.

The test is applicable in the presence of sulphide or sulphite; if thiocyanate is originally present, the cyanide must be isolated first by precipitation, e.g. as zinc cyanide.

CuS
↓
dissolves

7. *Copper sulphide test* solutions of cyanides readily dissolve copper(II) sulphide forming the colourless tetracyanocuprate(I) ions:

$$2CuS\downarrow + 10CN^- \rightarrow 2[Cu(CN)_4]^{3-} + 2S^{2-} + (CN)_2\uparrow$$

Note that the oxidation number of copper in the solution is +1. The test is best carried out on filter paper or drop-reaction paper and is applicable in the presence of chlorides, bromides, iodides, and hexacyanoferrate(II) and (III) ions.

Place a drop of a freshly prepared copper sulphide suspension on a filter paper (or on a spot plate) and add a drop of the test solution. The brown colour of copper sulphide disappears at once.

Sensitivity 2.5 µg CN⁻.

Concentration limit 1 in 20 000.

4.9 Cyanates, OCN⁻

Solubility the cyanates of the alkalis and of the alkaline earths are soluble in water. Those of silver, mercury(I), lead, and copper are insoluble. The free acid is a colourless liquid with an unpleasant odour; it is very unstable.

OCN⁻ reactions

To study these reactions use a 0.2M solution of potassium cyanate, KOCN.

H₂SO₄
↓
CO₂ gas

1. *Dilute sulphuric acid* vigorous effervescence, due largely to the evolution of carbon dioxide. The free cyanic acid, HOCN, which is liberated initially, is decomposed into carbon dioxide and ammonia, the latter combining with the sulphuric acid present to form ammonium sulphate. A little cyanic acid, however, escapes decomposition and may be recognized in the evolved gas by its penetrating odour. If the resulting solution is warmed with sodium hydroxide, ammonia is evolved (test with mercury(I) nitrate paper).

$$OCN^- + H^+ \rightarrow HOCN$$

$$HOCN + H^+ + H_2O \rightarrow CO_2\uparrow + NH_4^+$$

2. *Concentrated sulphuric acid* the reaction is similar to that with the dilute acid, but is somewhat more vigorous.

AgNO₃
↓
white AgOCN

3. *Silver nitrate solution* white, curdy precipitate of silver cyanate, AgOCN, soluble in ammonia solution and in dilute nitric acid. The precipitate appears instantaneously, without complex formation (difference from cyanide):

$$OCN^- + Ag^+ \rightarrow AgOCN\downarrow$$

The solution obtained after dissolving in ammonia should be discarded quickly to avoid a serious **explosion!** (Cf. Section 3.6, reaction 1.)

4. *Barium chloride solution* no precipitate.

Co(CH₃COO)₂
↓
blue complex

5. *Cobalt acetate solution* when the reagent is added to a concentrated solution of potassium cyanate, a blue colouration, due to tetracyanato-cobaltate(II) ions, $[Co(OCN)_4]^{2-}$, is produced. The colour is stabilized and intensified somewhat by the addition of ethanol.

$$4OCN^- + Co^{2+} \rightarrow [Co(OCN)_4]^{2-}$$

CuSO₄ + pyridine
↓
blue ppt

6. *Copper sulphate–pyridine test* when a cyanate is added to a dilute solution of a copper salt to which a few drops of pyridine have been previously added, a lilac-blue precipitate is formed of the compound $[Cu(C_5H_5N)_2](OCN)_2$; this is soluble in chloroform with the production of a sapphire-blue solution. Thiocyanates interfere; excess of copper solution should be avoided.

$$2OCN^- + Cu^{2+} + 2C_5H_5N \rightarrow [Cu(C_5H_5N)_2](OCN)_2$$

Add a few drops of pyridine to 2–3 drops of a 0.25M solution of copper sulphate then introduce about 2 ml chloroform followed by a few drops of the neutral cyanate solution. Shake the mixture briskly; the chloroform will acquire a blue colour.

Concentration limit 1 part cyanate in 20 000.

The blue complex is stable in the presence of a moderate excess of acetic acid; the reaction can therefore be applied to the detection of cyanates in alkaline solution. The solution to be tested is added to the copper-pyridine-chloroform mixture, acetic acid added slowly and the solution shaken vigorously after each addition. As soon as the solution is neutral, the chloroform will assume a blue colour.

4.10 Thiocyanates[2], SCN⁻

Solubility silver and copper(I) thiocyanates are practically insoluble in water, mercury(II) and lead thiocyanates are sparingly soluble; the solubilities, in g l⁻¹ at 20°C, are 0.0003, 0.0005, 0.7, and 0.45 respectively. The thiocyanates of most other metals are soluble.

---

[2] Raman spectra of thiocyanates appear to indicate that the ion is $-S-C\equiv N$ rather than $S=C=N-$, since a line assignable to the triple bond and none for the double bond was observed. Salts will be written as, e.g. KSCN and the ion as SCN⁻.

To study these reactions use a 0.1M solution of potassium thiocyanate, KSCN.

H$_2$SO$_4$
↓
COS gas

1. *Sulphuric acid* with the concentrated acid a yellow colouration is produced in the cold; upon warming a violent reaction occurs and carbonyl sulphide (burns with a blue flame) is formed.

$$SCN^- + H_2SO_4 + H_2O \rightarrow COS\uparrow + NH_4^+ + SO_4^{2-}$$

The reaction however is more complex than this, because sulphur dioxide (fuchsin solution decolourized) and carbon dioxide may also be detected in the gaseous decomposition products. In the solution some sulphur is precipitated, and formic acid can also be detected.

With the 3M acid no reaction occurs in the cold, but on boiling a yellow solution is formed and sulphur dioxide and a little carbonyl sulphide are evolved. A similar but slower reaction takes place with M sulphuric acid.

AgNO$_3$
↓
white AgSCN

2. *Silver nitrate solution* white, curdy precipitate of silver thiocyanate, AgSCN, soluble in ammonia solution but insoluble in dilute nitric acid.

$$SCN^- + Ag^+ \rightarrow AgSCN\downarrow$$

$$AgSCN\downarrow + 2NH_3 \rightarrow [Ag(NH_3)_2]^+ + SCN^-$$

Upon boiling with M sodium chloride solution, the precipitate is converted into silver chloride:

$$AgSCN\downarrow + Cl^- \rightarrow AgCl\downarrow + SCN^-$$

(distinction and method of separation from silver halides). After acidification with hydrochloric acid, thiocyanate ions can be detected with iron(III) chloride (cf. reaction 6).

The solution obtained after dissolving in ammonia should be discarded quickly to avoid a serious **explosion!** (Cf. Section 3.6, reaction 1.)

Silver thiocyanate also decomposes upon ignition or upon fusion with sodium carbonate.

CuSO$_4$
↓
black Cu(SCN)$_2$

3. *Copper sulphate solution* first a green colouration, then a black precipitate of copper(II) thiocyanate is observed:

$$2SCN^- + Cu^{2+} \rightarrow Cu(SCN)_2\downarrow$$

Adding sulphurous acid (a saturated solution of sulphur dioxide) the precipitate turns into white copper(I) thiocyanate:

$$2Cu(SCN)_2 + SO_2 + 2H_2O \rightarrow 2CuSCN\downarrow + 2SCN^- + SO_4^{2-} + 4H^+$$

Hg(NO$_3$)$_2$
↓
white Hg(SCN)$_2$

4. *Mercury(II) nitrate solution* (**POISON**) white precipitate of mercury(II) thiocyanate Hg(SCN)$_2$, readily soluble in excess of the thiocyanate solution. If the precipitate is heated, it swells up enormously forming 'Pharaoh's serpents', a polymerized cyanogen product.

$$2SCN^- + Hg^{2+} \rightarrow Hg(SCN)_2\downarrow$$

$$Hg(SCN)_2\downarrow + 2SCN^- \rightarrow [Hg(SCN)_4]^{2-}$$

Zn + HCl
↓
H₂S + HCN gases

5. *Zinc and dilute hydrochloric acid* hydrogen sulphide and hydrogen cyanide (**POISONOUS**) are evolved:

$$SCN^- + Zn + 3H^+ \rightarrow H_2S\uparrow + HCN\uparrow + Zn^{2+}$$

FeCl₃
↓
red complex

6. *Iron(III) chloride solution* blood-red colouration, due to the formation of a complex:

$$3SCN^- + Fe^{3+} \rightleftarrows Fe(SCN)_3$$

In fact there are a series of complex cations and anions formed. The (uncharged) complex can be extracted by shaking with ether. The red colour is removed by fluorides, mercury(II) ions, and oxalates; when colourless, more stable complexes are formed:

$$Fe(SCN)_3 + 6F^- \rightarrow [FeF_6]^{3-} + 3SCN^-$$

$$4Fe(SCN)_3 + 3Hg^{2+} \rightarrow 3[Hg(SCN)_4]^{2-} + 4Fe^{3+}$$

$$Fe(SCN)_3 + 3(COO)_2^{2-} \rightarrow [Fe\{(COO)_2\}_3]^{3-} + 3SCN^-$$

HNO₃
↓
HCN + NO gases

7. *Dilute nitric acid* decomposition upon warming, a red colouration is produced, and nitrogen oxide and hydrogen cyanide (**POISONOUS**) are evolved:

$$SCN^- + H^+ + 2NO_3^- \rightarrow 2NO\uparrow + HCN\uparrow + SO_4^{2-}$$

With concentrated nitric acid a more vigorous reaction takes place, with the formation of nitrogen oxide and carbon dioxide.

Test by distillation

8. *Distillation test* free thiocyanic acid, HSCN, can be liberated by hydrochloric acid, distilled into ammonia solution, where it can be identified with iron(III) chloride (cf. reaction 6). This test may be applied to separate thiocyanate from mixtures with ions which would interfere with reaction 6.

Place a few drops of the test solution in a semimicro test-tube, acidify with dilute hydrochloric acid, add a small fragment of broken porcelain and attach a gas absorption pipette (Fig. 2.28c) charged with a drop or two of ammonia solution. Boil the solution in the test-tube gently so as to distil any HSCN present into the ammonia solution. Rinse the ammonia solution into a clean semimicro test-tube, acidify slightly with dilute hydrochloric acid and add a drop of iron(III) chloride solution. A red colouration is obtained.

Co(NO₃)₂
↓
blue complex

9. *Cobalt nitrate solution* blue colouration, due to the formation of $[Co(SCN)_4]^{2-}$ (Section 3.26, reaction 6), but no precipitate [distinction from cyanide, hexacyanoferrate(II) and (III)]:

$$4SCN^- + Co^{2+} \rightarrow [Co(SCN)_4]^{2-}$$

The spot-test technique is as follows. Mix a drop of the test solution in a micro crucible with a very small drop (0.02 ml) of 0.5M solution of cobalt nitrate and evaporate to dryness. The residue, whether thiocyanate is present or not, is coloured violet and the colour slowly fades. Add a few drops of acetone. A blue-green or green colouration is obtained.

Sensitivity 1 µg SCN⁻.

Concentration limit 1 in 50 000.

Nitrites yield a red colour due to nitrosyl thiocyanate and therefore interfere with the test.

NaN₃ + I₃⁻
↓
reaction catalysed

10. *Catalysis of iodine–azide reaction test*   traces of thiocyanates act as powerful catalysts in the otherwise extremely slow reaction between tri-iodide (iodine) and sodium azide:

$$I_3^- + 2N_3^- \rightarrow 3I^- + 3N_2\uparrow$$

Sulphides (see Section 4.6, reaction 8) and thiosulphates (see Section 4.5, reaction 10) have a similar catalytic effect; these may be removed by precipitation with mercury(II) nitrate solution:

$$Hg^{2+} + S^{2-} \rightarrow HgS\downarrow$$

$$Hg^{2+} + S_2O_3^{2-} + H_2O \rightarrow HgS\downarrow + SO_4^{2-} + 2H^+$$

Considerable amounts of iodine retard the reaction; iodine should therefore be largely removed by the addition of an excess of mercury(II) nitrate solution whereupon the complex $[HgI_4]^{2-}$ ion, which does not affect the catalysis, is formed.

Mix a drop of the test solution with 1 drop of the iodine–azide reagent on a spot plate. Bubbles of gas (nitrogen) are evolved.

Sensitivity 1.5 µg KSCN.

Concentration limit 1 in 30 000.

4.11   Hexacyanoferrate(II) ions, $[Fe(CN)_6]^{4-}$

Solubility   the alkali and alkaline earth hexacyanoferrate(II)s are soluble in water; those of the other metals are insoluble in water and in cold dilute acids, but are decomposed by alkalis.

$[Fe(CN)_6]^{4-}$ reactions

Use a 0.025M solution of potassium hexacyanoferrate(II), (often called potassium ferrocyanide), $K_4[Fe(CN)_6]\cdot3H_2O$, to study these reactions.

cc H₂SO₄
↓
decomposition

1. *Concentrated sulphuric acid*   complete decomposition occurs on pro-longed boiling with the evolution of carbon monoxide, which burns with a blue flame:

$$[Fe(CN)_6]^{4-} + 6H_2SO_4 + 6H_2O \rightarrow Fe^{2+} + 6NH_4^+ + 6CO\uparrow + 6SO_4^{2-}$$

A little sulphur dioxide may also be produced, due to the oxidation of iron(II) with sulphuric acid:

$$2Fe^{2+} + 2H_2SO_4 \rightarrow 2Fe^{3+} + SO_2\uparrow + SO_4^{2-} + 2H_2O$$

With dilute sulphuric acid, little reaction occurs in the cold, but on boiling, a partial decomposition of hexacyanoferrate(II) occurs with the evolution of hydrogen cyanide (**POISON**):

$$[Fe(CN)_6]^{4-} + 6H^+ \rightarrow 6HCN\uparrow + Fe^{2+}$$

Iron(II) ions, formed in this reaction, react with some of the undecomposed hexacyanoferrate, yielding initially a white precipitate of potassium iron(II) hexacyanoferrate(II):

$$[Fe(CN)_6]^{4-} + Fe^{2+} + 2K^+ \rightarrow K_2Fe[Fe(CN)_6]\downarrow$$

This precipitate is gradually oxidized to Prussian blue by atmospheric oxygen (cf. Section 3.21, reaction 6).

AgNO$_3$
↓
white Ag$_4$[Fe(CN)$_6$]

2. *Silver nitrate solution*   white precipitate of silver hexacyanoferrate(II):

$$[Fe(CN)_6]^{4-} + 4Ag^+ \rightarrow Ag_4[Fe(CN)_6]\downarrow$$

The precipitate is insoluble in ammonia [distinction from hexacyanoferrate(III)] and nitric acid, but soluble in potassium cyanide and sodium thiosulphate:

$$Ag_4[Fe(CN)_6]\downarrow + 8CN^- \rightarrow 4[Ag(CN)_2]^- + [Fe(CN)_6]^{4-}$$

$$Ag_4[Fe(CN)_6]\downarrow + 8S_2O_3^{2-} \rightarrow 4[Ag(S_2O_3)_2]^{3-} + [Fe(CN)_6]^{4-}$$

Upon warming with concentrated nitric acid, the precipitate is converted to orange-red silver hexacyanoferrate(III), when it becomes soluble in ammonia:

$$3Ag_4[Fe(CN)_6]\downarrow + HNO_3 + 3H^+ \rightarrow$$

$$\rightarrow 3Ag_3[Fe(CN)_6]\downarrow + 3Ag^+ + NO\uparrow + 2H_2O$$

$$Ag_3[Fe(CN)_6]\downarrow + 6NH_3 \rightarrow 3[Ag(NH_3)_2]^+ + [Fe(CN)_6]^{3-}$$

The solution obtained after dissolving in ammonia should be discarded quickly to avoid a serious **explosion!** (Cf. Section 3.6, reaction 1.)

FeCl$_3$
↓
blue ppt
Prussian blue

3. *Iron(III) chloride solution*   precipitate of Prussian blue in neutral or acid solutions:

$$3[Fe(CN)_6]^{4-} + 4Fe^{3+} \rightarrow Fe_4[Fe(CN)_6]_3\downarrow$$

the precipitate is decomposed by solutions of alkali hydroxides, brown iron(III) hydroxide being formed (cf. Section 3.22, reaction 6).

The spot-test technique is as follows. Mix a drop of the test solution on a spot plate with a drop of iron(III) chloride solution. A blue precipitate or stain is formed.

Sensitivity 1 μg [Fe(CN)$_6$]$^{4-}$.

Concentration limit 1 in 400 000.

FeSO$_4$
↓
white ppt turns blue

4. *Iron(II) sulphate solution*   white precipitate of potassium iron(II) hexacyanoferrate(II), K$_2$Fe[Fe(CN)$_6$], which turns rapidly blue by oxidation (see reaction 1 and also Section 3.21, reaction 6).

CuSO$_4$
↓
brown Cu$_2$[Fe(CN)$_6$]

5. *Copper sulphate solution* brown precipitate of copper hexacyanoferrate(II):

$$[Fe(CN)_6]^{4-} + 2Cu^{2+} \rightarrow Cu_2[Fe(CN)_6]\downarrow$$

The precipitate is insoluble in dilute acetic acid, but decomposes in solutions of alkali hydroxides.

Th(NO$_3$)$_4$
↓
white Th[Fe(CN)$_6$]

6. *Thorium nitrate solution* white precipitate of thorium hexacyanoferrate(II):

$$[Fe(CN)_6]^{4-} + Th^{4+} \rightarrow Th[Fe(CN)_6]\downarrow$$

It is difficult to filter this precipitate, as it tends to form a colloid. The reaction can be used to distinguish hexacyanoferrate(II) ions from hexacyanoferrate(III) and thiocyanate, which do not react.

HCl
↓
free acid extractable

7. *Hydrochloric acid* if a concentrated solution of potassium hexacyanoferrate(II) is mixed with 6M hydrochloric acid, hydrogen hexacyanoferrate(II) is formed, which can be extracted by ether:

$$[Fe(CN)_6]^{4-} + 4H^+ \rightleftharpoons H_4[Fe(CN)_6]$$

By evaporating the ether, the substance is obtained as a white crystalline solid.

Dry test

8. *Dry test* all hexacyanoferrate(II) compounds are decomposed on heating, nitrogen and cyanogen being evolved:

$$3K_4[Fe(CN)_6] \rightarrow N_2\uparrow + 2(CN)_2\uparrow + 12KCN + Fe_3C + C$$

4.12 Hexacyanoferrate(III) ions, [Fe(CN)$_6$]$^{3-}$

Solubility alkali and alkaline earth hexacyanoferrate(III)s are soluble in water, as is iron(III) hexacyanoferrate(III). Those of most other metals are insoluble or sparingly soluble. Metal hexacyanoferrate(III)s are in general more soluble than metal hexacyanoferrate(II)s.

[Fe(CN)$_6$]$^{3-}$ reactions

To study these reactions use a 0.033M solution of potassium hexacyanoferrate(III) K$_3$[Fe(CN)$_6$].

cc H$_2$SO$_4$
↓
decomposition

1. *Concentrated sulphuric acid* on warming a solid hexacyanoferrate(III) with this acid, it is decomposed completely, carbon monoxide gas being evolved:

$$K_3[Fe(CN)_6] + 6H_2SO_4 + 6H_2O \rightarrow$$
$$\rightarrow 6CO\uparrow + Fe^{3+} + 3K^+ + 6NH_4^+ + 6SO_4^{2-}$$

With cold dilute sulphuric acid, no reaction occurs, but on boiling hydrocyanic acid (**POISON**) is evolved:

$$[Fe(CN)_6]^{3-} + 6H^+ \rightarrow Fe^{3+} + 6HCN\uparrow$$

This test must be carried out in a fume cupboard with good ventilation.

AgNO$_3$
↓
orange Ag$_3$[Fe(CN)$_6$]

2. *Silver nitrate solution* orange-red precipitate of silver hexacyano-ferrate(III):

$$[Fe(CN)_6]^{3-} + 3Ag^+ \rightarrow Ag_3[Fe(CN)_6]\downarrow$$

The precipitate is soluble in ammonia [distinction from hexacyano-ferrate(II)], but insoluble in nitric acid. The solution obtained after dissolving in ammonia should be discarded quickly to avoid a serious **explosion!** (Cf. Section 3.6, reaction 1.)

FeSO$_4$
↓
blue ppt

3. *Iron(II) sulphate solution* dark-blue precipitate of Prussian blue (formerly Turnbull's blue), in neutral or acid solution. (Section 3.21, reaction 7.)

FeCl$_3$
↓
brown complex

4. *Iron(III) chloride solution* brown colouration, owing to the formation of undissociated iron(III) hexacyanoferrate(III):

$$[Fe(CN)_6]^{3-} + Fe^{3+} \rightarrow Fe[Fe(CN)_6]$$

(cf. Section 3.22, reaction 7).

CuSO$_4$
↓
green Cu$_3$[Fe(CN)$_6$]$_2$

5. *Copper sulphate solution* green precipitate of copper(II) hexacyano-ferrate(III):

$$2[Fe(CN)_6]^{3-} + 3Cu^{2+} \rightarrow Cu_3[Fe(CN)_6]_2\downarrow$$

cc HCl
↓
brown H$_3$[Fe(CN)$_6$]

6. *Concentrated hydrochloric acid* adding concentrated hydrochloric acid to a saturated solution of potassium hexacyanoferrate(III) in cold, a brown precipitate of free hydrogen hexacyanoferrate(III) (hexacyanoferric acid) is obtained:

$$[Fe(CN)_6]^{3-} + 3HCl \rightarrow H_3[Fe(CN)_6]\downarrow + 3Cl^-$$

KI
↓
oxidized to I$_3^-$

7. *Potassium iodide solution* iodine is liberated in the presence of dilute hydrochloric acids and may be identified by the blue colour produced with starch solution:

$$2[Fe(CN)_6]^{3-} + 3I^- \rightarrow 2[Fe(CN)_6]^{4-} + I_3^-$$

The reaction is reversible; in neutral solution iodine oxidizes hexacyano-ferrate(II) ions.

Dry test

8. *Action of heat* alkali hexacyanoferrate(III)s decompose similarly to hexacyanoferrate(II)s (cf. Section 4.11, reaction 8):

$$6K_3[Fe(CN)_6] \rightarrow 6N_2\uparrow + 3(CN)_2\uparrow + 18KCN + 2Fe_3C + 10C$$

4.13 Hypochlorites, OCl$^-$

Solubility all hypochlorites are soluble in water. They give an alkaline reaction because of hydrolysis:

$$OCl^- + H_2O \rightleftarrows HOCl + OH^-$$

In solution hypochlorites disproportionate, slowly in the cold but quickly when hot, forming chlorate and chloride ions:

$$3OCl^- \rightarrow ClO_3^- + 2Cl^-$$

Thus, if these reactions are to be studied, the solution should be freshly prepared. By saturating 2M sodium hydroxide with chlorine gas, a M solution of sodium hypochlorite is obtained:

$$Cl_2 + 2OH^- \rightarrow ClO^- + Cl^- + H_2O$$

Chloride ions are invariably present and these will interfere with some ionic reactions.

HCl
↓
Cl₂ gas

1. *Dilute hydrochloric acid*   the solution first turns yellow and later effervesces as chlorine is evolved:

$$OCl^- + H^+ \rightarrow HOCl$$

$$HOCl + H^+ + Cl^- \rightarrow Cl_2\uparrow + H_2O$$

The gas may be identified (*a*) by its yellowish-green colour and irritating odour, (*b*) by its bleaching a wet litmus paper, and (*c*) by its action upon potassium iodide-starch paper, which it turns blue-black:

$$Cl_2\uparrow + 3I^- \rightarrow 2Cl^- + I_3^-$$

KI
↓
oxidized to I₃⁻

2. *Potassium iodide–starch paper*   a bluish-black colour is formed in neutral or weakly alkaline solution as the result of separation of iodine:

$$OCl^- + 3I^- + H_2O \rightarrow I_3^- + 2OH^- + Cl^-$$

If the solution is too alkaline the colour disappears because hypoiodite and iodide ions are formed:

$$I_3^- + 2OH^- \rightarrow OI^- + 2I^- + H_2O$$

Pb(CH₃COO)₂
↓
oxidized to PbO₂

3. *Lead acetate or lead nitrate solution*   brown lead dioxide is produced on boiling:

$$OCl^- + Pb^{2+} + H_2O \rightarrow PbO_2\downarrow + 2H^+ + Cl^-$$

Co(NO₃)₂
↓
oxidized to black
Co(OH)₃

4. *Cobalt nitrate solution*   adding a few drops of the reagent to a solution of the hypochlorite, a black precipitate of cobalt(III) hydroxide is obtained:

$$2Co^{2+} + OCl^- + 5H_2O \rightarrow 2Co(OH)_3\downarrow + Cl^- + 4H^+$$

Hydrogen ions, formed in the reaction, are neutralized by the excess alkali present. On warming, oxygen is liberated (identified by the rekindling of a glowing splint), the cobalt acting as a catalyst:

$$2OCl^- \rightarrow 2Cl^- + O_2\uparrow$$

5. *Mercury*   on shaking a slightly acidified (use sulphuric acid) solution of a hypochlorite with mercury, a brown precipitate of basic mercury(II) chloride $(HgCl)_2O$ is formed:

$$2Hg + 2H^+ + 2OCl^- \rightarrow O{\overset{\displaystyle HgCl}{\underset{\displaystyle HgCl\downarrow}{\big<}}} + H_2O$$

The precipitate is soluble in dilute hydrochloric acid, when undissociated mercury(II) chloride is formed:

$$O{\overset{\displaystyle HgCl}{\underset{\displaystyle HgCl}{\big<}}} + 2H^+ + 2Cl^- \rightarrow 2HgCl_2 + H_2O$$

If the precipitate is separated from the excess of mercury, washed and dissolved in hydrochloric acid, mercury can be detected in the solution by passing hydrogen sulphide into it (cf. Section 3.8).

Chlorine water, under similar conditions, produces a white precipitate of mercury(I) chloride Hg_2Cl_2.

4.14   Chlorides, Cl^-

Solubility   most chlorides are soluble in water. Mercury(I) chloride, Hg_2Cl_2, silver chloride, $AgCl$, lead chloride, $PbCl_2$ (this is sparingly soluble in cold but readily soluble in boiling water), copper(I) chloride, $CuCl$, bismuth oxychloride, $BiOCl$, antimony oxychloride, $SbOCl$, and mercury(II) oxychloride, Hg_2OCl_2, are insoluble in water.

To study these reactions use a 0.1M solution of sodium chloride, $NaCl$.

1. *Concentrated sulphuric acid*   considerable decomposition of the chloride occurs in the cold, becoming complete on warming, with the evolution of hydrogen chloride,

$$Cl^- + H_2SO_4 \rightarrow HCl\uparrow + HSO_4^-$$

The product is recognized (*a*) by its pungent odour and the production of white fumes, consisting of fine drops of hydrochloric acid, on blowing across the mouth of the tube, (*b*) by the formation of white clouds of ammonium chloride when a glass rod moistened with ammonia solution is held near the mouth of the vessel, and (*c*) by its turning blue litmus paper red.

2. *Manganese dioxide and concentrated sulphuric acid*   if the solid chloride is mixed with an equal quantity of precipitated manganese dioxide,[3] concentrated sulphuric acid added and the mixture gently warmed, chlorine is

---

[3] The commercial substance (pyrolusite) usually contains considerable quantities of chlorides.

evolved which is identified by its suffocating odour, yellowish-green colour, its bleaching of moistened litmus paper, and turning of potassium iodide–starch paper blue.

$$MnO(OH)_2 + 2H_2SO_4 + 2Cl^- \rightarrow Mn^{2+} + Cl_2\uparrow + 2SO_4^{2-} + 3H_2O$$

AgNO₃
↓
white AgCl

3. *Silver nitrate solution* white, curdy precipitate of silver chloride, AgCl, insoluble in water and in dilute nitric acid, but soluble in dilute ammonia solution, in potassium cyanide (**POISON**) and in sodium thiosulphate solutions (see under silver, Section 3.6, reaction 1):

$$Cl^- + Ag^+ \rightarrow AgCl\downarrow$$

$$AgCl\downarrow + 2NH_3 \rightarrow [Ag(NH_3)_2]^+ + Cl^-$$

$$[Ag(NH_3)_2]^+ + Cl^- + 2H^+ \rightarrow AgCl\downarrow + 2NH_4^+$$

The solution obtained after dissolving in ammonia should be discarded quickly to avoid a serious **explosion!** (Cf. Section 3.6, reaction 1.)

If the silver chloride precipitate is filtered off, washed with distilled water, and then shaken with sodium arsenite solution, it is converted into yellow silver arsenite (distinction from silver bromide and silver iodide, which are unaffected by this treatment). This may be used as a confirmatory test for a chloride.

$$3AgCl + AsO_3^{3-} \rightarrow Ag_3AsO_3\downarrow + 3Cl^-$$

Pb(CH₃COO)₂
↓
white PbCl₂

4. *Lead acetate solution* white precipitate of lead chloride, PbCl₂, from concentrated solutions (see under lead, Section 3.4, reaction 1):

$$2Cl^- + Pb^{2+} \rightarrow PbCl_2\downarrow$$

Chromyl chloride test

Selective for Cl⁻

5. *Potassium dichromate and sulphuric acid (chromyl chloride test)* assemble a small, preferably all-glass, distillation apparatus as shown in Fig. 4.1. Place 10 ml 2M sodium hydroxide in the recipient test-tube. The sample which contains chloride must be in solid form. Mix the sample thoroughly with about three times as much powdered potassium dichromate, and place the mixture in the distillation flask. Add an equal volume of concentrated sulphuric acid and warm the mixture gently.[4] Deep-red vapours of chromyl chloride, CrO_2Cl_2, are formed. When reacting with the sodium hydroxide in the recipient tube, chromyl chloride is converted to chromate ions, resulting in a yellow solution. The presence of chromate can be verified by appropriate reactions (cf. Section 4.33).

One simple test is the following: acidify 2 ml of the solution in the test tube with 4 ml M sulphuric acid, add 1 ml amyl alcohol[5] followed by 1 ml of 10% hydrogen peroxide. After gentle shaking the organic layer turns blue (see

---

[4] This test must not be carried out in the presence of chlorates because of the danger of forming explosive chlorine dioxide (Section 4.19, reaction 1).

[5] Diethyl ether may also be used, but owing to its highly inflammable character and the possible presence of peroxides (unless previously removed by special treatment), it is preferable to employ amyl alcohol or, less efficiently, amyl acetate.

Section 4.33, reaction 4). Alternatively, the diphenylcarbazide test (Section 4.33, reaction 9) may be applied.

The formation of chromate in the distillate indicates that chloride was present in the sample, as chromyl chloride is a volatile liquid (b.p. 116.5°C).

$$4Cl^- + Cr_2O_7^{2-} + 6H^+ \rightarrow 2CrO_2Cl_2\uparrow + 3H_2O$$

$$CrO_2Cl_2\uparrow + 4OH^- \rightarrow CrO_4^{2-} + 2Cl^- + 2H_2O$$

Some chlorine may also be liberated, owing to the reaction:

$$6Cl^- + Cr_2O_7^{2-} + 14H^+ \rightarrow 3Cl_2\uparrow + 2Cr^{3+} + 7H_2O$$

and this decreases the sensitivity of the test.

Bromides and iodides give rise to the free halogens, which yield colourless solutions with sodium hydroxide; if the ratio of iodide to chloride exceeds 1:15, the chromyl chloride formation is largely prevented and chlorine is evolved.[6] Fluorides give rise to the volatile chromyl fluoride, CrO_2F_2, which is decomposed by water, and hence should be absent or removed. Nitrites and nitrates interfere, as nitrosyl chloride may be formed. Chlorates must, of course, be absent.

The chlorides of mercury, owing to their slight ionization, do not respond to this test. Only partial conversion to CrO_2Cl_2 occurs with the chlorides of lead, silver, antimony, and tin.

The spot-test technique is as follows. Into the tube of Fig. 2.54 place a few milligrams of the solid sample (or evaporate a drop or two of the test solution in it), add a small quantity of powdered potassium dichromate and a drop of concentrated sulphuric acid. Place a column about 1 mm long of a 1% solution of diphenylcarbazide in alcohol into the capillary of the stopper and heat the apparatus for a few minutes. The chromyl chloride evolved causes the reagent to assume a violet colour.

Sensitivity 1.5 µg Cl^-.

Concentration limit 1 in 30 000.

4.15   Bromides, Br^-

Solubility   silver, mercury(I), and copper(I) bromides are insoluble in water. Lead bromide is sparingly soluble in cold, but more soluble in boiling water. All other bromides are soluble.

Br^- reactions

To study these reactions use a 0.1M solution of potassium bromide KBr.

cc H_2SO_4
↓
Br_2 gas

1. *Concentrated sulphuric acid*   if concentrated sulphuric acid is poured on some solid potassium bromide, first a reddish-brown solution is formed, then reddish-brown bromine vapour accompanies the hydrogen bromide

---

[6] The iodine reacts with the chromic acid yielding iodic acid: the latter, in the presence of concentrated sulphuric acid and especially on warming, liberates chlorine from chlorides, regenerating iodide. This explains the failure to form chromyl chloride.

(fuming in moist air) which is evolved:

$$KBr + H_2SO_4 \rightarrow HBr\uparrow + HSO_4^- + K^+$$

$$2KBr + 2H_2SO_4 \rightarrow Br_2\uparrow + SO_2\uparrow + SO_4^{2-} + 2K^+ + 2H_2O$$

These reactions are accelerated by warming. If concentrated phosphoric acid is substituted for the sulphuric acid and the mixture is warmed, only hydrogen bromide is formed:

$$KBr + H_3PO_4 \rightarrow HBr\uparrow + H_2PO_4^- + K^+$$

The properties of hydrogen bromide are similar to those of hydrogen chloride.

cc H_2SO_4 + MnO_2
↓
Br_2 gas

2. *Manganese dioxide and concentrated sulphuric acid*   when a mixture of a solid bromide, precipitated manganese dioxide, and concentrated sulphuric acid is warmed, reddish-brown vapour of bromine is evolved. Bromine is recognized (*a*) by its powerful irritating odour, (*b*) by its bleaching of litmus paper, (*c*) by its staining of starch paper orange-red and (*d*) by the red colouration produced upon filter paper impregnated with fluorescein (see reaction 8 below):

$$2KBr + MnO_2 + 2H_2SO_4 \rightarrow Br_2\uparrow + 2K^+ + Mn^{2+} + 2SO_4^{2-} + 2H_2O$$

$AgNO_3$
↓
pale yellow AgBr

3. *Silver nitrate solution*   curdy, pale-yellow precipitate of silver bromide, AgBr, sparingly soluble in dilute, but readily soluble in concentrated ammonia solution. The precipitate is also soluble in potassium cyanide and sodium thiosulphate solutions, but insoluble in dilute nitric acid.

$$Br^- + Ag^+ \rightarrow AgBr\downarrow$$

The solution obtained after dissolving in ammonia should be discarded quickly to avoid a serious **explosion!** (Cf. Section 3.6, reaction 1.)

$Pb(CH_3COO)_2$
↓
white $PbBr_2$

4. *Lead acetate solution*   white crystalline precipitate of lead bromide:

$$2Br^- + Pb^{2+} \rightarrow PbBr_2\downarrow$$

The precipitate is soluble in boiling water.

Cl_2
↓
Br_2

5. *Chlorine water*[7]   the addition of this reagent dropwise to a solution of a bromide liberates free bromine, which colours the solution orange-red; if chloroform (2 ml) is added and the liquid shaken, the bromine dissolves in the solvent and, after allowing to stand, forms a reddish-brown solution below the colourless aqueous layer. With excess chlorine water, the bromine is converted into yellow bromine monochloride

---

[7] In practice, it is more convenient to use dilute sodium hypochlorite solution, acidified with dilute hydrochloric acid.

and a pale yellow solution results (difference from iodide).

$$2Br^- + Cl_2\uparrow \rightarrow Br_2\uparrow + 2Cl^-$$

$$Br_2\uparrow + Cl_2\uparrow \rightleftharpoons 2BrCl$$

cc $H_2SO_4 + K_2Cr_2O_7$
↓
Br_2 gas

6. *Potassium dichromate and concentrated sulphuric acid*   on gently warming a mixture of a solid bromide, concentrated sulphuric acid, and potassium dichromate (see chlorides, Section 4.14, reaction 5) and passing the evolved vapours into water, a yellowish-brown solution, containing free bromine but no chromium, is produced. A colourless (or sometimes a pale yellow) solution is obtained on treatment with sodium hydroxide solution; this does not give the chromate reaction with dilute sulphuric acid, hydrogen peroxide and amyl alcohol, or with the diphenyl carbazide reagent (distinction from chloride).

$$6KBr + K_2Cr_2O_7 + 7H_2SO_4 \rightarrow 3Br_2\uparrow + 2Cr^{3+} + 8K^+ + 7SO_4^{2-} + 7H_2O$$

Hot HNO_3
↓
Br_2

7. *Nitric acid*   hot, fairly concentrated (8M) nitric acid oxidizes bromides to bromine:

$$6Br^- + 8HNO_3 \rightarrow 3Br_2\uparrow + 2NO\uparrow + 6NO_3^- + 4H_2O$$

Fluorescein test
↓
red eosin

8. *Fluorescein test*   free bromine converts the yellow dyestuff fluorescein (1) into the red tetrabromofluorescein or eosin (2). Filter paper impregnated with fluorescein solution is therefore a valuable reagent for bromine vapour since the paper acquires a red colour.

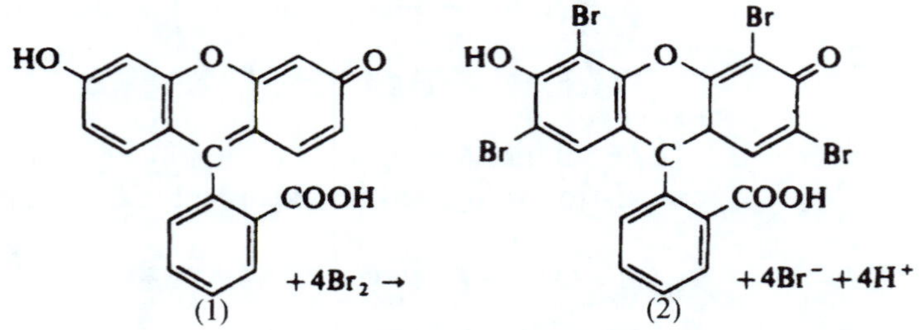

Chlorine tends to bleach the reagent. Iodine forms the red-violet coloured iodoeosin and hence must be absent. If the bromide is oxidized to free bromine by heating with lead dioxide and acetic acid, practically no chlorine is simultaneously evolved from chlorides, and hence the test may be conducted in the presence of chlorides.

$$2Br^- + PbO_2 + 4CH_3COOH \rightarrow Br_2\uparrow + Pb^{2+} + 4CH_3COO^- + 2H_2O$$

Place a drop of the test solution together with a few milligrams of lead dioxide and acetic acid in the apparatus of Fig. 2.53 and close the tube with the funnel stopper carrying a piece of filter paper which has been impregnated with the reagent and dried. Warm the apparatus gently. A circular red spot is formed on the yellow test paper.

Alternatively, the apparatus of Fig. 2.54 may be used; a column, about 1 mm long, of the reagent, is employed.

Sensitivity 2 µg Br$_2$.

Concentration limit 1 in 25 000.

Fuchsin test
↓
violet colour

9. *Fuschin (or Magenta) test* the dyestuff fuschin

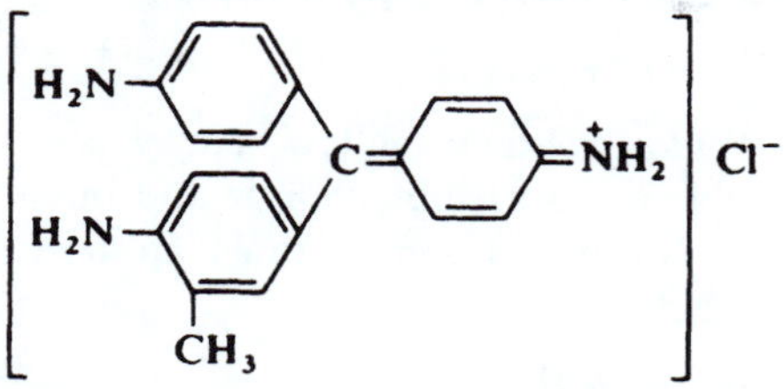

forms a colourless addition compound with hydrogen sulphite. Free bromine converts the thus decolourized fuchsin into a blue or violet brominated dyestuff. Neither free chlorine nor free iodine affect the colourless fuchsin hydrogen sulphite compound, hence the reaction may be employed for the detection of bromides in the presence of chlorides and iodides.

Place a drop of the test solution (or a few milligrams of the test solid) in the tube of the apparatus shown in Fig. 2.54, add 2–4 drops 25% chromic acid solution and close the apparatus with the 'head' which contains 1–2 drops of the reagent solution in the capillary. Warm the apparatus gently (do not allow it to boil). In a short time the liquid in the capillary assumes a violet colour.

Sensitivity 3 µg Br$^-$.

Concentration limit 1 in 50 000.

The reagent is prepared from 0.1% fuchsin solution by just decolourizing it with a few drops of 0.5M sodium hydrogen sulphite solution.

4.16 Iodides, I$^-$

Solubility the solubilities of the iodides are similar to the chlorides and bromides. Silver, mercury(I), mercury(II), copper(I), and lead iodides are the least soluble salts. These reactions can be studied with a 0.1M solution of potassium iodide, KI.

I$^-$ reactions

cc H$_2$SO$_4$
↓
oxidized to brown I$_3^-$

1. *Concentrated sulphuric acid* with a solid iodide, iodine is liberated; on warming, violet vapour is evolved, which turns starch paper blue. Some hydrogen iodide is formed – this can be seen by blowing across the mouth of the vessel, when white fumes are produced – but most of it reduces the sulphuric acid to sulphur dioxide, hydrogen sulphide, and sulphur, the relative proportions of which depend upon the concentrations of the reagents.

$$3I^- + 2H_2SO_4 \rightarrow I_3^- \uparrow + SO_4^{2-} + 2H_2O + SO_2\uparrow$$

$$I^- + H_2SO_4 \rightarrow HI\uparrow + HSO_4^-$$

$$9I^- + 4H_2SO_4 \rightarrow 3I_3^- \uparrow + S\downarrow + 3SO_4^{2-} + 4H_2O$$

$$12I^- + 5H_2SO_4 \rightarrow 4I_3^- \uparrow + H_2S\uparrow + 4SO_4^{2-} + 4H_2O$$

Pure hydrogen iodide is formed on warming with concentrated phosphoric acid:

$$I^- + H_3PO_4 \rightarrow HI\uparrow + H_2PO_4^-$$

If manganese dioxide is added to the mixture, only iodine is formed and the sulphuric acid does not get reduced:

$$3I^- + MnO_2 + 2H_2SO_4 \rightarrow I_3^-\uparrow + Mn^{2+} + 2SO_4^{2-} + 2H_2O$$

AgNO₃
↓
yellow AgI

2. *Silver nitrate solution* yellow, curdy precipitate of silver iodide AgI, readily soluble in potassium cyanide and in sodium thiosulphate solutions, very slightly soluble in concentrated ammonia solution, and insoluble in dilute nitric acid.

$$I^- + Ag^+ \rightarrow AgI$$

Pb(CH₃COO)₂
↓
yellow PbI₂

3. *Lead acetate solution* yellow precipitate of lead iodide, PbI₂, soluble in much hot water forming a colourless solution, and yielding golden-yellow plates ('spangles') on cooling.

$$2I^- + Pb^{2+} \rightarrow PbI_2\downarrow$$

Cl₂
↓
brown colour, I₃⁻

4. *Chlorine water*[8] when this reagent is added dropwise to a solution of an iodide, iodine is liberated, which colours the solution brown; on shaking with 1–2 ml chloroform it dissolves forming a violet solution, which settles out below the aqueous layer. The free iodine may also be identified by the characteristic blue colour it forms with starch solution. If excess chlorine water is added, the iodine is oxidized to colourless iodic acid.

$$3I^- + Cl_2\uparrow \rightarrow I_3^- + 2Cl^-$$

$$I_3^- + 8Cl_2\uparrow + 9H_2O \rightarrow 3IO_3^- + 16Cl^- + 18H^+$$

K₂Cr₂O₇ or NaNO₂
↓
brown colour, I₃⁻

5. *Potassium dichromate and concentrated sulphuric acid* only iodine is liberated, and no chromate is present in the distillate (see chlorides, Section 4.14, reaction 5) (difference from chloride).

$$6I^- + Cr_2O_7^{2-} + 7H_2SO_4 \rightarrow 3I_2\uparrow + 2Cr^{3+} + 7SO_4^{2-} + 7H_2O$$

6. *Potassium nitrite solution* iodine is liberated when this reagent is added to an iodide solution acidified with dilute acetic or sulphuric acid (difference from bromide and chloride). The iodine may be identified by colouring starch paste blue, or chloroform violet.

$$3I^- + 2NO_2^- + 4H^+ \rightarrow I_3^- + 2NO\uparrow + 2H_2O$$

CuSO₄
↓
white CuI
+
brown colour, I₃⁻

7. *Copper sulphate solution* brown precipitate consisting of a mixture of copper(I) iodide, CuI, and iodine. The iodine may be removed by the addition of sodium thiosulphate solution or sulphurous acid, and a nearly white precipitate of copper(I) iodide obtained.

$$5I^- + 2Cu^{2+} \rightarrow 2CuI\downarrow + I_3^-$$

$$I_3^- + 2S_2O_3^{2-} \rightarrow 3I^- + S_4O_6^{2-}$$

---

[8] In practice it is more convenient to use dilute sodium hypochlorite solution acidified with dilute hydrochloric acid.

HgCl₂
↓
scarlet HgI₂

8. *Mercury(II) chloride solution* (**POISON**)   scarlet precipitate of mercury(II) iodide:

$$2I^- + HgCl_2 \rightarrow HgI_2\downarrow + 2Cl^-$$

(note that mercury(II) chloride is practically undissociated in solution). The precipitate dissolves in excess potassium iodide, forming a tetraiodomercurate(II) complex:

$$HgI_2\downarrow + 2I^- \rightarrow [HgI_4]^{2-}$$

Starch
+
oxidizing agent
↓
blue complex

9. *Starch test*   iodides are readily oxidized in acid solution to free iodine by a number of oxidizing agents; the free iodine may then be identified by the deep-blue colouration produced with starch solution. The best oxidizing agent to employ in the spot test reaction is acidified potassium nitrite solution (cf. reaction 6):

$$3I^- + 2NO_2^- + 4H^+ \rightarrow I_3^- + 2NO\uparrow + 2H_2O$$

Cyanides interfere because of the formation of cyanogen iodide: they are therefore removed before the test either by heating with sodium hydrogen carbonate solution or by acidifying and heating:

$$I_3^- + CN^- \rightarrow ICN\uparrow + 2I^-$$

Mix a drop of the acid test solution on a spot plate with a drop of the reagent and add a drop of 50% potassium nitrite solution. A blue colouration is obtained.
Sensitivity 2.5 μg I₂.
Concentration limit 1 in 20 000.

Ce(IV) + As(III)
reaction
↓
catalysed

10. *Catalytic reduction of cerium(IV) salts test*   the reduction of cerium(IV) salts in acid solution by arsenites takes place very slowly:

$$2Ce^{4+} + AsO_3^{3-} + H_2O \rightarrow 2Ce^{3+} + AsO_4^{3-} + 2H^+$$

Iodides accelerate this change, possibly owing to iodine liberated in the instantaneous reaction (i):

$$2Ce^{4+} + 3I^- \rightarrow 2Ce^{3+} + I_3^- \tag{i}$$

reacting further according to (ii):

$$AsO_3^{3-} + I_3^- + H_2O \rightarrow AsO_4^{3-} + 3I^- + 2H^+ \tag{ii}$$

the iodide ion reacting again as in (i). The completion of the reduction is indicated by the disappearance of the yellow colour of the cerium(IV) solution. Osmium and ruthenium salts have a similar catalytic effect. Moderate amounts of chlorides, bromides, sulphates, and nitrates have no influence, but cyanides and also mercury(II), silver, and manganese salts interfere.

Place a drop of the test solution together with a drop each of neutral or slightly acid 0.1M sodium arsenite solution and 0.1M cerium(IV) sulphate solution on a spot plate. The yellow colour soon disappears.
Sensitivity 0.03 μg I⁻.
Concentration limit 1 in 1 000 000.

4.17 Fluorides, F⁻

Solubility the fluorides of the common alkali metals and of silver, mercury, aluminium, and nickel are readily soluble in water, those of lead, copper, iron(III), barium, and lithium are slightly soluble, and those of the other alkaline earth metals are insoluble.

F⁻ reactions

To study these reactions use a 0.1M solution of sodium fluoride, NaF.

cc H_2SO_4
↓
H_2F_2

1. *Concentrated sulphuric acid* with the solid fluoride, a colourless, corrosive gas, hydrogen fluoride, H_2F_2, is evolved on warming; the gas fumes in moist air, and the test-tube acquires a greasy appearance as a result of the corrosive action of the vapour on the silica in the glass, which liberates the gas, silicon tetrafluoride, SiF_4. By holding a moistened glass rod in the vapour, gelatinous silicic acid H_2SiO_3 is deposited on the rod; this is a product of the decomposition of the silicon tetrafluoride.

$$2F^- + H_2SO_4 \rightarrow H_2F_2\uparrow + SO_4^{2-}$$

$$SiO_2 + 2H_2F_2 \rightarrow SiF_4\uparrow + 2H_2O$$

$$3SiF_4\uparrow + 3H_2O \rightarrow 2[SiF_6]^{2-} + H_2SiO_3\downarrow + 4H^+$$

Note that at room temperature hydrogen fluoride gas is almost completely dimerized, therefore its formula has been written as H_2F_2. At elevated temperatures (say 90°C) it dissociates completely to monomer hydrogen fluoride:

$$H_2F_2 \rightleftharpoons 2HF$$

The same result is more readily attained by mixing the solid fluoride with an equal bulk of silica, making into a paste with concentrated sulphuric acid and warming gently; silicon tetrafluoride is quickly evolved.

Etching glass

2. *The etching test* a clean watch glass is coated on the convex side with paraffin wax, and part of the glass is exposed by scratching a design on the wax with a nail or wire. A mixture of about 0.3 g fluoride and 1 ml concentrated sulphuric acid is placed in a small lead or platinum crucible, and the latter immediately covered with the watch glass, convex side down. A little water should be poured in the upper (concave) side of the watch glass to prevent the wax from melting. The crucible is very gently warmed (best on a boiling water bath). After 5–10 minutes, the hydrogen fluoride will have etched the glass. This is readily seen after removing the paraffin wax by holding above a flame or with hot water, and then breathing upon the surface of the glass.

The test may also be conducted in a small lead capsule, provided with a close-fitting lid made from lead foil. A small hole of about 3 mm diameter is pierced in the lid. About 0.1 g suspected fluoride and a few drops concentrated sulphuric acid are placed in the clean capsule, and a small piece of glass (e.g. a microscope slide) is placed over the hole in the lid. Upon warming very gently (best on a water bath) it will be found that an etched spot appears on the glass where it covers the hole.

Chlorates, silicates, and borates interfere and should therefore be absent.

3. *Silver nitrate solution* no precipitate, since silver fluoride is soluble in water.

CaCl$_2$
↓
white CaF$_2$

4. *Calcium chloride solution* white, slimy precipitate of calcium fluoride, CaF$_2$, sparingly soluble in acetic acid, but slightly more soluble in dilute hydrochloric acid.

$$2F^- + Ca^{2+} \rightarrow CaF_2\downarrow$$

FeCl$_3$
↓
white Na$_3$[FeF$_6$]

5. *Iron(III) chloride solution* white crystalline precipitate of sodium hexafluoroferrate(III) from concentrated solutions of fluorides, sparingly soluble in water. The precipitate does not give the reactions of iron (e.g. with ammonium thiocyanate), except upon acidification.

$$6F^- + Fe^{3+} + 3Na^+ \rightarrow Na_3[FeF_6]\downarrow$$

Zr + alizarin
↓
lake decolourizes

6. *Zirconium–alizarin lake test* hydrochloric acid solutions of zirconium salts are coloured reddish-violet by alizarin red S or by alizarin (see under aluminium, Section 3.23, reactions 8 and 9 and under zirconium, Section 6.18, reaction 12); upon adding a solution of a fluoride the colour of such solutions changes immediately to a pale yellow (that of the liberated alizarin sulphonic acid or alizarin) because of the formation of the colourless hexafluorozirconate(IV) ion [ZrF$_6$]$^{2-}$. The test may be performed on a spot plate.

Mix together on a spot plate 2 drops each (equal volumes) of a 0.1% alizarin red S (sodium alizarin sulphonate) and 0.1% zirconyl chloride solution. Upon the addition of a drop or two of the fluoride solution the zirconium lake is decolourized to a clear yellow solution.

4.18 Nitrates, NO$_3^-$

Solubility all nitrates are soluble in water. The nitrates of mercury and bismuth yield basic salts on treatment with water; these are soluble in dilute nitric acid.

NO$_3^-$ reactions

The reactions can be studied with a 0.1M solution of potassium nitrate KNO$_3$.

cc H$_2$SO$_4$
↓
red NO$_2$ gas

1. *Concentrated sulphuric acid* reddish-brown vapour of nitrogen dioxide, and pungent acid vapour of nitric acid which fumes in the air, are formed on heating the solid nitrate with the reagent. Dilute sulphuric acid has no effect (difference from nitrite):

$$4NO_3^- + 2H_2SO_4 \rightarrow 4NO_2\uparrow + O_2\uparrow + 2SO_4^{2-} + 2H_2O$$

cc H$_2$SO$_4$ + Cu
↓
blue colour (Cu^{2+})
or red fumes (NO$_2$)

2. *Concentrated sulphuric acid and bright copper turnings* on heating these with the solid nitrate, reddish-brown fumes of nitrogen dioxide are evolved, and the solution acquires a blue colour owing to the formation of copper(II) ions. A solution of the nitrate may also be used; the sulphuric

acid is then added very cautiously.

$$2NO_3^- + 4H_2SO_4 + 3Cu \rightarrow 3Cu^{2+} + 2NO\uparrow + 4SO_4^{2-} + 4H_2O$$

$$2NO\uparrow + O_2\uparrow \rightarrow 2NO_2\uparrow$$

cc H_2SO_4 + $FeSO_4$
↓
brown ring

3. *Iron(II) sulphate solution and concentrated sulphuric acid (brown ring test)* this test is carried out in either of two ways: (*a*) Add 3 ml freshly prepared saturated solution of iron(II) sulphate to 2 ml nitrate solution, and pour 3–5 ml concentrated sulphuric acid slowly down the side of the test-tube so that the acid forms a layer beneath the mixture. A brown ring will form where the liquids meet. (*b*) Add 4 ml concentrated sulphuric acid slowly to 2 ml nitrate solution, mix the liquids thoroughly and cool the mixture under a stream of cold water from the tap. Pour a saturated solution of iron(II) sulphate slowly down the side of the tube so that it forms a layer on top of the liquid. A brown ring will form at the zone of contact of the two liquids.

The brown ring is due to the formation of the $[Fe(NO)]^{2+}$. On shaking and warming the mixture the brown colour disappears, nitrogen oxide is evolved, and a yellow solution of iron(III) ions remains. The test is unreliable in the presence of bromide, iodide, nitrite, chlorate, and chromate (see Section 5.9, reactions 3 and 4).

$$2NO_3^- + 4H_2SO_4 + 6Fe^{2+} \rightarrow 6Fe^{3+} + 2NO\uparrow + 4SO_4^{2-} + 4H_2O$$

$$Fe^{2+} + NO\uparrow \rightarrow [Fe(NO)]^{2+}$$

Bromides and iodides interfere because of the liberated halogen; the test is not trustworthy in the presence of chromates, sulphites, thiosulphates, iodates, cyanides, thiocyanates, hexacyanoferrate(II) and (III) ions. All of these anions may be removed by adding an excess of nitrate-free Ag_2SO_4 to an aqueous solution (or sodium carbonate extract), shaking vigorously for 3–4 minutes, and filtering the insoluble silver salts, etc.

Nitrites react similarly to nitrates. They are best removed by adding a little sulphamic acid (compare Section 4.7, reaction 10). The following reaction takes place in the cold:

$$H_2N-HSO_3 + NO_2^- \rightarrow N_2\uparrow + SO_4^{2-} + H^+ + H_2O$$

The spot-test technique is as follows. Place a crystal of iron(II) sulphate about as large as a pin head on a spot plate. Add a drop of the test solution and allow a drop of concentrated sulphuric acid to run in at the side of the drop. A brown ring forms round the crystal of iron(II) sulphate.

Sensitivity 2.5 µg NO_3^-.

Concentration limit 1 in 25 000.

Reduction of nitrates

4. *Reduction of nitrates in alkaline medium* ammonia is evolved (detected by its odour; by its action upon red litmus paper and upon mercury(I) nitrate paper; or by the tannic acid–silver nitrate test, Section 3.38, reaction 7) when a solution of a nitrate is boiled with zinc dust or gently warmed with aluminium powder and sodium hydroxide solution. Excellent results are obtained by the use of Devarda's alloy (45% Al, 5% Zn, 50% Cu).

Ammonium ions must, of course, be removed by boiling the solution with sodium hydroxide solution (and, preferably, evaporating almost to dryness) before the addition of the metal.

$$NO_3^- + 4Zn + 7OH^- + 6H_2O \rightarrow NH_3\uparrow + 4[Zn(OH)_4]^{2-}$$

$$3NO_3^- + 8Al + 5OH^- + 18H_2O \rightarrow 3NH_3\uparrow + 8[Al(OH)_4]^-$$

Nitrites give a similar reaction and may be removed most simply with the aid of sulphamic acid (see reaction 3). Another, but more expensive, procedure involves the addition of sodium azide to the acid solution; the solution is allowed to stand for a short time and then boiled in order to complete the reaction and to expel the readily volatile hydrogen azide:

$$NO_2^- + N_3^- + 2H^+ \rightarrow N_2\uparrow + N_2O\uparrow + H_2O$$

$$N_3^- + H^+ \rightarrow HN_3\uparrow$$

Other nitrogen compounds which evolve ammonia under the above conditions are cyanides, thiocyanates, hexacyanoferrate(II) and (III) ions. These may be removed by treating the aqueous solution (or a sodium carbonate extract) with excess of nitrate-free silver sulphate, warming the mixture to about 60°C, shaking vigorously for 3–4 minutes and filtering from the silver salts of the interfering anions and excess of precipitant. The excess silver ions are removed from the filtrate by adding an excess of sodium hydroxide solution, and filtering from the precipitated silver oxide. The filtrate is concentrated and tested with zinc, aluminium, or Devarda's alloy.

Attention is directed to the fact that arsenites are reduced in alkaline solution by aluminium, Devarda's alloy, etc., to arsine, which blackens mercury(I) nitrate paper and also gives a positive tannic acid–silver nitrate test. Hence neither the mercury(I) nitrate test nor the tannic acid–silver nitrate test for ammonia is applicable if arsenites are present.

Diphenylamine
↓
blue ring

5. *Diphenylamine reagent* $(C_6H_5 \cdot NH \cdot C_6H_5)$ pour the nitrate solution carefully down the side of the test-tube so that it forms a layer above the solution of the reagent; a blue ring is formed at the zone of contact of the two liquids. The test is a very sensitive one, but unfortunately is also given by a number of oxidizing agents, such as nitrites, chlorates, bromates, iodates, permanganates, chromates, vanadates, molybdates, and iron(III) salts.

Nitron
↓
white ppt

6. *Nitron reagent (diphenyl-endo-anilo-dihydrotriazole $C_{20}H_{16}N_4$)*

white crystalline precipitate of nitron nitrate $C_{20}H_{16}N_4$,HNO_3 with solutions of nitrates. Bromides, iodides, nitrites, chromates, chlorates, perchlorates, thiocyanates, oxalates, and picrates also yield insoluble compounds, and hence the reaction is not very characteristic.

Heating
↓
decomposition

7. *Action of heat*   the result varies with the metal. The nitrates of sodium and potassium evolve oxygen (test with glowing splint) and leave solid nitrites (brown fumes with dilute acid); ammonium nitrate yields dinitrogen oxide and steam; the nitrates of the noble metals leave a residue of the metal, and a mixture of nitrogen dioxide and oxygen is evolved; the nitrates of the other metals, such as those of lead and copper, evolve oxygen and nitrogen dioxide, and leave a residue of the oxide.

$$2NaNO_3 \rightarrow 2NaNO_2 + O_2\uparrow$$

$$NH_4NO_3 \rightarrow N_2O\uparrow + 2H_2O$$

$$2AgNO_3 \rightarrow 2Ag + 2NO_2\uparrow + O_2\uparrow$$

$$2Pb(NO_3)_2 \rightarrow 2PbO + 4NO_2\uparrow + O_2\uparrow$$

Zn
↓
reduction to NO_2^-

8. *Reduction to nitrite test*   nitrates are reduced to nitrites by metallic zinc in acetic acid solution; the nitrite can be readily detected by means of the sulphanilic-acid-1-naphthylamine reagent (see under nitrites, Section 4.7, reaction 11). Nitrites, of course, interfere and are best removed with sulphamic acid (see reaction 3 above).

Mix on a spot plate a drop of the neutral or acetic acid test solution with a drop of the sulphanilic acid reagent and a drop of the 1-naphthylamine reagent, and add a few milligrams of zinc dust. A red colouration develops.

Sensitivity $0.05\,\mu g\ NO_3^-$.
Concentration limit 1 in 1 000 000.

4.19   Chlorates, ClO_3^-

Solubility   all chlorates are soluble in water; potassium chlorate is one of the least soluble ($66\,g\,l^{-1}$ at 18°C) and lithium chlorate one of the most soluble ($3150\,g\,l^{-1}$ at 18°C).

ClO_3^- reactions

To study these reactions use a 0.1M solution of potassium chlorate.

cc H_2SO_4
↓
ClO_2 gas
(DANGER)

1. *Concentrated sulphuric acid* (**DANGER**)   all chlorates are decomposed with the formation of the greenish-yellow gas, chlorine dioxide, ClO_2, which dissolves in the sulphuric acid to give an orange-yellow solution. On warming gently (**DANGER**) an explosive crackling occurs, which may develop into a violent explosion. In carrying out this test one or two small crystals of potassium chlorate (weighing not more than 0.1 g) are treated with 1 ml concentrated sulphuric acid in the cold; the yellow explosive chlorine dioxide can be seen on shaking the solution. The test-tube should not be warmed, and its mouth should be directed away from the experimenter.

$$3KClO_3 + 3H_2SO_4 \rightarrow 2ClO_2\uparrow + ClO_4^- + 3SO_4^{2-} + 4H^+ + 3K^+ + H_2O$$

cc HCl
↓
$ClO_2 + Cl_2$ gases
(DANGER)

2. *Concentrated hydrochloric acid*   all chlorates are decomposed by this acid, and chlorine, together with varying quantities of the explosive chlorine dioxide, is evolved; chlorine dioxide imparts a yellow colour to the acid. The mixture of gases is sometimes known as 'euchlorine'. The experiment should be conducted on a very small scale, and not more than 0.1 g potassium chlorate should be used. The following two chemical reactions probably occur simultaneously:

$$2KClO_3 + 4HCl \rightarrow 2ClO_2 + Cl_2\uparrow + 2K^+ + 2Cl^- + 2H_2O$$

$$KClO_3 + 6HCl \rightarrow 3Cl_2\uparrow + K^+ + Cl^- + 3H_2O$$

Reducing agents
↓
reduction to Cl^-

3. *Potassium nitrite solution*   on warming this reagent with a solution of the chlorate, the latter is reduced to a chloride, which may be identified by adding silver nitrate solution after acidification with dilute nitric acid. The nitrite must, of course, be free from chloride. A solution of sulphurous acid or of formaldehyde (10%) acts similarly. Excellent results are obtained with zinc, aluminium or Devarda's alloy and sodium hydroxide solution (see under nitrates, Section 4.18, reaction 4); the solution is acidified with dilute nitric acid after several minutes boiling[9] and silver nitrate solution added.

$$ClO_3^- + 3NO_2^- \rightarrow Cl^- + 3NO_3^-$$

$$ClO_3^- + 3H_2SO_3 \rightarrow Cl^- + 3SO_4^{2-} + 6H^+$$

$$ClO_3^- + 3HCHO \rightarrow Cl^- + 3HCOOH$$

$$ClO_3^- + 3Zn + 6OH^- + 3H_2O \rightarrow Cl^- + 3[Zn(OH)_4]^{2-}$$

$$ClO_3^- + 2Al + 2OH^- + 3H_2O \rightarrow Cl^- + 2[Al(OH)_4]^-$$

4. *Silver nitrate solution*   no precipitate in neutral solution or in the presence of dilute nitric acid. Upon the addition of a little pure (chloride-free) sodium nitrite to the dilute nitric acid solution, a white precipitate of silver chloride is obtained because of the reduction of the chlorate to chloride (see reaction 3 above).

No precipitate is obtained with barium chloride solution.

KI
↓
brown colour, I_3^-

5. *Potassium iodide solution*   iodine is liberated if a mineral acid is present. If acetic acid is used, no iodine separates even on long standing (difference from iodate).

$$ClO_3^- + 9I^- + 6H^+ \rightarrow 3I_3^- + Cl^- + 3H_2O$$

$FeSO_4$
↓
reduction to Cl^-

6. *Iron(II) sulphate solution*   reduction to chloride upon boiling in the presence of mineral acid (difference from perchlorate).

$$ClO_3^- + 6Fe^{2+} + 6H^+ \rightarrow Cl^- + 6Fe^{3+} + 3H_2O$$

Indigo + SO_3^{2-}
↓
colour bleaches

7. *Indigo test*   a dilute solution of indigo in concentrated sulphuric acid is added to the chlorate solution until the latter has a pale-blue colour. Dilute sulphurous acid or sodium sulphite solution is then added drop by

---

[9] It is best to filter off the excess metal before adding the silver nitrate solution.

drop; the blue colour disappears. The chlorate is reduced by the sulphurous acid to chlorine or to hypochlorite, and the latter bleaches the indigo.

Aniline + cc H_2SO_4
↓
blue colour

8. *Aniline sulphate test, $(C_6H_5 \cdot NH_2)_2H_2SO_4$* a small quantity of the solid chlorate (say, 0.05 g) (**DANGER**) is mixed with 1 ml concentrated sulphuric acid, and 2–3 ml aqueous aniline sulphate solution added; a deep-blue colour is obtained (distinction from nitrate).

$MnSO_4 + H_3PO_4$
↓
violet complex

9. *Manganese(II) sulphate–phosphoric acid test* manganese(II) sulphate in concentrated phosphoric acid solution reacts with chlorates to form the violet-coloured diphosphatomanganate(III) ion:

$$ClO_3^- + 6Mn^{2+} + 12PO_4^{3-} + 6H^+ \rightarrow 6[Mn(PO_4)_2]^{3-} + Cl^- + 3H_2O$$

Peroxodisulphates, nitrites, bromates, iodates, and also periodates react similarly. The first-named may be decomposed by evaporating the sulphuric acid solution with a little silver nitrate as catalyst.

Place a drop of the test solution in a micro crucible and add a drop of the reagent. Warm rapidly over a micro burner and allow to cool. A violet colouration appears. Very pale colourations may be intensified by adding a drop of 1% alcoholic diphenylcarbazide solution when a deep-violet colour, due to an oxidation product of the diphenylcarbazide, is obtained.

Sensitivity 0.05 μg ClO_3^-.

Concentration limit 1 in 1 000 000.

The reagent is prepared by mixing equal volumes of saturated manganese(II) sulphate solution and concentrated phosphoric acid.

Heating
↓
O_2 gas

10. *Action of heat* all chlorates are decomposed by heat into chlorides and oxygen. Some perchlorate is usually formed as an intermediate product. The chloride is identified in the residue by extracting with water and adding dilute nitric acid and silver nitrate solution. An insoluble chlorate should be mixed with sodium carbonate before ignition.

$$2KClO_3 \rightarrow 2KCl + 3O_2\uparrow$$

$$2KClO_3 \rightarrow KClO_4 + KCl + O_2\uparrow$$

4.20 Bromates, BrO_3^-

Solubility silver, barium, and lead bromates are slightly soluble in water, the solubilities being respectively 2.0 g l^{-1}, 7.0 g l^{-1} and 13.5 g l^{-1} at 20°C; mercury(I) bromate is also sparingly soluble. Most of the other metallic bromates are readily soluble in water.

BrO_3^- reactions

To study these reactions use a 0.1M solution of potassium bromate.

1. *Concentrated sulphuric acid* add 2 ml acid to 0.5 g solid bromate; bromine and oxygen are evolved in the cold.

$$4KBrO_3 + 2H_2SO_4 \rightarrow 2Br_2\uparrow + 5O_2\uparrow + 4K^+ + 2SO_4^{2-} + 2H_2O$$

AgNO₃
↓
white AgBrO₃

2. *Silver nitrate solution* a white crystalline precipitate of silver bromate, $AgBrO_3$, is produced with a concentrated solution of a bromate. The precipitate is soluble in hot water, readily soluble in dilute ammonia solution forming a complex salt, and sparingly soluble in dilute nitric acid.

$$BrO_3^- + Ag^+ \rightarrow AgBrO_3\downarrow$$

The solution obtained after dissolving in ammonia should be discarded quickly to avoid a serious **explosion!** (Cf. Section 3.6, reaction 1.)

Precipitates of the corresponding bromates are also produced by the addition of solutions of barium chloride, lead acetate or mercury(I) nitrate to a concentrated solution of a bromate.

$$2BrO_3^- + Ba^{2+} \rightarrow Ba(BrO_3)_2\downarrow$$

$$2BrO_3^- + Pb^{2+} \rightarrow Pb(BrO_3)_2\downarrow$$

$$2BrO_3^- + Hg_2^{2+} \rightarrow Hg_2(BrO_3)_2\downarrow$$

If the solution of silver bromate in dilute ammonia solution is treated dropwise with sulphurous acid solution, silver bromide separates: the latter dissolves in concentrated ammonia solution (difference from iodate).

SO₂ gas
↓
reduction to Br⁻

3. *Sulphur dioxide* if the gas is bubbled through a solution of a bromate, the latter is reduced to a bromide (see bromides, Section 4.15). A similar result is obtained with hydrogen sulphide and with sodium nitrite solution (see under chlorates, Section 4.19, reaction 3).

$$BrO_3^- + 3H_2SO_3 \rightarrow Br^- + 3SO_4^{2-} + 6H^+$$

$$BrO_3^- + 3H_2S \rightarrow Br^- + 3S\downarrow + 3H_2O$$

$$BrO_3^- + 3NO_2^- \rightarrow Br^- + 3NO_3^-$$

HBr
↓
brown colour, Br₂

4. *Hydrobromic acid* mix together solutions of potassium bromate and bromide, and acidify with dilute sulphuric acid; bromine is liberated as a result of interaction between the bromic and hydrobromic acids set free. The bromine may be extracted by adding a little chloroform.

$$BrO_3^- + 5Br^- + 6H^+ \rightarrow 3Br_2\uparrow + 3H_2O$$

Heating
↓
O₂

5. *Action of heat* potassium bromate on heating evolves oxygen and a bromide remains. Sodium and calcium bromates behave similarly, but cobalt, zinc, and other similar metallic bromates evolve oxygen and bromine, and leave an oxide.

$$2KBrO_3 \rightarrow 2KBr + 3O_2\uparrow$$

MnSO₄
↓
brown MnO₂

6. *Manganese(II) sulphate test* if a bromate solution is treated with a little of a $1+1$ mixture of saturated manganese(II) sulphate solution and M sulphuric acid, a transient red colouration (due to manganese(III) ions) is observed. Upon concentrating the solution rapidly, brown, hydrated manganese dioxide separates. The latter is insoluble in dilute sulphuric acid, but dissolves in a mixture of dilute sulphuric and oxalic acids (difference

from chlorates and iodates, which neither give the colouration nor yield the brown precipitate).

$$BrO_3^- + 6Mn^{2+} + 6H^+ \rightarrow 6Mn^{3+} + Br^- + 3H_2O$$

$$BrO_3^- + 6Mn^{3+} + 9H_2O \rightarrow 6MnO_2\downarrow + Br^- + 18H^+$$

$$MnO_2\downarrow + (COO)_2^{2-} + 4H^+ \rightarrow Mn^{2+} + 2CO_2\uparrow + 2H_2O$$

4.21 Iodates, IO_3^-

Solubility the iodates of the alkali metals are soluble in water; those of the other metals are sparingly soluble and, in general, less soluble than the corresponding chlorates and bromates. Some solubilities in $g\,l^{-1}$ at 20°C are: lead iodate 0.03 (25°C), silver iodate 0.06, barium iodate 0.22, calcium iodate 3.7, potassium iodate 81.3, and sodium iodate 90.0. Iodic acid is a crystalline solid, and has a solubility of $2330\,g\,l^{-1}$ at 20°C.

To study these reactions use a 0.1M solution of potassium iodate, KIO_3.

IO_3^- reactions

cc H_2SO_4 + $FeSO_4$
↓
brown ppt

1. *Concentrated sulphuric acid* no action in the absence of reducing agents; readily converted to iodide in the presence of iron(II) sulphate:

$$IO_3^- + 6Fe^{2+} + 6H^+ \rightarrow I^- + 6Fe^{3+} + 3H_2O$$

I_2
(with excess IO_3^-)

If the iodate is in excess, iodine is formed because of the interaction of iodate and iodide (cf. reaction 6):

$$IO_3^- + 5I^- + 6H^+ \rightarrow 3I_2\downarrow + 3H_2O$$

$AgNO_3$
↓
white $AgIO_3$

2. *Silver nitrate solution* white, curdy precipitate of silver iodate, readily soluble in dilute ammonia solution, but sparingly soluble in dilute nitric acid.

$$IO_3^- + Ag^+ \rightarrow AgIO_3\downarrow$$

$$AgIO_3\downarrow + 2NH_3 \rightarrow [Ag(NH_3)_2]^+ + IO_3^-$$

The solution obtained after dissolving in ammonia should be discarded quickly to avoid a serious **explosion!** (Cf. Section 3.6, reaction 1.)

If the ammoniacal solution of the precipitate is treated dropwise with sulphurous acid solution, silver iodide is precipitated; the latter is not dissolved by concentrated ammonia solution (difference from bromate).

$$[Ag(NH_3)_2]^+ + IO_3^- + 3H_2SO_3 \rightarrow AgI\downarrow + 3SO_4^{2-} + 2NH_4^+ + 4H^+$$

$BaCl_2$
↓
white $Ba(IO_3)_2$

3. *Barium chloride solution* white precipitate of barium iodate (difference from chlorate), sparingly soluble in hot water and in dilute nitric acid, but insoluble in alcohol (difference from iodide). If the precipitate of barium iodate is well washed, treated with a little sulphurous acid solution and 1–2 ml chloroform, the latter is coloured violet by the liberated iodine, and barium sulphate is precipitated:

$$2IO_3^- + Ba^{2+} \rightarrow Ba(IO_3)_2\downarrow$$

$$Ba(IO_3)_2\downarrow + 5H_2SO_3 \rightarrow I_2\downarrow + BaSO_4\downarrow + 4SO_4^{2-} + 8H^+ + H_2O$$

$Hg(NO_3)_2$
↓
white $Hg(IO_3)_2$

4. *Mercury(II) nitrate solution* (**POISON**)   white precipitate of mercury(II) iodate (difference from chlorate and bromate). Lead acetate solution similarly gives a precipitate of lead iodate. Mercury(II) chloride solution, which is practically un-ionized (as mercuric chloride is covalent) gives no precipitate.

$$2IO_3^- + Hg^{2+} \rightarrow Hg(IO_3)_2\downarrow$$

$$2IO_3^- + Pb^{2+} \rightarrow Pb(IO_3)_2\downarrow$$

SO_2 or H_2S
↓
first brown ppt, I_2

In excess
↓
colourless
(white ppt
with H_2S)

5. *Sulphur dioxide or hydrogen sulphide*   passage of sulphur dioxide or of hydrogen sulphide into a solution of an iodate, acidified with dilute hydrochloric acid, liberates iodine as a brown precipitate, which may be recognized by the addition of starch solution or chloroform. With an excess of either reagent, the iodine is further reduced to iodide.

$$2IO_3^- + 5H_2SO_3 \rightarrow I_2\downarrow + 5SO_4^{2-} + 8H^+ + H_2O$$

$$I_2\downarrow + H_2SO_3 + H_2O \rightarrow 2I^- + SO_4^{2-} + 4H^+$$

$$2IO_3^- + 5H_2S + 2H^+ \rightarrow I_2\downarrow + 5S\downarrow + 6H_2O$$

$$I_2\downarrow + H_2S \rightarrow 2I^- + 2H^+ + S\downarrow$$

KI
↓
brown colour, I_3^-

6. *Potassium iodide solution*   mix together solutions of potassium iodide and potassium iodate, and acidify with hydrochloric acid, acetic acid or with tartaric acid solution; iodine is immediately liberated (use the chloroform test).

$$IO_3^- + 8I^- + 6H^+ \rightarrow 3I_3^- + 3H_2O$$

Heating
↓
decomposition

7. *Action of heat*   the alkali iodates decompose into oxygen and an iodide. Most iodates of the bivalent metals yield iodine and oxygen and leave a residue of oxide; barium iodate, exceptionally, gives the periodate (more precisely, hexoxoperiodate):

$$2KIO_3 \rightarrow 2KI + 3O_2\uparrow$$

$$2Pb(IO_3)_2 \rightarrow 2I_2\uparrow + 5O_2\uparrow + 2PbO$$

$$5Ba(IO_3)_2 \rightarrow Ba_5(IO_6)_2 + 4I_2\uparrow + 9O_2\uparrow$$

4.22   Perchlorates, ClO_4^-

Solubility   the perchlorates are generally soluble in water. Potassium perchlorate is one of the least soluble ($7.5\,g\,l^{-1}$ and $218\,g\,l^{-1}$ at 0°C and 100°C respectively), and sodium perchlorate is one of the most soluble ($2.096\,g\,l^{-1}$ at 25°C). Note that operations with perchlorates can be **DANGEROUS**.

ClO_4^- reactions

To study these reactions use a 2M solution of perchloric acid or solid sodium perchlorate, $NaClO_4$.

cc H$_2$SO$_4$
↓
white fumes, HClO$_4$

1. *Concentrated sulphuric acid*   no visible action with the solid salt in cold, but on strong heating, white fumes of perchloric acid monohydrate are produced:

$$NaClO_4 + H_2SO_4 + H_2O \rightarrow HClO_4 \cdot H_2O\uparrow + HSO_4^- + Na^+$$

Note that hot, concentrated perchloric acid **EXPLODES** if particles of dust, coal or filter paper, etc. fall into it.

KCl
↓
white KClO$_4$

2. *Potassium chloride solution*   white precipitate of KClO$_4$, insoluble in alcohol (see Section 3.36, reaction 3). Ammonium chloride solution gives a similar reaction:

$$ClO_4^- + K^+ \rightarrow KClO_4\downarrow$$

$$ClO_4^- + NH_4^+ \rightarrow NH_4ClO_4\downarrow$$

3. *Barium chloride solution*   no precipitate. A similar result is obtained with silver nitrate solution.

4. *Indigo test*   no decolourization even in the presence of acid (difference from hypochlorite and chlorate).

5. *Sulphur dioxide, hydrogen sulphide, or iron(II) salts*   no reduction (difference from chlorate).

Ti$_2$(SO$_4$)$_3$
↓
reduction to Cl$^-$

6. *Titanium(III) sulphate solution*   in the presence of sulphuric acid, perchlorates are reduced to chlorides:

$$ClO_4^- + 8Ti^{3+} + 8H^+ \rightarrow Cl^- + 8Ti^{4+} + 4H_2O$$

Chloride ions can be identified in the solution with the usual tests (see Section 4.14).

CdSO$_4$ + NH$_3$
↓
white ppt

7. *Tetramminecadmium perchlorate test*   when a neutral solution of perchlorate is treated with a saturated solution of cadmium sulphate in concentrated ammonia solution, a white crystalline precipitate of tetramminecadmium perchlorate is obtained:

$$2ClO_4^- + [Cd(NH_3)_4]^{2+} \rightarrow [Cd(NH_3)_4](ClO_4)_2\downarrow$$

Sulphides interfere by precipitating cadmium as cadmium sulphide, and should therefore be absent.

Heating
↓
O$_2$ gas

8. *Action of heat*   oxygen is evolved from a solid perchlorate, and in the residue chloride ions can be tested (see Section 4.14):

$$NaClO_4 \rightarrow NaCl + 2O_2\uparrow$$

4.23   Borates, BO$_3^{3-}$, B$_4$O$_7^{2-}$, BO$_2^-$

The borates are derived from the three boric acids: orthoboric acid, H$_3$BO$_3$, pyroboric acid, H$_2$B$_4$O$_7$, and metaboric acid, HBO$_2$. Orthoboric acid is a white, crystalline solid, sparingly soluble in cold but more soluble in hot water; very few salts of this acid are definitely known. On heating orthoboric acid at 100°C, it is converted into metaboric acid; at 140°C pyroboric acid is produced. Most of the salts are derived from the meta- and pyro-acids. Owing to the weakness of boric acid, the soluble salts are hydrolysed in solution and therefore give an alkaline reaction.

$$BO_3^{3-} + 3H_2O \rightleftharpoons H_3BO_3 + 3OH^-$$

$$B_4O_7^{2-} + 7H_2O \rightleftharpoons 4H_3BO_3 + 2OH^-$$

$$BO_2^- + 2H_2O \rightleftharpoons H_3BO_3 + OH^-$$

Solubility the borates of the alkali metals are readily soluble in water. The borates of the other metals are, in general, sparingly soluble in water, but fairly soluble in acids and in ammonium chloride solution.

Borate reactions

To study these reactions use a 0.1M solution of sodium tetraborate (sodium pyroborate, borax) $Na_2B_4O_7 \cdot 10H_2O$.

cc H_2SO_4
↓
white fumes, H_3BO_3

1. *Concentrated sulphuric acid* no visible action in the cold, although ortho-boric acid, H_3BO_3, is set free. On heating, however, white fumes of boric acid are evolved. If concentrated hydrochloric acid is added to a concentrated solution of borax, boric acid is precipitated:

$$Na_2B_4O_7 + H_2SO_4 + 5H_2O \rightarrow 4H_3BO_3\uparrow + 2Na^+ + SO_4^{2-}$$

$$Na_2B_4O_7 + 2HCl + 5H_2O \rightarrow 4H_3BO_3\downarrow + 2Na^+ + 2Cl^-$$

Methyl borate
flame test
↓
green flames

2. *Concentrated sulphuric acid and alcohol (flame test)* if a little borax is mixed with 1 ml concentrated sulphuric acid and 5 ml methanol or ethanol (the former is to be preferred owing to its greater volatility) in a small porcelain basin, and the alcohol ignited, the latter will burn with a green-edged flame, due to the formation of methyl borate $B(OCH_3)_3$ or of ethyl borate $B(OC_2H_5)_3$. Both these esters are **POISONOUS**. Copper and barium salts may give a similar green flame. The following modification of the test, which depends upon the greater volatility of boron trifluoride, BF_3, can be used in the presence of copper and barium compounds; these do not form volatile compounds under the experimental conditions given below. Thoroughly mix the borate with powdered calcium fluoride and a little concentrated sulphuric acid, and bring a little of the paste thus formed on the loop of a platinum wire, or upon the end of a glass rod, very close to the edge of the base of a Bunsen flame without actually touching it; volatile boron trifluoride is formed and colours the flame green.

$$H_3BO_3 + 3CH_3OH \rightarrow B(OCH_3)_3\uparrow + 3H_2O$$

$$Na_2B_4O_7 + 6CaF_2 + 7H_2SO_4 \rightarrow$$

$$\rightarrow 4BF_3\uparrow + 6CaSO_4 + 2Na^+ + SO_4^{2-} + 7H_2O$$

Turmeric paper test
↓
red or black colour

3. *Turmeric (curcumin) paper test* dip a piece of turmeric paper into a solution of borate, acidified with 2M hydrochloric acid. Dry the paper at 100°C in a drying oven; it turns reddish-brown. Moisten the paper with a few drops of 2M sodium hydroxide; it now turns bluish-black or

greenish-black. This is a selective and sensitive test for borate. Oxidizing agents (like chromates, chlorates, nitrites) as well as iodides bleach the reagent paper and therefore should be absent.

Turmeric paper can be prepared by immersing dry strips of filter paper in a saturated alcoholic solution of curcumin, followed by drying in air.

AgNO₃
↓
white AgBO₂

4. *Silver nitrate solution* white precipitate of silver metaborate, $AgBO_2$, from fairly concentrated borax solution, soluble in both dilute ammonia solution and in acetic acid. On boiling the precipitate with water, it is completely hydrolysed and a brown precipitate of silver oxide is obtained. A brown precipitate of silver oxide is produced directly in very dilute solutions.

$$B_4O_7^{2-} + 4Ag^+ + H_2O \rightarrow 4AgBO_2\downarrow + 2H^+$$

$$2AgBO_2\downarrow + 3H_2O \rightarrow Ag_2O\downarrow + 2H_3BO_3$$

The solution obtained after dissolving in ammonia should be discarded quickly to avoid a serious **explosion!** (Cf. Section 3.6, reaction 1.)

The boric acid, formed in this reaction, is practically undissociated.

BaCl₂
↓
white Ba(BO₂)₂

5. *Barium chloride solution* white precipitate of barium metaborate, $Ba(BO_2)_2$, from fairly concentrated solutions; the precipitate is soluble in excess reagent, in dilute acids, and in solutions of ammonium salts. Solutions of calcium and strontium chloride behave similarly.

$$B_4O_7^{2-} + 2Ba^{2+} + H_2O \rightarrow 2Ba(BO_2)_2\downarrow + 2H^+$$

Heating
↓
swelling

6. *Action of heat* powdered borax when heated in an ignition tube, or upon a platinum wire, swells up considerably, and then subsides, leaving a colourless glass of the anhydrous salt. The glass possesses the property of dissolving many oxides on heating, forming metaborates, which often have characteristic colours. This is the basis of the borax bead test for various metals (see Section 2.2, 5).

Chromotrope 2B
↓
colour changes to
greenish blue

7. *4-Nitrobenzene-azo-chromotropic acid reagent (Chromotrope 2B)*

Borates cause the blue-violet reagent to assume a greenish-blue colour.

Evaporate a drop of the slightly alkaline solution to dryness in a semi-micro crucible. Stir the warm residue with 2–3 drops of the reagent. A greenish-blue colouration is obtained on cooling. A blank test should be performed simultaneously.

Sensitivity 0.1 µg B.

Concentration limit 1 in 500 000.

Oxidizing agents and fluorides interfere, the latter because of the formation of borofluorides. Oxidizing agents, including nitrates and chlorates, are rendered innocuous by evaporating with solid hydrazine sulphate, whilst fluorides may be removed as silicon tetrafluoride by evaporation with silicic acid and sulphuric acid.

The experimental details are as follows. Treat 2 drops of the test solution in a small porcelain crucible either with a little solid hydrazine sulphate or with a few specks of precipitated silica and 1–2 drops concentrated sulphuric acid, and heat cautiously until fumes of sulphuric acid appear. Add 3–4 drops of the reagent whilst the residue is still warm and observe the colour on cooling.

Sensitivity 0.25 µg **B** in the presence of 12 000 times that amount of $KClO_3$ or KNO_3; 0.5 µg **B** in the presence of 2500 times that amount of NaF.

Mannitol
↓
increased
acid strength

8. *Mannitol–bromothymol blue test* boric acid acts as a very weak monobasic acid ($K_a = 5.8 \times 10^{-10}$), but on the addition of certain organic polyhydroxy compounds, such as mannitol (mannite), glycerol, dextrose or invert sugar, it is transformed into a relatively strong acid, probably of the type:

$$-\underset{|}{\overset{|}{C}}-OH \quad HO \diagdown \qquad -\underset{|}{\overset{|}{C}}-O \diagdown$$
$$\qquad \qquad + \qquad B-OH \; = \qquad \qquad B-O^- + H^+ + 2H_2O$$
$$-\underset{|}{\overset{|}{C}}-OH \quad HO \diagup \qquad -\underset{|}{\overset{|}{C}}-O \diagup$$

The *p*H of the solution therefore decreases. Hence if the solution is initially almost neutral to bromothymol blue (green colour), then upon the addition of mannitol the colour becomes yellow. (It is advisable when testing for minute quantities of borates to recrystallize the mannitol from a solution neutralized to bromothymol blue, wash with pure acetone and dry at 100°C.) Only periodate interferes with the test; it can be decomposed by heating on charcoal.

Render the test solution almost neutral to bromothymol blue by treating it with dilute acid or alkali (as necessary) until the indicator turns green. Place a few drops of the test solution in a micro test-tube, and add a few drops of the reagent solution. A yellow colouration is obtained in the presence of a borate. It is advisable to carry out a blank test with distilled water simultaneously.

Sensitivity 0.001 µg **B**.
Concentration limit 1 in 30 000 000.

4.24 Sulphates, SO_4^{2-}

Solubility the sulphates of barium, strontium, and lead are practically insoluble in water,[10] those of calcium and mercury(II) are slightly soluble, and most of the remaining metallic sulphates are soluble. Some basic sulphates such as those of mercury, bismuth, and chromium are also insoluble in water, but these dissolve in dilute hydrochloric or nitric acid.

---

[10] Of these three sulphates, that of strontium is the most soluble.

Sulphuric acid is a colourless, oily, and hygroscopic liquid, of specific gravity 1.838. The pure, commercial, concentrated acid is a constant boiling point mixture, boiling point 338°C and containing c. 98% acid. It is miscible with water in all proportions with the evolution of considerable heat; on mixing the two, the acid should always be poured in a thin stream into the water (if the water is poured into the heavier acid, steam may be suddenly generated which will carry with it some of the acid and may therefore cause considerable damage).

SO_4^{2-} reactions

To study these reactions use a 0.1M solution of sodium sulphate, $Na_2SO_4 \cdot 10H_2O$.

$BaCl_2$
↓
white $BaSO_4$

1. *Barium chloride solution* white precipitate of barium sulphate, $BaSO_4$ (see under barium, Section 3.31), insoluble in warm dilute hydrochloric acid and in dilute nitric acid, but moderately soluble in boiling, concentrated hydrochloric acid:

$$SO_4^{2-} + Ba^{2+} \rightarrow BaSO_4\downarrow$$

The test is usually carried out by adding the reagent to the solution acidified with dilute hydrochloric acid; carbonates, sulphites, and phosphates are not precipitated under these conditions. Concentrated hydrochloric acid or concentrated nitric acid should not be used, as a precipitate of barium chloride or of barium nitrate may form; these dissolve, however, upon dilution with water. The barium sulphate precipitate may be filtered from the hot solution and fused on charcoal with sodium carbonate, when sodium sulphide will be formed. The latter may be extracted with water, and the extract filtered into a freshly prepared solution of sodium nitroprusside, when a transient, purple colouration is obtained (see under sulphides, Section 4.6, reaction 6). An alternative method is to add a few drops of very dilute hydrochloric acid to the fused mass, and to cover the latter with lead acetate paper; a black stain of lead sulphide is produced on the paper. The so-called Hepar reaction, which is less sensitive than the above two tests, consists of placing the fusion product on silver foil and moistening with a little water; a brownish-black stain of silver sulphide results.

$$BaSO_4 + 4C + Na_2CO_3 \rightarrow Na_2S + BaCO_3 + 4CO\uparrow$$

$$Na_2S \rightarrow 2Na^+ + S^{2-}$$

$$2S^{2-} + 4Ag + O_2 + 2H_2O \rightarrow 2Ag_2S\downarrow + 4OH^-$$

A more efficient method for decomposing most sulphur compounds is to heat them with sodium or potassium, and then test the solution of the product for sulphide. The test is rendered sensitive by heating the substance with potassium in an ignition tube, dissolving the melt in water, and testing for sulphide by the nitroprusside or methylene blue reactions (see under sulphides, Section 4.6, reactions 6 and 7). Heating substances with potassium metal is a **DANGEROUS** operation and must be carried out with appropriate equipment under supervision.

The above tests (depending upon the formation of a sulphide) are not exclusive to sulphates but are given by most sulphur compounds. If, however, the barium sulphate precipitated in the presence of hydrochloric acid is employed, then the reaction may be used as a confirmatory test for sulphates.

Pb(CH₃COO)₂
↓
white PbSO₄

2. *Lead acetate solution* white precipitate of lead sulphate, PbSO₄, soluble in hot concentrated sulphuric acid, in solutions of ammonium acetate and of ammonium tartrate (see under lead, Section 3.4, reaction 5), and in sodium hydroxide solution. In the last case sodium tetrahydroxoplumbate(II) is formed, and on acidification with hydrochloric acid, the lead crystallizes out as the chloride. If any of the aqueous solutions of the precipitate are acid-ified with acetic acid and potassium chromate solution added, yellow lead chromate is precipitated (see under lead, Section 3.4, reaction 6).

$$SO_4^{2-} + Pb^{2+} \rightarrow PbSO_4\downarrow$$

AgNO₃
↓
white Ag₂SO₄

3. *Silver nitrate solution* white, crystalline precipitate of silver sulphate, Ag₂SO₄ (solubility 5.8 g l⁻¹ at 18°C), from concentrated solutions.

$$SO_4^{2-} + 2Ag^{+} \rightarrow Ag_2SO_4\downarrow$$

Sodium
rhodizonate test
↓
removal of colour

4. *Sodium rhodizonate test* barium salts yield a reddish-brown precipitate with sodium rhodizonate (see under barium, Section 3.31, reaction 7). Sulphates and sulphuric acid cause immediate decolourization because of the formation of insoluble barium sulphate. This test is specific for sulphates.

Place a drop of barium chloride solution upon filter or drop-reaction paper, followed by a drop of a freshly prepared 0.5% aqueous solution of sodium rhodizonate. Treat the reddish-brown spot with a drop of the acid or alkaline test solution. The coloured spot disappears.
Sensitivity 4 µg SO_4^{2-}.
Concentration limit 1 in 10 000.

KMnO₄ on BaSO₄
↓
cannot be reduced

5. *Potassium permanganate–barium sulphate test* if barium sulphate is precipitated in a solution containing potassium permanganate, it is coloured pink (violet) by adsorption of some of the permanganate. The permanganate which has been adsorbed on the precipitate cannot be reduced by the common reducing agents (including hydrogen peroxide); the excess of potassium permanganate in the mother liquor reacts readily with reducing agents, thus rendering the pink barium sulphate clearly visible in the colour-less solution.

Place 3 drops test solution in a semimicro centrifuge tube, add 2 drops 0.02M potassium permanganate solution and 1 drop barium chloride solution. A pink precipitate is obtained. Add a few drops 3% hydrogen peroxide solution or 0.5M oxalic acid solution (in the latter case it will be necessary to warm on a water bath until decolourization is complete). Centrifuge. The coloured precipitate is clearly visible.
Sensitivity 2.5 µg SO_4^{2-}.
Concentration limit 1 in 20 000.

Hg(NO₃)₂
↓
yellow ppt

6. *Mercury(II) nitrate solution* (**POISON**) yellow precipitate of basic mercury(II) sulphate:

$$SO_4^{2-} + 3Hg^{2+} + 2H_2O \rightarrow HgSO_4 \cdot 2HgO\downarrow + 4H^{+}$$

This is a sensitive test, given even by suspensions of barium or lead sulphates.

4.25 Peroxodisulphates, $S_2O_8^{2-}$

Solubility   the best known peroxodisulphates, those of sodium, potassium, ammonium, and barium are soluble in water, the potassium salt being the least soluble ($17.7 \, \text{g} \, \text{l}^{-1}$ at $0°C$).

$S_2O_8^{2-}$ reactions

To study these reactions use a freshly prepared 0.1M solution of ammonium peroxodisulphate $(NH_4)_2S_2O_8$.

Boiling
↓
O_2 + some O_3 gas

1. *Water*   all peroxodisulphates are decomposed on boiling with water into the sulphate, free sulphuric acid, and oxygen. The oxygen contains appreciable quantities of ozone, which may be detected by its odour or by its property of turning starch–iodide paper blue. A similar result is obtained with dilute sulphuric or nitric acid. With dilute hydrochloric acid, chlorine is evolved (cf. reaction 4 below). By dissolving the solid peroxodisulphate in concentrated sulphuric acid at $0°C$, peroxomonosulphuric acid (Caro's acid), H_2SO_5, is formed in solution; this possesses strong oxidizing properties.

$$2S_2O_8^{2-} + 2H_2O \rightarrow 4SO_4^{2-} + 4H^+ + O_2\uparrow$$

$$3S_2O_8^{2-} + 3H_2O \rightarrow 6SO_4^{2-} + 6H^+ + O_3\uparrow$$

$$O_3\uparrow + 3I^- + 2H^+ \rightarrow I_3^- + O_2\uparrow + H_2O$$

AgNO₃
↓
black Ag₂O₂

2. *Silver nitrate solution*   black precipitate of silver peroxide, Ag_2O_2, from concentrated solutions. If only a little silver nitrate solution is added followed by dilute ammonia solution, the silver peroxide, or the silver ion, acts catalytically leading to the evolution of nitrogen and the liberation of considerable heat.

$$S_2O_8^{2-} + 2Ag^+ + 2H_2O \longrightarrow Ag_2O_2\downarrow + 2SO_4^{2-} + 4H^+$$

$$3S_2O_8^{2-} + 8NH_3 \xrightarrow{(Ag^+)} N_2\uparrow + 6SO_4^{2-} + 6NH_4^+$$

3. *Barium chloride solution*   no immediate precipitate in the cold with a solution of a pure peroxodisulphate; on standing for some time or on boiling, a precipitate of barium sulphate is obtained, due to the decomposition of the peroxodisulphate.

KI⁻
↓
slowly brown colour,
I₃⁻

4. *Potassium iodide solution*   iodine is slowly liberated in the cold and rapidly on warming (test with starch solution) (distinction from perborate and percarbonate, which liberate iodine immediately). Traces of copper catalyse the reaction.

$$S_2O_8^{2-} + 3I^- \rightarrow I_3^- + 2SO_4^{2-}$$

MnSO₄ .
↓
brown MnO(OH)₂

5. *Manganese(II) sulphate solution*   brown precipitate in neutral or preferably alkaline solution. The precipitate is manganese(IV) dioxide hydrate with a composition approximating $MnO(OH)_2$.

$$S_2O_8^{2-} + Mn^{2+} + 4OH^- \rightarrow MnO(OH)_2\downarrow + 2SO_4^{2-} + H_2O$$

In the presence of nitric or sulphuric acid and small amounts of silver nitrate, permanganate ions are formed on warming, and the solution turns to violet:

$$5S_2O_8^{2-} + 2Mn^{2+} + 8H_2O \rightarrow 2MnO_4^- + 10SO_4^{2-} + 16H^+$$

Silver ions act as catalysts (cf. Section 3.28, reaction 6).

6. *Potassium permanganate solution* unaffected (distinction from hydrogen peroxide). Peroxodisulphates are also unaffected by a solution of titanium(IV) sulphate.

4.26 Silicates, SiO_3^{2-}

The silicic acids may be represented by the general formula $xSiO_2 \cdot yH_2O$. Salts corresponding to orthosilicic acid H_4SiO_4 ($SiO_2 \cdot 2H_2O$), metasilicic acid H_2SiO_3 ($SiO_2 \cdot H_2O$), and disilicic acid $H_2Si_2O_5$ ($2SiO_2 \cdot H_2O$) are definitely known. The metasilicates are sometimes designated simply as silicates.

Solubility only the silicates of the alkali metals are soluble in water; they are hydrolysed in aqueous solution and therefore give an alkaline reaction.

$$SiO_3^{2-} + 2H_2O \rightleftharpoons H_2SiO_3 + 2OH^-$$

SiO_3^{2-} reactions

To study these reactions use a M solution of sodium silicate, Na_2SiO_3. Commercial (30%) water-glass solutions should be diluted with a fivefold amount of water.

HCl
↓
white H_2SiO_3

1. *Dilute hydrochloric acid* add dilute hydrochloric acid to the solution of the silicate; a gelatinous precipitate of metasilicic acid is obtained, particularly on boiling. The precipitate is insoluble in concentrated acids. The freshly precipitated substance is appreciably soluble in water and in dilute acids. It is converted by repeated evaporation with concentrated hydrochloric acid on the water bath into a white insoluble powder (silica, SiO_2).

If a dilute solution (say, 1–10%) of water glass is quickly added to moderately concentrated hydrochloric acid, no precipitation of silicic acid takes place; it remains in colloidal solution (sol).

$$SiO_3^{2-} + 2H^+ \rightarrow H_2SiO_3\downarrow$$

NH_4Cl
↓
white H_2SiO_3

2. *Ammonium chloride or ammonium carbonate solution* gelatinous precipitate of silicic acid. This reaction is important in routine qualitative analysis since silicates, unless previously removed, will be precipitated by ammonium chloride solution in Group IIIA.

$$SiO_3^{2-} + 2NH_4^+ \rightarrow H_2SiO_3\downarrow + 2NH_3$$

$AgNO_3$
↓
yellow Ag_2SiO_3

3. *Silver nitrate solution* yellow precipitate of silver silicate, soluble in dilute acids and in ammonia solution:

$$SiO_3^{2-} + 2Ag^+ \rightarrow Ag_2SiO_3\downarrow$$

The solution obtained after dissolving in ammonia should be discarded quickly to avoid a serious **explosion!** (Cf. Section 3.6, reaction 1.)

BaCl₂
↓
white BaSiO₃

4. *Barium chloride solution*   white precipitate of barium silicate, soluble in dilute nitric acid.

$$SiO_3^{2-} + Ba^{2+} \rightarrow BaSiO_3\downarrow$$

Calcium chloride solution gives a similar precipitate of calcium silicate.

Microcosmic salt
↓
no fusion

5. *Microcosmic salt bead test*   most silicates, and also silica, when fused in a bead of microcosmic salt, $Na(NH_4)HPO_4 \cdot 4H_2O$, in a loop of platinum wire give this test. The microcosmic salt first fuses to a transparent bead consisting largely of sodium metaphosphate (see Section 2.2, 6); when a minute quantity of the solid silicate or even of the solution is introduced into the bead (best achieved by dipping the hot bead into the substance) and the whole again heated, the silica produced will not dissolve in the bead, but will move in the fused mass, and is visible as white opaque masses or 'skeletons' in both the fused and the cold bead.

$$CaSiO_3 + NaPO_3 \rightarrow CaNaPO_4 + SiO_2$$

Fusion with
Na₂CO₃
↓
soluble Na₂SiO₃

Insoluble silicates are best brought into solution by fusing the powdered solid, mixed with six times its weight of anhydrous sodium carbonate, in a platinum crucible or upon platinum foil; the alkali carbonate reacts with the silicate yielding sodium silicate. The cold mass is then evaporated to dryness on the water bath with excess 6M hydrochloric acid; the alkali silicate is thereby decomposed yielding first gelatinous silicic acid and, ultimately, white, amorphous silica, whilst the metallic oxides derived from the insoluble silicate are converted into chlorides. The residue is extracted with boiling 6M hydrochloric acid; this removes the metals as chlorides and insoluble silica remains behind. A simpler, but not quantitative, method is to extract the melt with boiling water: sufficient sodium silicate passes into solution to give any of the reactions referred to above.

$$SiO_2 + Na_2CO_3 \rightarrow CO_2\uparrow + Na_2SiO_3$$

$$SiO_3^{2-} + 2H^+ \rightarrow SiO_2\downarrow + H_2O$$

SiF₄ test
↓
specific for SiO₃²⁻

6. *Silicon tetrafluoride test*   this test depends upon the fact that when silica (isolated from a silicate by treatment with ammonium chloride solution or with hydrochloric acid, etc.) is heated with less than an equivalent amount of calcium fluoride and some concentrated sulphuric acid, silicon tetrafluoride is evolved. The latter is identified by its action upon a drop of water held in a loop of platinum wire, when a turbidity (due to silica) is produced.

$$SiO_2 + 2CaF_2 + 2H_2SO_4 \rightarrow 2CaSO_4 + SiF_4\uparrow + 2H_2O;$$

$$3SiF_4\uparrow + 2H_2O \rightarrow SiO_2\downarrow + 2[SiF_6]^{2-} + 4H^+$$

An excess of calcium fluoride should be avoided since a mixture of H_2F_2 and SiF_4 will be formed and interfere with the test.

Mix the solid substance (or, preferably, silicic acid isolated by treatment of the silicate with ammonium chloride solution) with one-third of its weight of calcium fluoride in a small lead (or platinum) capsule, and add sufficient concentrated sulphuric acid to form a thin paste: mix the contents of the capsule with a stout platinum wire. Warm gently (FUME CUPBOARD) and hold close above the mixture a loop of platinum wire supporting a drop of water. The drop of water will become turbid, due to the hydrolysis of the silicon tetrafluoride absorbed.

$(NH_4)_2MoO_4 + SnCl_2$
$\downarrow$
blue colour
(molybdenum blue)

7. *Ammonium molybdate–tin(II) chloride test*   the silicate is separated by volatilization as silicon tetrafluoride and the latter collected in a little sodium hydroxide solution. The resulting silicate is treated with ammonium molybdate solution and the ammonium salt of silicomolybdic acid, $H_4[SiMo_{12}O_{40}]$, is reduced by tin(II) chloride solution to 'molybdenum blue'. Tin(II) chloride does not reduce ammonium molybdate solution.

Phosphates and arsenates give the same reaction, but do not interfere under the conditions of the test: large amounts of borates should be absent, but may be removed by warming with methanol and sulphuric acid.

Place a little of the solid silicate in a small lead or platinum crucible, add a little sodium fluoride and a few drops of concentrated sulphuric acid. Cover the crucible with a small sheet of cellophane from which is suspended a drop of 2M sodium hydroxide solution (freshly prepared from the analytical grade solid). Warm gently for 3–5 minutes over a micro burner with the crucible about 8 cm from the flame. Transfer the drop of sodium hydroxide solution to a porcelain micro crucible, add 2 drops ammonium molybdate reagent and then 2M acetic acid until slightly acidic. Then add a few drops of 0.25M tin(II) chloride followed by sufficient sodium hydroxide solution to dissolve the tin(II) hydroxide. A blue colouration is obtained.

Concentration limit 1 in 10 000.

The reaction may be applied to aqueous solutions of silicates, but phosphates and arsenates must be absent.

4.27   Hexafluorosilicates (silicofluorides), $[SiF_6]^{2-}$

Solubility   most metallic hexafluorosilicates (with the exception of the barium and potassium salts which are sparingly soluble) are soluble in water. A solution of the acid (hydrofluorosilicic acid $H_2[SiF_6]$) is one of the products of the action of water upon silicon tetrafluoride, and is also formed by dissolving silica in hydrofluoric acid.

$$SiO_2 + 3H_2F_2 \rightarrow H_2[SiF_6] + 2H_2O$$

$[SiF_6]^{2-}$ reactions

To study these reactions use a freshly prepared 0.1M solution of sodium hexafluorosilicate.

cc H_2SO_4
$\downarrow$
$SiF_4 + H_2F_2$ gases

1. *Concentrated sulphuric acid*   silicon tetrafluoride and hydrogen fluoride are evolved on warming the reagent with the solid salt. If the reaction is carried out in a platinum or lead capsule or crucible, the escaping gas will

etch glass and will cause a drop of water to become turbid (see under silicates, Section 4.26, reaction 6).

$$Na_2[SiF_6] + H_2SO_4 \rightarrow SiF_4\uparrow + H_2F_2\uparrow + 2Na^+ + SO_4^{2-}$$

BaCl₂
↓
white Ba[SiF₆]

2. *Barium chloride solution* white, crystalline precipitate of barium hexa-fluorosilicate, $Ba[SiF_6]$, sparingly soluble in water ($0.25\,g\,l^{-1}$ at 25°C) and insoluble in dilute hydrochloric acid. The precipitate is distinguished from barium sulphate by the evolution of hydrogen fluoride and silicon fluoride, which etch glass, on heating with concentrated sulphuric acid in a lead crucible.

$$[SiF_6]^{2-} + Ba^{2+} \rightarrow Ba[SiF_6]\downarrow$$

KCl
↓
white K₂[SiF₆]

3. *Potassium chloride solution* white, gelatinous precipitate of potassium hexafluorosilicate, $K_2[SiF_6]$, from concentrated solutions. The precipitate is slightly soluble in water ($1.77\,g\,l^{-1}$ at 25°C), less soluble in excess of the reagent and in 50% alcohol.

$$[SiF_6]^{2-} + 2K^+ \rightarrow K_2[SiF_6]\downarrow$$

NH₃
↓
white H₂SiO₃

4. *Ammonia solution* decomposition occurs with the separation of gelatinous silicic acid.

$$[SiF_6]^{2-} + 4NH_3 + 3H_2O \rightarrow H_2SiO_3\downarrow + 6F^- + 4NH_4^+$$

Heating
↓
decomposition

5. *Action of heat* decomposition occurs into silicon tetrafluoride, which renders a drop of water turbid, and the metallic fluoride, which can be tested for in the usual manner (see under fluorides, Section 4.17).

$$Na_2[SiF_6] \rightarrow SiF_4\uparrow + 2Na^+ + 2F^-$$

4.28 Orthophosphates, PO_4^{3-}

Three phosphoric acids are known: orthophosphoric, H_3PO_4; pyro-phosphoric, $H_4P_2O_7$; and metaphosphoric, HPO_3. Salts of the three acids exist: the orthophosphates are the most stable and the most important;[11] solutions of pyro- and metaphosphates convert into orthophosphates slowly at room temperature, and more rapidly on boiling. Meta-phosphates, unless prepared by special methods, are usually polymeric, i.e. are derived from $(HPO_3)_n$.

Orthophosphoric acid is a tribasic acid giving rise to three series of salts: primary orthophosphates, e.g. NaH_2PO_4; secondary orthophosphates, e.g. Na_2HPO_4; and tertiary orthophosphates, e.g. Na_3PO_4. If a solution of orthophosphoric acid is neutralized with sodium hydroxide solution using methyl orange as indicator, the neutral point is reached when the acid is converted into the primary phosphate; with phenolphthalein as indicator, the solution will react neutral when the secondary phosphate is formed;

---

[11] They are often referred to simply as phosphates.

with 3 moles of alkali, the tertiary or normal phosphate is formed. NaH_2PO_4 is neutral to methyl orange and acid to phenolphthalein, Na_2HPO_4 is neutral to phenolphthalein and alkaline to methyl orange, Na_3PO_4 is alkaline to most indicators because of its extended hydrolysis. Ordinary 'sodium phosphate' is disodium hydrogen phosphate, $Na_2HPO_4 \cdot 12H_2O$.

Solubility the phosphates of the alkali metals, with the exception of lithium and of ammonium, are soluble in water; the primary phosphates of the alkaline earth metals are also soluble. All the phosphates of the other metals, and also the secondary and tertiary phosphates of the alkaline earth metals, are sparingly soluble or insoluble in water.

PO_4^{3-} reactions

To study these reactions use a 0.033M solution of disodium hydrogen phosphate $Na_2HPO_4 \cdot 12H_2O$.

$AgNO_3$
↓
yellow Ag_3PO_4

1. *Silver nitrate solution* yellow precipitate of normal silver orthophosphate, Ag_3PO_4 (distinction from meta- and pyrophosphate), soluble in dilute ammonia solution and in dilute nitric acid.

$$HPO_4^{2-} + 3Ag^+ \rightarrow Ag_3PO_4\downarrow + H^+$$

$$Ag_3PO_4\downarrow + 2H^+ \rightarrow H_2PO_4^- + 3Ag^+$$

$$Ag_3PO_4\downarrow + 6NH_3 \rightarrow 3[Ag(NH_3)_2]^+ + PO_4^{3-}$$

The solution obtained after dissolving in ammonia should be discarded quickly to avoid a serious **explosion!** (Cf. Section 3.6, reaction 1.)

$BaCl_2$
↓
white $BaHPO_4$

2. *Barium chloride solution* white, amorphous precipitate of secondary barium phosphate, $BaHPO_4$, from neutral solutions, soluble in dilute mineral acids and in acetic acid. In the presence of dilute ammonia solution, the less soluble tertiary phosphate, $Ba_3(PO_4)_2$, is precipitated.

$$HPO_4^{2-} + Ba^{2+} \rightarrow BaHPO_4\downarrow$$

$$2HPO_4^{2-} + 3Ba^{2+} + 2NH_3 \rightarrow Ba_3(PO_4)_2\downarrow + 2NH_4^+$$

$Mg(NO_3)_2 + NH_4NO_3$
↓
white $MgNH_4PO_4$

3. *Magnesium nitrate reagent or magnesia mixture* the former is a solution containing $Mg(NO_3)_2$, NH_4NO_3, and a little aqueous NH_3, and the latter is a solution containing $MgCl_2$, NH_4Cl, and a little aqueous NH_3. The magnesium nitrate reagent is generally preferred since it may be employed in any subsequent test with silver nitrate solution. With either reagent a white, crystalline precipitate of magnesium ammonium phosphate, $Mg(NH_4)PO_4 \cdot 6H_2O$, is produced: this precipitate is soluble in acetic acid and in mineral acids, but practically insoluble in 2.5% ammonia solution (see under magnesium, Section 3.35, reaction 5; also under arsenic(V), Section 3.13, reaction 3).

$$HPO_4^{2-} + Mg^{2+} + NH_3 \rightarrow MgNH_4PO_4\downarrow$$

Arsenates give a similar precipitate $Mg(NH_4)AsO_4 \cdot 6H_2O$ with either reagent. They are most simply distinguished from one another by treating the washed precipitate with silver nitrate solution containing a few drops of dilute acetic acid: the phosphate turns yellow (Ag_3PO_4) whilst the arsenate assumes a brownish-red colour (Ag_3AsO_4).

$(NH_4)_2MoO_4$
↓
yellow ppt

4. *Ammonium molybdate reagent* the addition of a large excess (2–3 ml) of this reagent to a small volume (0.5 ml) of a phosphate solution produces a yellow, crystalline precipitate of ammonium phosphomolybdate (ammonium dodecamolybdatophosphate) $(NH_4)_3[PMo_{12}O_{40}]$ or $(NH_4)_3[P(Mo_3O_{10})_4]$, the Mo_3O_{10} group replacing each oxygen atom in phosphate. The resulting solution should be strongly acid with nitric acid; the latter is usually present in the reagent and addition is therefore unnecessary. Precipitation is accelerated by warming to a temperature not exceeding 40°C, and by the addition of ammonium nitrate solution.

$$HPO_4^{2-} + 3NH_4^+ + 12MoO_4^{2-} + 23H^+ \rightarrow$$

$$\rightarrow (NH_4)_3[P(Mo_3O_{10})_4]\downarrow + 12H_2O$$

The precipitate is soluble in ammonia solution and in solutions of caustic alkalis. Large quantities of hydrochloric acid interfere with the test and should preferably be removed by evaporation to a small volume with excess concentrated nitric acid. Reducing agents, such as sulphides, sulphites, hexacyanoferrate(II)s, and tartrates, seriously affect the reaction, and should be destroyed before carrying out the test.

Arsenates give a similar reaction on boiling (see under arsenic, Section 3.13, reaction 4). Both ammonium phosphomolybdate and ammonium arsenomolybdate dissolve on boiling with ammonium acetate solution, but only the latter yields a white precipitate on cooling.

Note: Commercial ammonium molybdate has the formula $(NH_4)_6Mo_7O_{24}\cdot4H_2O$ (a heptamolybdate) and not $(NH_4)_2MoO_4$; but the ion MoO_4^{2-} is employed in the equations for purposes of simplicity. This ion may exist under the experimental conditions of the reaction.

$FeCl_3$
↓
yellow $FePO_4$

5. *Iron(III) chloride solution* yellowish-white precipitate of iron(III) phosphate, $FePO_4$, soluble in dilute mineral acids, but insoluble in dilute acetic acid.

$$HPO_4^{2-} + Fe^{3+} \rightarrow FePO_4\downarrow + H^+$$

Precipitation is incomplete owing to the free mineral acid produced. If the hydrogen ions, arising from the complete ionization of the mineral acid, are removed by the addition of the salt of a weak acid, such as ammonium or sodium acetate, precipitation is almost complete. This is the basis of one of the methods for the removal of phosphates, which interfere with the precipitation of Group IIIA metals, in qualitative analysis.

$ZrO(NO_3)_2$
↓
white $ZrO(HPO_4)$

6. *Zirconium nitrate reagent* when the reagent is added to a solution of a phosphate containing hydrochloric acid not exceeding M in concentration, a white gelatinous precipitate of zirconyl phosphate $ZrO(HPO_4)$ is obtained. This reaction forms the basis of a simple method for the removal of phosphate prior to the precipitation of Group IIIA.

$$HPO_4^{2-} + ZrO^{2+} \rightarrow ZrO(HPO_4)\downarrow$$

In fact the precipitate is of variable composition, depending on the concentrations of zirconium, phosphate, and hydrogen ions. Species like $ZrO(H_2PO_4)_2$ and $ZrPO_4$ may also be formed.

7. *Cobalt nitrate test* phosphates when heated on charcoal and then moistened with a few drops of cobalt nitrate solution give a blue mass of the phosphate, $NaCoPO_4$. This must not be confused with the blue mass produced with aluminium compounds (see Section 3.23).

$Co(NO_3)_2$
on charcoal
↓
blue mass

$$Na(NH_4)HPO_4 \rightarrow NaPO_3 + NH_3\uparrow + H_2O;$$

$$NaPO_3 + CoO \rightarrow NaCoPO_4$$

8. *Ammonium molybdate–quinine sulphate reagent* phosphates give a yellow precipitate with this reagent: the exact composition appears to be unknown.

$(NH_4)_2MoO_4$
+
quinine
↓
yellow ppt

Reducing agents (sulphides, thiosulphates, etc.) interfere since they yield 'molybdenum blue'; hexacyanoferrate(II) ions give a red colouration. Arsenates (warming is usually required), arsenites, chromates, oxalates, tartrates, and silicates give a similar reaction with some variation in the colour of the precipitate. All should be removed before applying the test.

Place 1 ml test solution in a semimicro test-tube and add 1 ml reagent. A yellow precipitate is produced within a few minutes; gentle warming (water bath) is sometimes necessary.

Concentration limit 1 in 20 000.

4.29 Pyrophosphates, $P_2O_7^{4-}$, and metaphosphates, PO_3^-

Sodium pyrophosphate is prepared by heating disodium hydrogen phosphate:

$$2Na_2HPO_4 \rightarrow Na_4P_2O_7 + H_2O\uparrow$$

This is the normal salt. Acid salts, e.g. $Na_2H_2P_2O_7$, are known.

Sodium metaphosphate (polymeric) may be prepared by heating microcosmic salt or sodium dihydrogen phosphate:

$$Na(NH_4)HPO_4 \rightarrow NaPO_3 + NH_3\uparrow + H_2O\uparrow$$

$$NaH_2PO_4 \rightarrow NaPO_3 + H_2O\uparrow$$

$P_2O_7^{4-}$, PO_3^-
reactions
Cf. Table 4.1

A number of metaphosphates are known and these may be regarded as derived from the polymeric acid $(HPO_3)_n$, i.e. polymetaphosphoric acid. *Calgon*, used for water softening, is probably $(NaPO_3)_6$ or $Na_2[Na_4(PO_3)_6]$.

Pyro- and metaphosphates give the ammonium molybdate test on warming for some time; this is doubtless due to their initial conversion in solution into orthophosphates. The chief differences between ortho-, pyro- and metaphosphates are incorporated in Table 4.1. To study these reactions use freshly prepared solutions of sodium pyrophosphate and sodium metaphosphate.

4.30 Phosphites, HPO_3^{2-}

Solubility the phosphites of the alkali metals are soluble in water; all other metallic phosphites are insoluble in water.

HPO_3^{2-} reactions

To study these reactions use a freshly prepared 0.1M solution of sodium phosphite $Na_2HPO_3 \cdot 5H_2O$.

Table 4.1   **Reactions of ortho-, pyro- and metaphosphates**

Reagent	Orthophosphate	Pyrophosphate	Metaphosphate
1. Silver nitrate solution.	Yellow ppt., soluble in dilute HNO_3 and in dilute NH_3 solution.[a]	White ppt., soluble in dilute HNO_3 and in dilute NH_3 solution:[a] sparingly soluble in dilute acetic acid.	White ppt. (separates slowly), soluble in dilute HNO_3, in dilute NH_3 solution[a] and in dilute acetic acid.
2. Albumin and dilute acetic acid.	No coagulation.	No coagulation.	Coagulation.
3. Copper sulphate solution.	Pale blue ppt.	Very pale blue ppt.	No ppt.
4. Magnesia mixture or $Mg(NO_3)_2$ reagent.	White ppt., insoluble in excess of the reagent.	White ppt., soluble in excess of reagent, but reprecipitated on boiling.	No ppt., even on boiling.
5. Cadmium acetate solution and dilute acetic acid.	No ppt.	White ppt.	No ppt.
6. Zinc sulphate solution.	White ppt., soluble in dilute acetic acid.	White ppt., insoluble in dilute acetic acid; soluble in dilute NH_3 solution,[a] yielding a white ppt. on boiling.	White ppt. on warming; soluble in dilute acetic acid.

[a] The solution obtained after dissolving in ammonia should be discarded quickly to avoid a serious **explosion!**

$AgNO_3$
↓
white Ag_2HPO_3

1. *Silver nitrate solution*   white precipitate of silver phosphite, Ag_2HPO_3, which soon passes in the cold into black metallic silver. Warming is necessary with dilute solutions. Upon adding the reagent to a warm solution of a phosphite, a black precipitate of metallic silver is obtained immediately.

$$HPO_3^{2-} + 2Ag^+ \rightarrow Ag_2HPO_3\downarrow$$

$$Ag_2HPO_3 + H_2O \rightarrow 2Ag\downarrow + H_3PO_4$$

$$HPO_3^{2-} + 2Ag^+ + H_2O \rightarrow 2Ag\downarrow + H_3PO_4$$

$BaCl_2$
↓
white $BaHPO_3$

2. *Barium chloride solution*   white precipitate of barium phosphite, $BaHPO_3$, soluble in dilute acids.

$$HPO_3^{2-} + Ba^{2+} \rightarrow BaHPO_3\downarrow$$

$HgCl_2$
↓
white Hg_2Cl_2

3. *Mercury(II) chloride solution* (**POISON**)   white precipitate of calomel in the cold; on warming with excess of the phosphite solution, grey, metallic mercury is produced.

$$HPO_3^{2-} + 2HgCl_2 + H_2O \rightarrow Hg_2Cl_2\downarrow + H_3PO_4 + 2Cl^-$$

$$Hg_2Cl_2\downarrow + HPO_3^{2-} + H_2O \rightarrow 2Hg\downarrow + H_3PO_4 + 2Cl^-$$

Note that $HgCl_2$ and H_3PO_4 are virtually undissociated.

4. *Potassium permanganate solution*   no action with a cold solution acidified with acetic acid, but decolourized on warming.

cc H_2SO_4
↓
SO_2 gas

5. *Concentrated sulphuric acid*   no reaction in the cold with the solid salt, but on warming sulphur dioxide is evolved.

$$Na_2HPO_3 + H_2SO_4 \rightarrow SO_2\uparrow + 2Na^+ + HPO_4^{2-} + H_2O$$

$$H_3PO_3 + H_2SO_4 \rightarrow SO_2\uparrow + H_3PO_4 + H_2O$$

$Zn + H_2SO_4$
↓
PH_3 gas

6. *Zinc and dilute sulphuric acid*   phosphites are reduced by the nascent hydrogen to phosphine, PH_3, which may be identified as described in the Gutzeit test under arsenic (Section 3.14); the silver nitrate paper is stained first yellow and then black.

$$HPO_3^{2-} + 3Zn + 8H^+ \rightarrow PH_3\uparrow + 3Zn^{2+} + 3H_2O$$

$$PH_3\uparrow + 6Ag^+ + 3NO_3^- \rightarrow Ag_3P\cdot3AgNO_3\downarrow(\text{yellow}) + 3H^+$$

$$Ag_3P\cdot3AgNO_3\downarrow + 3H_2O \rightarrow 6Ag\downarrow(\text{black}) + H_3PO_3 + 3H^+ + 3NO_3^-$$

$CuSO_4$
↓
blue $CuHPO_3$

7. *Copper sulphate solution*   light-blue precipitate of copper phosphite, $CuHPO_3$; the precipitate merely dissolves when it is boiled with acetic acid (cf. hypophosphites, Section 4.31).

$$HPO_3^{2-} + Cu^{2+} \rightarrow CuHPO_3\downarrow$$

$Pb(CH_3COO)_2$
↓
white $PbHPO_3$

8. *Lead acetate solution*   white precipitate, insoluble in acetic acid.

$$HPO_3^{2-} + Pb^{2+} \rightarrow PbHPO_3\downarrow$$

Heating
↓
PH_3 gas

9. *Action of heat*   inflammable phosphine is evolved, and a mixture of phosphates is produced.

$$8Na_2HPO_3 \rightarrow 2PH_3\uparrow + 4Na_3PO_4 + Na_4P_2O_7 + H_2O$$

4.31   Hypophosphites, $H_2PO_2^-$

Solubility   all hypophosphites are soluble in water.

$H_2PO_2^-$ reactions

To study these reactions use a freshly prepared 0.1M solution of sodium hypophosphite $NaH_2PO_2\cdot H_2O$.

$AgNO_3$
↓
white AgH_2PO_2

1. *Silver nitrate solution*   white precipitate of silver hypophosphite, AgH_2PO_2, which is slowly reduced to silver at room temperature, but more rapidly on warming. Hydrogen is simultaneously evolved.

$$H_2PO_2^- + Ag^+ \rightarrow AgH_2PO_2\downarrow$$

$$2AgH_2PO_2\downarrow + 4H_2O \rightarrow 2Ag\downarrow + 2H_3PO_4 + 3H_2\uparrow$$

2. *Barium chloride solution*   no precipitate.

$HgCl_2$
↓
white Hg_2Cl_2

3. *Mercury(II) chloride solution* (**POISON**)   white precipitate of calomel in the cold, converted by warming into grey, metallic mercury.

$$H_2PO_2^- + 4HgCl_2 + 2H_2O \rightarrow 2Hg_2Cl_2\downarrow + H_3PO_4 + 3H^+ + 4Cl^-$$

$$H_2PO_2^- + 2Hg_2Cl_2\downarrow + 2H_2O \rightarrow 4Hg\downarrow + H_3PO_4 + 3H^+ + 4Cl^-$$

$CuSO_4$ + heat
↓
red CuH

4. *Copper sulphate solution*   no precipitate in the cold, but on warming red copper(I) hydride is precipitated.

$$3H_2PO_2^- + 4Cu^{2+} + 6H_2O \rightarrow 4CuH\downarrow + 3H_3PO_4 + 5H^+$$

When heated with concentrated hydrochloric acid, the precipitate decomposes to white copper(I) chloride precipitate and hydrogen gas.

$$CuH\downarrow + HCl \rightarrow CuCl\downarrow + H_2\uparrow$$

$KMnO_4$
↓
reduced to Mn^{2+}

5. *Potassium permanganate solution*   reduced immediately in the cold to a colourless solution.

$$5H_2PO_2^- + 4MnO_4^- + 17H^+ \rightarrow 5H_3PO_4 + 4Mn^{2+} + 6H_2O$$

cc H_2SO_4
↓
SO_2 gas

6. *Concentrated sulphuric acid*   reduced to sulphur dioxide by the solid salt on warming.

$$NaH_2PO_2 + 4H_2SO_4 \rightarrow 2SO_2\uparrow + H_3PO_4 + 3H^+ + Na^+ + 2SO_4^{2-} + 2H_2O$$

Some elemental sulphur may also be formed during the process.

cc NaOH
↓
H_2 gas

7. *Concentrated sodium hydroxide solution*   hydrogen gas is evolved and phosphate ions formed on warming.

$$H_2PO_2^- + 2OH^- \rightarrow 2H_2\uparrow + PO_4^{3-}$$

Zn + H_2SO_4
↓
PH_3 gas

8. *Zinc and dilute sulphuric acid*   inflammable phosphine is evolved (see under phosphites, Section 4.30, reaction 6).

$(NH_4)_2MoO_4$
↓
molybdenum blue

9. *Ammonium molybdate solution*   reduced to 'molybdenum blue' in solution acidified with dilute sulphuric acid (difference from phosphite).

Heating
↓
PH_3 gas

10. *Action of heat*   phosphine is evolved, and a pyrophosphate is produced.

$$4NaH_2PO_2 \rightarrow Na_4P_2O_7 + 2PH_3\uparrow + H_2O$$

4.32   **Arsenites, AsO_3^{3-}, and arsenates, AsO_4^{3-}**

See under arsenic, Sections 3.12 and 3.13.

4.33   **Chromates, CrO_4^{2-}, and dichromates, $Cr_2O_7^{2-}$**

The metallic chromates are usually coloured solids, yielding yellow solutions when soluble in water. In the presence of dilute mineral acids, i.e. of hydrogen ions, chromates are converted into dichromates; the latter yield orange-red aqueous solutions. The change is reversed by alkalis, i.e. by hydroxyl ions.

$$2CrO_4^{2-} + 2H^+ \rightleftarrows Cr_2O_7^{2-} + H_2O$$

or

$$Cr_2O_7^{2-} + 2OH^- \rightleftarrows 2CrO_4^{2-} + H_2O$$

The reactions may also be expressed as:

$$2CrO_4^{2-} + 2H^+ \rightleftarrows 2HCrO_4^- \rightleftarrows Cr_2O_7^{2-} + H_2O$$

Solubility the chromates of the alkali metals and of calcium and magnesium are soluble in water; strontium chromate is sparingly soluble. Most other metallic chromates are insoluble in water. Sodium, potassium, and ammonium dichromates are soluble in water.

To study these reactions use a 0.1M solution of potassium chromate, K_2CrO_4, or potassium dichromate, $K_2Cr_2O_7$.

CrO_4^{2-}, $Cr_2O_7^{2-}$ reactions

1. *Barium chloride solution* pale-yellow precipitate of barium chromate, $BaCrO_4$, insoluble in water and in acetic acid, but soluble in dilute mineral acids.

• $BaCl_2$
↓
yellow $BaCrO_4$

$$CrO_4^{2-} + Ba^{2+} \rightarrow BaCrO_4\downarrow$$

Dichromate ions produce the same precipitate, but as a strong acid is formed, precipitation is only partial:

$$Cr_2O_7^{2-} + 2Ba^{2+} + H_2O \rightleftarrows 2BaCrO_4\downarrow + 2H^+$$

If sodium hydroxide or sodium acetate is added, precipitation becomes quantitative.

$AgNO_3$
↓
red Ag_2CrO_4

2. *Silver nitrate solution* brownish-red precipitate of silver chromate, Ag_2CrO_4, with a solution of a chromate. The precipitate is soluble in dilute nitric acid and in ammonia solution, but is insoluble in acetic acid. Hydrochloric acid converts the precipitate into silver chloride (white).

$$CrO_4^{2-} + 2Ag^+ \rightarrow Ag_2CrO_4\downarrow$$

$$2Ag_2CrO_4\downarrow + 2H^+ \rightarrow 4Ag^+ + Cr_2O_7^{2-} + H_2O$$

$$Ag_2CrO_4\downarrow + 4NH_3 \rightarrow 2[Ag(NH_3)_2]^+ + CrO_4^{2-}$$

$$Ag_2CrO_4\downarrow + 2Cl^- \rightarrow 2AgCl\downarrow + CrO_4^{2-}$$

The solution obtained after dissolving in ammonia should be discarded quickly to avoid a serious **explosion!** (Cf. Section 3.6, reaction 1.)

A reddish-brown precipitate of silver dichromate, $Ag_2Cr_2O_7$, is formed with a concentrated solution of a dichromate; on boiling with water, this is converted into the less soluble silver chromate.

$$Cr_2O_7^{2-} + 2Ag^+ \rightarrow Ag_2Cr_2O_7\downarrow$$

$$Ag_2Cr_2O_7\downarrow + H_2O \rightarrow Ag_2CrO_4\downarrow + CrO_4^{2-} + 2H^+$$

$Pb(CH_3COO)_2$
↓
yellow $PbCrO_4$

3. *Lead acetate solution*   yellow precipitate of lead chromate, $PbCrO_4$, insoluble in acetic acid, but soluble in dilute nitric acid.

$$CrO_4^{2-} + Pb^{2+} \rightarrow PbCrO_4\downarrow$$

$$2PbCrO_4\downarrow + 2H^+ \rightleftarrows 2Pb^{2+} + Cr_2O_7^{2-} + H_2O$$

The precipitate is soluble in sodium hydroxide solution; acetic acid reprecipitates lead chromate from such solutions.

The solubility in sodium hydroxide solution is due to the formation of the soluble complex tetrahydroxoplumbate(II) ion, which suppresses the Pb^{2+} ion concentration to such an extent that the solubility product of lead chromate is no longer exceeded, and consequently the latter dissolves.

$$PbCrO_4\downarrow + 4OH^- \rightleftarrows [Pb(OH)_4]^{2-} + CrO_4^{2-}$$

H_2O_2
↓
blue CrO_5
↓
decomposes to
green Cr^{3+}

4. *Hydrogen peroxide*   if an acid solution of a chromate is treated with hydrogen peroxide, a deep-blue solution of chromium pentoxide is obtained (cf. chromium, Section 3.24, reaction 6b). The blue solution is very unstable and soon decomposes, yielding oxygen and a green solution of a chromium(III) salt. The blue compound is soluble in amyl alcohol and also in amyl acetate and in diethyl ether, and can be extracted from aqueous solutions by these solvents to yield somewhat more stable solutions. Amyl alcohol is recommended. Diethyl ether is not recommended for general use owing to its highly inflammable character and also because it frequently contains peroxides after storage for comparatively short periods, and these may interfere. Peroxides may be removed from diethyl ether by shaking with a concentrated solution of iron(II) salt or with sodium sulphite.

The blue colouration is attributed to the presence of chromium pentoxide, CrO_5 (cf. Section 3.24, reaction 6b).

$$CrO_4^{2-} + 2H^+ + 2H_2O_2 \rightarrow CrO_5 + 3H_2O$$

The enhanced stability in solutions of amyl alcohol, ether etc. is due to the formation of complexes with these oxygen-containing compounds.

Carefully acidify a cold solution of a chromate with dilute sulphuric acid or dilute nitric acid, add 1–2 ml amyl alcohol, then 1 ml of 3% hydrogen peroxide solution dropwise and with shaking after each addition: the organic layer is coloured blue. The chromium pentoxide is more stable below 0°C than at laboratory temperature.

H_2S
↓
white S + green
colour, Cr^{3+}

5. *Hydrogen sulphide*   an acid solution of a chromate is reduced by this reagent to a green solution of chromium(III) ions, accompanied by the separation of sulphur.[12]

$$2CrO_4^{2-} + 3H_2S + 10H^+ \rightarrow 2Cr^{3+} + 3S\downarrow + 8H_2O$$

[12] In qualitative analysis the production of sulphur in the reduction of chromates may be troublesome, because the precipitate is almost colloidal and is difficult to filter. On adding a few strips of filter paper to the mixture and boiling, the precipitate coagulates and can be filtered easily. Alternatively, before the use of hydrogen sulphide, chromates can be reduced by heating the solid substance with concentrated hydrochloric acid, evaporating most of the acid and then diluting with water (see reaction 7).

Reducing agents
↓
green solution.
Cr^{3+}

6. *Reduction of chromates and dichromates*

(a) Sulphur dioxide, in the presence of dilute mineral acid, reduces chromates or dichromates:

$$2CrO_4^{2-} + 3SO_2 + 4H^+ \rightarrow 2Cr^{3+} + 3SO_4^{2-} + 2H_2O$$

The solution becomes green owing to the formation of chromium(III) ions.

(b) Potassium iodide, in the presence of dilute mineral acid, can be used as a reductant. The colour of the solution becomes brown or bluish, according to the amounts of iodine and chromium(III) ions, which are formed in the reaction:

$$2CrO_4^{2-} + 9I^- + 16H^+ \rightarrow 2Cr^{3+} + 3I_3^- + 8H_2O$$

(c) Iron(II) sulphate, in the presence of mineral acid, reduces chromates or dichromates smoothly:

$$CrO_4^{2-} + 3Fe^{2+} + 8H^+ \rightarrow Cr^{3+} + 3Fe^{3+} + 4H_2O$$

(d) Ethanol, again in the presence of mineral acid, reduces chromates or dichromates slowly in cold, but rapidly if the solution is heated:

$$2CrO_4^{2-} + 3C_2H_5OH + 10H^+ \rightarrow 2Cr^{3+} + 3CH_3CHO\uparrow + 8H_2O$$

The smell of acetaldehyde (CH_3CHO) is easily observed if the mixture is heated.

HCl on solid chromates
↓
Cl_2 gas + Cr^{3+} ions

7. *Reduction of solid chromates or dichromates with concentrated hydrochloric acid*   On heating a solid chromate or dichromate with concentrated hydrochloric acid, chlorine is evolved, and a solution containing chromium(III) ions is produced:

$$2K_2CrO_4 + 16HCl \rightarrow 2Cr^{3+} + 3Cl_2\uparrow + 4K^+ + 10Cl^- + 8H_2O$$

$$K_2Cr_2O_7 + 14HCl \rightarrow 2Cr^{3+} + 3Cl_2\uparrow + 2K^+ + 8Cl^- + 7H_2O$$

If the solution is evaporated almost to dryness (to remove chlorine), after dilution chromium(III) ions can be tested for in the solution (cf. Section 3.24).

Chromyl chloride test

8. *Concentrated sulphuric acid and a chloride*   see chromyl chloride test under chlorides, Section 4.14, reaction 5.

Diphenylcarbazide
↓
red colour

9. *1,5-Diphenylcarbazide reagent (1% in ethanol)*   The solution is acidified with dilute sulphuric acid or with dilute acetic acid, and 1–2 ml of the reagent added. A deep-red colouration is produced. With small quantities of chromates, the solution is coloured violet.

Full details of the use of the reagent as a spot test are given under chromium, Section 3.24, reaction 6c.

Chromotropic acid test

10. *Chromotropic acid test*   a red colouration, best seen by transmitted light, is given by chromates. For details, see under chromium, Section 3.24, reaction 6d.

4.34   Permanganates, MnO_4^-

Solubility   all permanganates are soluble in water, forming purple (reddish-violet) solutions.

To study these reactions use a 0.02M solution of potassium permanganate, $KMnO_4$.

$H_2O_2 + H^+$
↓
reduction to Mn^{2+}

1. *Hydrogen peroxide*   the addition of this reagent to a solution of potassium permanganate, acidified with dilute sulphuric acid, results in decolourization and the evolution of pure but moist oxygen.

$$2MnO_4^- + 5H_2O_2 + 6H^+ \rightarrow 5O_2\uparrow + 2Mn^{2+} + 8H_2O$$

Reducing agents + H^+
↓
reduction to
colourless Mn^{2+}

2. *Reduction of permanganates*   in acid solutions the reduction proceeds to the formation of colourless manganese(II) ions. The following reducing agents may be used:

(a) Hydrogen sulphide: in the presence of dilute sulphuric acid the solution decolourizes and sulphur is precipitated.[13]

$$2MnO_4^- + 5H_2S + 6H^+ \rightarrow 5S\downarrow + 2Mn^{2+} + 8H_2O$$

(b) Sulphur dioxide, in the presence of sulphuric acid, reduces permanganate instantaneously:

$$2MnO_4^- + 5SO_2 + 2H_2O \rightarrow 5SO_4^{2-} + 2Mn^{2+} + 4H^+$$

Some dithionate ($S_2O_6^{2-}$) may also be formed in this reaction, the quantity being dependent upon the experimental conditions:

$$2MnO_4^- + 6SO_2 + 2H_2O \rightarrow 2Mn^{2+} + S_2O_6^{2-} + 4SO_4^{2-} + 4H^+$$

(c) Iron(II) sulphate, in the presence of sulphuric acid, reduces permanganate to manganese(II). The solution becomes yellow because of the formation of iron(III) ions.

$$MnO_4^- + 5Fe^{2+} + 8H^+ \rightarrow 5Fe^{3+} + Mn^{2+} + 4H_2O$$

The yellow colour disappears if phosphoric acid or potassium fluoride is added; they form colourless complexes with iron(III).

(d) Potassium iodide, in the presence of sulphuric acid, reduces permanganate with the formation of iodine:

$$2MnO_4^- + 15I^- + 16H^+ \rightarrow 5I_3^- + 2Mn^{2+} + 8H_2O$$

(e) Sodium nitrite, in the presence of sulphuric acid, reduces permanganate with formation of nitrate ions:

$$2MnO_4^- + 5NO_2^- + 6H^+ \rightarrow 2Mn^{2+} + 5NO_3^- + 3H_2O$$

---

[13] To avoid the production of sulphur by the reduction of potassium permanganate by hydrogen sulphide in systematic qualitative analysis, the solution may be reduced with formaldehyde solution, or the solid substance may be boiled with concentrated hydrochloric acid.

$$2MnO_4^- + 5HCHO + 6H^+ \rightarrow 2Mn^{2+} + 5HCOOH + 3H_2O$$

Some side reactions also take place, and the mixture has the smell of nitrogen oxide gas.

(f) Oxalic acid, in the presence of sulphuric acid, produces carbon dioxide gas:

$$2MnO_4^- + 5(COO)_2^{2-} + 16H^+ \rightarrow 10CO_2\uparrow + 2Mn^{2+} + 8H_2O$$

This reaction is slow at room temperature, but is rapid at 60°C. Manganese(II) ions catalyse the reaction; thus, the reaction is **autocatalytic**; once manganese(II) ions are formed, it becomes faster and faster.

In alkaline solution the permanganate is decolourized, but (hydrated) manganese dioxide is precipitated. In the presence of sodium hydroxide solution, potassium iodide is converted into potassium iodate, and sodium sulphite solution into sodium sulphate on boiling.

$$2MnO_4^- + I^- + 3H_2O \rightarrow 2MnO(OH)_2\downarrow + IO_3^- + 2OH^-$$

$$2MnO_4^- + 3SO_3^{2-} + 3H_2O \rightarrow 2MnO(OH)_2\downarrow + 3SO_4^{2-} + 2OH^-$$

cc HCl
↓
Cl_2 gas + Mn^{2+}

3. *Concentrated hydrochloric acid* all permanganates on boiling with concentrated hydrochloric acid evolve chlorine:

$$2MnO_4^- + 16HCl \rightarrow 5Cl_2\uparrow + 2Mn^{2+} + 6Cl^- + 8H_2O$$

cc NaOH
↓
reduction to
green MnO_4^{2-}

4. *Sodium hydroxide solution* upon warming a concentrated solution of potassium permanganate with concentrated sodium hydroxide solution, a green solution of potassium manganate is produced and oxygen is evolved. When the manganate solution is poured into a large volume of water or is acidified with dilute sulphuric acid, the purple colour of the potassium permanganate is restored, and manganese dioxide is precipitated.

$$4MnO_4^- + 4OH^- \rightarrow 4MnO_4^{2-} + O_2\uparrow + 2H_2O$$

$$3MnO_4^{2-} + 3H_2O \rightarrow 2MnO_4^- + MnO(OH)_2\downarrow + 4OH^-$$

The latter reaction is in fact a **disproportionation**, when manganese(VI) is partly oxidized to manganese(VII) and partly reduced to manganese(IV).

A manganate is produced when a manganese compound is fused with potassium nitrate and sodium carbonate (see Section 3.28, dry tests).

Heating
solid $KMnO_4$
↓
O_2 gas

5. *Action of heat* when potassium permanganate is heated in a test-tube, pure oxygen is evolved, and a black residue of potassium manganate K_2MnO_4 and manganese dioxide remains behind. Upon extracting with a little water and filtering, a green solution of potassium manganate is obtained.

$$2KMnO_4 \rightarrow K_2MnO_4 + MnO_2 + O_2\uparrow$$

4.35 Acetates, CH_3COO^-

Solubility all normal acetates, with the exception of silver and mercury(I) acetates which are sparingly soluble, are readily soluble in water. Some basic acetates, e.g. those of iron, aluminium, and chromium, are insoluble in water. The free acid, CH_3COOH, is a colourless liquid with a pungent odour (boiling point 117°C, melting point 17°C) and is miscible with water in all proportions; it has a corrosive action on the skin.

CH_3COO^- reactions

To study these reactions use a 2M solution of sodium acetate $CH_3COONa \cdot 3H_2O$.

H_2SO_4
↓
smell of vinegar

1. *Dilute sulphuric acid* acetic acid, easily recognized by its vinegar-like odour, is evolved on warming.

$$CH_3COO^- + H^+ \rightarrow CH_3COOH\uparrow$$

2. *Concentrated sulphuric acid* acetic acid is evolved on heating, together with sulphur dioxide, the latter tending to mask the penetrating odour of the concentrated acetic acid vapour. The test with dilute sulphuric acid, in which the acetic acid vapour is diluted with steam, is therefore to be preferred as a test for an acetate.

$C_2H_5OH +$
cc H_2SO_4
↓
ethyl acetate,
pleasant odour

3. *Ethanol and concentrated sulphuric acid* 1 g of the solid acetate is treated with 1 ml concentrated sulphuric acid and 2–3 ml 96% ethanol in a test-tube, and the whole gently warmed for several minutes; ethyl acetate $CH_3COO \cdot C_2H_5$ is formed, which is recognized by its pleasant, fruity odour. On cooling and dilution with water on a clock glass, the fragrant odour will be more readily detected.

$$CH_3COONa + H_2SO_4 \rightarrow CH_3COOH + Na^+ + HSO_4^-$$

$$CH_3COOH + C_2H_5OH \rightarrow CH_3COO \cdot C_2H_5\uparrow + H_2O$$

(In the second reaction the sulphuric acid acts as a dehydrating agent.)

It is preferable to use amyl alcohol because the odour of the resulting amyl acetate is more readily distinguished from the alcohol itself than is the case with ethanol. It is as well to run a parallel test with a known acetate and to compare the odours of the two products.

$AgNO_3$
↓
white $AgCH_3COO$

4. *Silver nitrate solution* a white, crystalline precipitate of silver acetate is produced in concentrated solutions in the cold. The precipitate is more soluble in boiling water ($10.4 \, g \, l^{-1}$ at 30°C and $25.2 \, g \, l^{-1}$ at 80°C) and readily soluble in dilute ammonia solution.

$$CH_3COO^- + Ag^+ \rightleftharpoons CH_3COOAg\downarrow$$

The solution obtained after dissolving in ammonia should be discarded quickly to avoid a serious **explosion!** (Cf. Section 3.6, reaction 1.)

When heating the mixture, the precipitate redissolves without the formation of black precipitate of silver metal (distinction from formate ions).

5. *Barium chloride, calcium chloride or mercury(II) chloride solution* no change in the presence of acetates (distinction from oxalates and formates).

FeCl₃
↓
red ppt

6. *Iron(III) chloride solution* deep-red colouration, owing to the formation of a complex ion, $[Fe_3(OH)_2(CH_3COO)_6]^+$. On boiling the red solution it decomposes and a brownish-red precipitate of basic iron(III) acetate is formed:

$$6CH_3COO^- + 3Fe^{3+} + 2H_2O \rightarrow [Fe_3(OH)_2(CH_3COO)_6]^+ + 2H^+$$

$$[Fe_3(OH)_2(CH_3COO)_6]^+ + 4H_2O \rightarrow$$

$$\rightarrow 3Fe(OH)_2CH_3COO\downarrow + 3CH_3COOH + H^+$$

Heating with
As₂O₃
↓
cacodyl oxide

7. *Cacodyl oxide reaction* if a dry acetate, preferably that of sodium or potassium, is heated in an ignition tube or test-tube with a small quantity of solid arsenic(III) oxide, an extremely nauseating odour of cacodyl oxide is produced. All cacodyl compounds are extremely **POISONOUS**; the experiment must therefore be performed on a very small scale, and preferably in the fume cupboard. Mix not more than 0.2 g sodium acetate with 0.2 g arsenic(III) oxide in an ignition tube and warm; observe the extremely unpleasant odour that is produced.

$$4CH_3COONa + As_2O_3 \rightarrow \begin{matrix} CH_3 & & CH_3 \\ \diagdown & & \diagup \\ & As-O-As & \\ \diagup & & \diagdown \\ CH_3 & & CH_3 \end{matrix} \quad 2Na_2CO_3 + 2CO_2\uparrow$$

La(NO₃)₃ + I₂
↓
blue colour on ppt

8. *Lanthanum nitrate test* treat 0.5 ml of the acetate solution with 0.5 ml 0.1M lanthanum nitrate solution, add 0.5 ml 0.05M iodine solution and a few drops of dilute ammonia solution, and heat slowly to the boiling point. A blue colour is produced; this is probably due to the adsorption of the iodine by the basic lanthanum acetate. This reaction provides an extremely sensitive test for an acetate.

Sulphates and phosphates interfere, but can be removed by precipitation with barium nitrate solution before applying the test. Propionates give a similar reaction.

The spot-test technique is as follows. Mix a drop of the test solution on a spot plate with a drop of 0.1M lanthanum nitrate solution and a drop of 0.005M iodine. Add a drop of M ammonia solution. Within a few minutes a blue to blue-brown ring will develop round the drop of ammonia solution.

Sensitivity 50 µg CH_3COO^-.

Concentration limit 1 in 2000.

Indigo test

9. *Formation of indigo test* the test depends upon the conversion of acetone, formed by the dry distillation of acetates (see reaction 10), into indigo. No other fatty acids give this test, but the sensitivity is reduced in their presence.

Mix the solid test sample with calcium carbonate or, alternatively, evaporate a drop of the test solution to dryness with calcium carbonate; both operations may be carried out in the hard glass tube of Fig. 2.55. Cover the open end of the tube with a strip of quantitative filter paper moistened with a freshly prepared solution of 2-nitrobenzaldehyde in 2M sodium hydroxide, and hold the paper in position with a small glass cap or a small watch glass. Insert the tube into a hole in Teflon sheet and heat the tube gently. Acetone is evolved which colours the paper blue or bluish-green. For minute amounts of acetates, it is best to remove the filter paper after the reaction and treat it with a drop of dilute hydrochloric acid; the original yellow colour of the paper is thus bleached and the blue colour of the indigo is more readily apparent.

Sensitivity 60 μg CH_3COO^-.

Heating
↓
acetone vapours

10. *Action of heat* all acetates decompose upon strong ignition, yielding the highly inflammable acetone, $CH_3 \cdot CO \cdot CH_3$, and a residue which consists of the carbonates for the alkali acetates, of the oxides for the acetates of the alkaline earth and heavy metals, and of the metal for the acetates of silver and the noble metals. Carry out the experiment in an ignition tube with sodium acetate and lead acetate.

$$2CH_3COONa \rightarrow CH_3 \cdot CO \cdot CH_3\uparrow + Na_2CO_3$$

$$(CH_3COO)_2Pb \rightarrow CH_3 \cdot CO \cdot CH_3\uparrow + PbO + CO_2$$

4.36 Formates, $HCOO^-$

Solubility with the exception of the lead, silver, and mercury(I) salts which are sparingly soluble, most formates are soluble in water. The free acid, HCOOH, is a pungent smelling liquid (boiling point 100.5°C, melting point 8°C), miscible with water in all proportions, and producing blisters when allowed to come into contact with the skin.

$HCOO^-$ reactions

To study these reactions use a M solution of sodium formate, HCOONa.

H_2SO_4
↓
HCOOH vapour

1. *Dilute sulphuric acid* formic acid is liberated, the pungent odour of which can be detected on warming the mixture.

$$HCOO^- + H^+ \rightarrow HCOOH\uparrow$$

cc H_2SO_4
↓
CO gas

2. *Concentrated sulphuric acid* carbon monoxide (**HIGHLY POISONOUS**) is evolved on warming; the gas should be ignited and the characteristic blue flame observed. The test can be successfully carried out with solid sodium formate.

$$HCOONa + H_2SO_4 \rightarrow CO\uparrow + Na^+ + HSO_4^- + H_2O$$

$C_2H_5OH + $ cc H_2SO_4
↓
ethyl formate,
pleasant odour

3. *Ethanol and concentrated sulphuric acid* a pleasant odour, owing to ethyl formate, is apparent on warming (for details, see under acetates, Section 4.35, reaction 3).

$$HCOONa + H_2SO_4 \rightarrow HCOOH + Na^+ + HSO_4^-$$

$$HCOOH + C_2H_5OH \rightarrow HCOO \cdot C_2H_5\uparrow + H_2O$$

$AgNO_3$
↓
white HCOOAg
↓
decomposes to Ag

4. *Silver nitrate solution* white precipitate of silver formate in neutral solutions, slowly reduced at room temperature and more rapidly on warming. A black precipitate of silver is formed (distinction from acetate). With very dilute solutions, the silver may be deposited in the form of a mirror on the walls of the tube.

$$HCOO^- + Ag^+ \rightarrow HCOOAg\downarrow$$

$$2HCOOAg\downarrow \rightarrow 2Ag\downarrow + HCOOH + CO_2\uparrow$$

5. *Barium chloride or calcium chloride solution* no change (distinction from oxalates).

FeCl₃
↓
red colour

6. *Iron(III) chloride* a red colouration, due to the complex $[Fe_3(HCOO)_6]^{3+}$, is produced in practically neutral solutions; the colour is discharged by hydrochloric acid. If the red solution is diluted and boiled, a brown precipitate of basic iron(III) formate is formed.

$$6HCOO^- + 3Fe^{3+} \rightarrow [Fe_3(HCOO)_6]^{3+}$$

$$[Fe_3(HCOO)_6]^{3+} + 4H_2O \rightarrow 2Fe(OH)_2HCOO\downarrow + 4HCOOH\uparrow + Fe^{3+}$$

HgCl₂
↓
white Hg₂Cl₂
↓
decomposes

7. *Mercury(II) chloride solution* (**POISON**) white precipitate of mercury(I) chloride (calomel) is produced on warming; this passes into grey, metallic mercury, in the presence of excess formate solution (distinction from acetate).

$$2HCOO^- + 2HgCl_2 \rightarrow Hg_2Cl_2\downarrow + 2Cl^- + CO\uparrow + CO_2\uparrow + H_2O$$

$$2HCOO^- + Hg_2Cl_2\downarrow \rightarrow 2Hg\downarrow + 2Cl^- + CO\uparrow + CO_2\uparrow + H_2O$$

Mercury formate test

8. *Mercury(II) formate test* free formic acid is necessary for this test. The solution of sodium formate must be acidified with dilute sulphuric acid and shaken vigorously with a little mercury(II) oxide. The solution must then be filtered. The filtrate, when boiled, gives momentarily a white precipitate of mercury(I) formate, which rapidly changes to a grey precipitate of metallic mercury.

$$2HCOOH + HgO \rightarrow (HCOO)_2Hg + H_2O$$

$$2(HCOO)_2Hg \rightarrow (HCOO)_2Hg_2\downarrow + HCOOH + CO_2\uparrow$$

$$(HCOO)_2Hg_2\downarrow \rightarrow 2Hg\downarrow + HCOOH + CO_2\uparrow$$

Formic acid and mercury(II) formate are almost completely undissociated under these circumstances. Note that all compounds of mercury are **POISONOUS**.

Mg + HCl
↓
reduced to HCHO

9. *Formaldehyde–chromotropic acid test* formic acid, $H\cdot COOH$, is reduced to formaldehyde $H\cdot CHO$ by magnesium and hydrochloric acid. The formaldehyde is identified by its reaction with chromotropic acid (see Section 3.24, reaction 6d) in strong sulphuric acid when a violet-pink colouration appears. Other aliphatic aldehydes do not give the violet colouration.

Place a drop or two of the test solution in a semimicro test-tube, add a drop or two of dilute hydrochloric acid, followed by magnesium powder until the evolution of gas ceases. Introduce 3 ml 12M sulphuric acid and a little solid chromotropic acid, and warm to 60°C. A violet-pink colouration appears within a few minutes.

Sensitivity 1.5 μg $H\cdot COOH$.

Concentration limit 1 in 20 000.

Heating
↓
oxalate

10. *Action of heat* cautious ignition of the formates of the alkali metals yields the corresponding oxalates (for tests, see Section 4.37) and hydrogen.

$$2HCOONa \rightarrow (COO)_2Na_2 + H_2\uparrow$$

4.37 Oxalates, $(COO)_2^{2-}$

Solubility the oxalates of the alkali metals and of iron(II) are soluble in water; all other oxalates are either insoluble or sparingly soluble in water. They are all soluble in dilute acids. Some of the oxalates dissolve in a concentrated solution of oxalic acid by virtue of the formation of soluble acid or complex oxalates. Oxalic acid (a dibasic acid) is a colourless, crystalline solid, $(COOH)_2 \cdot 2H_2O$, and becomes anhydrous on heating to $110°C$; it is readily soluble in water ($111\,g\,l^{-1}$ at $20°C$).

$(COO)_2^{2-}$ reactions

To study these reactions a 0.1M solution of sodium oxalate, $(COONa)_2$, or ammonium oxalate, $(COONH_4)_2 \cdot H_2O$, should be used. Note that all oxalates are **POISONOUS**.

cc H_2SO_4
↓
$CO + CO_2$ gases

1. *Concentrated sulphuric acid* decomposition of all solid oxalates occurs with the evolution of carbon monoxide (**POISON**) and carbon dioxide; the latter can be detected by passing the escaping gases through lime water (distinction from formate) and the former by burning it at the mouth of the tube. With dilute sulphuric acid, there is no visible action; in the presence of manganese dioxide, however, carbon dioxide is evolved.

$$(COOH)_2 \rightarrow H_2O + CO\uparrow + CO_2\uparrow$$

$$(COO)_2^{2-} + MnO_2 + 4H^+ \rightarrow Mn^{2+} + 2CO_2\uparrow + 2H_2O$$

(the concentrated sulphuric acid acts as a dehydrating agent).

$AgNO_3$
↓
white $Ag_2(COO)_2$

2. *Silver nitrate solution* white, curdy precipitate of silver oxalate sparingly soluble in water, soluble in ammonia solution and in dilute nitric acid.

$$(COO)_2^{2-} + 2Ag^+ \rightarrow (COOAg)_2\downarrow$$

$$(COOAg)_2\downarrow + 4NH_3 \rightarrow 2[Ag(NH_3)_2]^+ + (COO)_2^{2-}$$

$$(COOAg)_2\downarrow + 2H^+ \rightarrow 2Ag^+ + (COO)_2^{2-} + 2H^+$$

The solution obtained after dissolving in ammonia should be discarded quickly to avoid a serious **explosion!** (Cf. Section 3.6, reaction 1.)

$CaCl_2$
↓
white $Ca(COO)_2$

3. *Calcium chloride solution* white, crystalline precipitate of calcium oxalate from neutral solutions, insoluble in dilute acetic acid, oxalic acid, and in ammonium oxalate solution, but soluble in dilute hydrochloric acid and in dilute nitric acid. It is the least soluble of all oxalates ($0.0067\,g\,l^{-1}$ at $13°C$) and is even precipitated by calcium sulphate solution and acetic acid. Barium chloride solution similarly gives a white precipitate of barium oxalate sparingly soluble in water ($0.016\,g\,l^{-1}$ at $8°C$), but soluble in solutions of acetic and of oxalic acids.

$$(COO)_2^{2-} + Ca^{2+} \rightarrow (COO)_2Ca\downarrow$$

$$(COO)_2^{2-} + Ba^{2+} \rightarrow (COO)_2Ba\downarrow$$

$KMnO_4 + H^+$
↓
reduced to Mn^{2+}

4. *Potassium permanganate solution* decolourized when warmed in acid solution to $60°C$ (cf. Section 4.34, reaction 2f). The bleaching of permanganate solution is also effected by many other organic compounds but if the

evolved carbon dioxide is tested for by the lime water reaction (Section 4.2, reaction 1), the test becomes specific for oxalates.

$$5(COO)_2^{2-} + 2MnO_4^- + 16H^+ \rightarrow 10CO_2\uparrow + 2Mn^{2+} + 8H_2O$$

Resorcinol test
↓
blue colour

5. *Resorcinol test* place a few drops of the test solution in a test-tube: add several drops of dilute sulphuric acid and a speck or two of magnesium powder. When the metal has dissolved, add about 0.1 g resorcinol, and shake until dissolved. Cool. Carefully pour down the side of the tube 3–4 ml concentrated sulphuric acid. A blue ring will form at the junction of the two liquids. Upon warming the sulphuric acid layer at the bottom of the tube very gently (**CAUTION**), the blue colour spreads downwards from the interface and eventually colours the whole of the sulphuric acid layer.

Citrates do not interfere. In the presence of tartrates, a blue ring is obtained in the cold or upon very gently warming (compare similar test under tartrates in Section 4.38).

MnO_2 + acid
↓
red complex

6. *Manganese(II) sulphate test* a solution of manganese(II) sulphate is treated with sodium hydroxide and the resulting mixture warmed gently to oxidize the precipitate by atmospheric oxygen to hydrated manganese dioxide (cf. Section 3.28, reaction 1). After cooling, the solution of the oxalate, acidified with dilute sulphuric acid, is added. The precipitate dissolves and a red colour is produced, which is probably due to the formation of trioxalato-manganate(III) complex ion:

$$7(COO)_2^{2-} + 2MnO(OH)_2\downarrow + 8H^+ \rightarrow$$
$$\rightarrow 2\{Mn[(COO)_2]_3\}^{3-} + 2CO_2\uparrow + 6H_2O$$

Heating
↓
decomposition

7. *Action of heat* all oxalates decompose upon ignition. Those of the alkali metals and of the alkaline earths yield chiefly the carbonates and carbon monoxide; a little carbon is also formed. The oxalates of the metals whose carbonates are easily decomposed into stable oxides, are converted into carbon monoxide, carbon dioxide, and the oxide, e.g. magnesium and zinc oxalates. Silver oxalate yields silver and carbon dioxide; silver oxide decomposes on heating. Oxalic acid decomposes into carbon dioxide and formic acid, the latter being further partially decomposed into carbon monoxide and water. Note that carbon monoxide is **poisonous**!

$$7(COONa)_2 \rightarrow 7Na_2CO_3 + 3CO\uparrow + 2CO_2\uparrow + 2C$$
$$(COO)_2Ba \rightarrow BaCO_3 + CO\uparrow$$
$$(COO)_2Mg \rightarrow MgO + CO\uparrow + CO_2\uparrow$$
$$(COOAg)_2 \rightarrow 2Ag + 2CO_2\uparrow$$
$$(COOH)_2 \rightarrow HCOOH + CO_2\uparrow$$
$$HCOOH \rightarrow CO\uparrow + H_2O\uparrow$$

Aniline blue test

8. *Formation of aniline blue test* upon heating insoluble oxalates with concentrated phosphoric acid and diphenylamine, or upon heating together

oxalic acid and diphenylamine, the dyestuff aniline blue (or diphenylamine blue) is formed. Formates, acetates, tartrates, citrates, succinates, benzoates, and salts of other organic acids do not react under these experimental conditions. In the presence of other anions which are precipitated by calcium chloride solution, e.g. tartrate, sulphate, sulphite, phosphate, and fluoride, it is best to heat the precipitate formed by calcium chloride with phosphoric acid as detailed below.

Place a few milligrams of the test sample (or, alternatively, use the residue obtained by evaporating 2 drops of the test solution to dryness) in a micro test-tube, add a little pure diphenylamine and melt over a free flame. When cold, take up the melt in a drop or two of alcohol when the latter will be coloured blue.

Sensitivity 5 µg $(COOH)_2$.

Concentration limit 1 in 10 000.

4.38 Tartrates, $C_4H_4O_6^{2-}$

Solubility tartaric acid, $HOOC \cdot [CH(OH)]_2COOH$ or $H_2 \cdot C_4H_4O_6$, is a crystalline solid, which is extremely soluble in water; it is a dibasic acid. Potassium and ammonium hydrogen tartrates are sparingly soluble in water; those of the other alkali metals are readily soluble. The normal tartrates of the alkali metals are easily soluble (those of the other metals being sparingly soluble in water) but dissolve in solutions of alkali tartrates forming complex salts, which often do not give the typical reactions of the metals present.

$C_4H_4O_6^{2-}$ reactions

To study these reactions use a 0.1M solution of tartaric acid, $H_2 \cdot C_4H_4O_6$, or a 0.1M solution of potassium sodium tartrate, $KNa \cdot C_4H_4O_6 \cdot 4H_2O$. The latter solution is neutral.

cc H_2SO_4
↓
decomposition

1. *Concentrated sulphuric acid* when a solid tartrate is heated with concentrated sulphuric acid, it is decomposed in a complex manner. Charring occurs almost immediately (owing to the separation of carbon), carbon dioxide and carbon monoxide are evolved, together with some sulphur dioxide; the last-named probably arises from the reduction of the sulphuric acid by the carbon. An odour, reminiscent of burnt sugar, can be detected in the evolved gases. Dilute sulphuric acid has no visible action.

$$H_2 \cdot C_4H_4O_6 \rightarrow CO\uparrow + CO_2\uparrow + 2C + 3H_2O\uparrow$$

$$C + 2H_2SO_4 \rightarrow 2SO_2\uparrow + CO_2\uparrow + 2H_2O\uparrow$$

$AgNO_3$
↓
white $Ag_2C_4H_4O_6$

2. *Silver nitrate solution* white, curdy precipitate of silver tartrate, $Ag_2 \cdot C_4H_4O_6$, from neutral solutions of tartrates, soluble in excess of the tartrate solution, in dilute ammonia solution and in dilute nitric acid. On warming the ammoniacal solution, metallic silver is precipitated; this can be deposited in the form of a mirror under suitable conditions.

$$C_4H_4O_6^{2-} + 2Ag^+ \rightarrow Ag_2 \cdot C_4H_4O_6\downarrow$$

$$Ag_2 \cdot C_4H_4O_6\downarrow + 4NH_3 \rightarrow 2[Ag(NH_3)_2]^+ + C_4H_4O_6^{2-}$$

The silver mirror test is best performed as follows. The solution of the tartrate is acidified with dilute nitric acid, excess of silver nitrate solution added and any precipitate present filtered off. Very dilute ammonia solution (approximately 0.02M) is then added to the solution until the precipitate at first formed is nearly redissolved. The solution is filtered, and the filtrate collected in a clean test-tube; the latter is then placed in a beaker of boiling water. A brilliant mirror is formed on the sides of the tube after a few minutes. The mirror from the test-tube may be removed by dissolving it with 8M nitric acid.

The solution obtained after dissolving in ammonia should be discarded quickly to avoid a serious **explosion!** (Cf. Section 3.6, reaction 1.)

$CaCl_2$
↓
white $CaC_4H_4O_6$

3. *Calcium chloride solution* white, crystalline precipitate of calcium tartrate, $Ca \cdot C_4H_4O_6$, with a concentrated neutral solution. The precipitate is soluble in dilute acetic acid (difference from oxalate), in dilute mineral acids, and in cold alkali solutions. An excess of the reagent must be added because calcium tartrate dissolves in an excess of an alkali tartrate solution forming a complex tartrate ion, $[Ca(C_4H_4O_6)_2]^{2-}$. Precipitation is slow in dilute solutions, but may be accelerated by vigorous shaking or by rubbing the walls of the test-tube with a glass rod.

$$C_4H_4O_6^{2-} + Ca^{2+} \rightarrow Ca \cdot C_4H_4O_6\downarrow$$

KCl
↓
white $KHC_4H_4O_6$

4. *Potassium chloride solution* when a concentrated neutral solution of a tartrate is treated with a solution of a potassium salt (e.g. M potassium chloride or potassium acetate) and then acidified with acetic acid, a colourless crystalline precipitate of potassium hydrogen tartrate (winestone), $KH \cdot C_4H_4O_6$, is obtained. The precipitate forms slowly in dilute solutions; crystallization is induced by vigorous shaking or by rubbing the walls of the vessel with a glass rod.

$$C_4H_4O_6^{2-} + K^+ + CH_3COOH \rightarrow KH \cdot C_4H_4O_6\downarrow + CH_3COO^-$$

$FeSO_4 + H_2O_2$
↓
violet colour

5. *Fenton's test* add 1 drop of a saturated solution of iron(II) sulphate to about 5 ml of the neutral or acid solution of the tartrate, and then 2–3 drops of hydrogen peroxide solution (3%). A deep-violet or blue colouration is developed on adding excess sodium hydroxide solution. The colour becomes more intense upon the addition of a drop of iron(III) chloride solution.

The violet colour is due to the formation of a salt of dihydroxymaleic acid, $CO_2H \cdot C(OH){=}C(OH) \cdot CO_2H$.

The test is not given by citrates, malates or succinates.

Resorcinol test

6. *Resorcinol test* place a few drops of the test solution in a test-tube; add several drops of dilute sulphuric acid and a speck or two of magnesium powder. When the metal has dissolved, add about 0.1 g resorcinol, and shake until dissolved. Cool. Carefully pour down the side of the tube 3–4 ml concentrated sulphuric acid. Upon warming the sulphuric acid layer at the bottom of the tube very gently (**CAUTION**), a red layer (or ring) forms at the junction of the two liquids. With continued gentle heating,

the red colour spreads downwards from the interface and eventually colours the whole of the sulphuric acid layer.

Citrates do not interfere. In the presence of oxalate, a blue ring is formed in the cold and on gentle warming of the sulphuric acid layer at the bottom of the tube the blue colouration spreads downwards into the concentrated acid layer and a red ring forms at the interface.

The colour is due to the formation of a condensation product of resorcinol, $C_6H_4(OH)_2$, and glycollic aldehyde, $CH_2OH \cdot CHO$, the latter arising from the action of the sulphuric acid upon the tartaric acid. The formula of the condensation product is $CH_2OH \cdot CH[C_6H_3(OH)_2]_2$.

$Cu(OH)_2$ + alkali
↓
dissolution
+
dark blue colour

7. *Copper hydroxide test* tartrates dissolve copper hydroxide in the presence of excess alkali hydroxide solution to form the dark-blue ditartrato-cuprate(II) ion, $[Cu(C_4H_4O_6)_2]^{2-}$, which is best detected by filtering the solution. If only small quantities of tartrate are present, the filtrate should be acidified with acetic acid and tested for copper by the potassium hexacyanoferrate(II) test. Arsenites, ammonium salts, and organic compounds convertible into tartaric acid also give a blue colouration, and should therefore be absent or be removed.

The experimental details are as follows. Treat the test solution with an equal volume of 2M sodium or potassium hydroxide (or treat the test solid directly with a few ml of the alkali hydroxide solution, warm for a few minutes with stirring and then cool). Add a few drops of 0.25M copper sulphate solution (i.e. just sufficient to yield a visible precipitate of copper(II) hydroxide), shake the mixture vigorously for 5 minutes, and filter. If the filtrate is not clear, warm to coagulate the colloidal copper hydroxide and filter again. A distinct blue colouration indicates the presence of a tartrate. If a pale colouration is obtained, it is advisable to add concentrated ammonia solution dropwise when the blue colour intensifies; it is perhaps better to acidify with 2M acetic acid and add potassium hexacyanoferrate(II) solution to the clear solution whereupon a reddish-brown precipitate, or (with a trace of a tartrate) a red colouration, is obtained.

1,1-Bi-2-naphthol
↓
green colour

8. *1,1-Bi-2-naphthol test*

when a solution of 1,1-bi-2-naphthol in concentrated sulphuric acid is heated with tartaric acid, a green colouration is obtained. Oxalic, citric, succinic, and cinnamic acids do not interfere.

Heat a few milligrams of the solid test sample or a drop of the test solution with 1–2 ml of the dinaphthol reagent in a water bath at 85°C for 30 minutes. A luminous green fluorescence appears during the heating; it intensifies on cooling, and at the same time the violet fluorescence of the reagent disappears.

Sensitivity 10 µg tartaric acid.
Concentration limit 1 in 5000.

9. *Action of heat* the tartrates and also tartaric acid decompose in a complex manner on heating; charring takes place, a smell of burnt sugar is apparent and inflammable vapours are evolved.

4.39 Citrates, $C_6H_5O_7^{3-}$

Solubility citric acid, $HOOC \cdot CH_2 \cdot C(OH)CO_2H \cdot CH_2 \cdot COOH \cdot H_2O$ or $H_3 \cdot C_6H_5O_7 \cdot H_2O$, is a crystalline solid which is very soluble in water; it becomes anhydrous at 55°C and melts at 160°C. It is a tribasic acid, and therefore gives rise to three series of salts. The normal citrates of the alkali metals dissolve readily in water; other metallic citrates are sparingly soluble. The acid citrates are more soluble than the acid tartrates.

$C_6H_5O_7^{3-}$ reactions

To study these reactions use a 0.5M solution of sodium citrate, $Na_3 \cdot C_6H_5O_7 \cdot 2H_2O$.

cc H_2SO_4 + heating
↓
decomposition

1. *Concentrated sulphuric acid* when a solid citrate is heated with concentrated sulphuric acid, carbon monoxide (**POISON**) and carbon dioxide are evolved; the solution slowly darkens owing to the separation of carbon, and sulphur dioxide is evolved (compare tartrates, Section 4.38, which char almost immediately).

The first products of the reaction are carbon monoxide and acetone dicarboxylic acid (1); the latter undergoes further partial decomposition into acetone (2) and carbon dioxide.

$$HOOC \cdot CH_2 \cdot C(OH)CO_2H \cdot CH_2 \cdot COOH \rightarrow$$

$$\rightarrow CO\uparrow + H_2O\uparrow + HOOC \cdot CH_2 \cdot CO \cdot CH_2 \cdot COOH;$$
(1)

$$HOOC \cdot CH_2CO \cdot CH_2 \cdot COOH \rightarrow$$

$$\rightarrow CH_3 \cdot CO \cdot CH_3\uparrow + 2CO_2\uparrow$$
(2)

Use may be made of the intermediate formation of acetone dicarboxylic acid and of the interaction of the latter with sodium nitroprusside solution to yield a red colouration as a test for citrates. When about 0.5 g of a citrate or of citric acid is treated with 1 ml concentrated sulphuric acid for 1 minute, the mixture cooled, cautiously diluted with water, rendered alkaline with sodium hydroxide solution and then a few millilitres of a freshly prepared solution of sodium nitroprusside added, an intense red colouration results.

AgNO₃
↓
white Ag₃C₆H₅O₇

2. *Silver nitrate solution* white, curdy precipitate of silver citrate, $Ag_3 \cdot C_6H_5O_7$, from neutral solutions. The precipitate is soluble in dilute ammonia solution, and this solution undergoes only very slight reduction to silver on boiling (distinction from tartrate).

$$C_6H_5O_7^{3-} + 3Ag^+ \rightarrow Ag_3 \cdot C_6H_5O_7\downarrow$$

$$Ag_3 \cdot C_6H_5O_7\downarrow + 6NH_3 \rightarrow 3[Ag(NH_3)_2]^+ + C_6H_5O_7^{3-}$$

The solution obtained after dissolving in ammonia should be discarded quickly to avoid a serious **explosion!** (Cf. Section 3.6, reaction 1.)

$CaCl_2$ + heat
↓
white ppt

3. *Calcium chloride solution* no precipitate with neutral solutions in the cold (difference from tartrate), but on boiling for several minutes a crystalline precipitate of calcium citrate, $Ca_3(C_6H_5O_7)_2 \cdot 4H_2O$, is produced. If sodium hydroxide solution is added to the cold solution containing excess calcium chloride, there results an immediate precipitation of amorphous calcium citrate, insoluble in solutions of caustic alkalis, but soluble in ammonium chloride solution. On boiling the ammonium chloride solution, crystalline calcium citrate is precipitated, which is now insoluble in ammonium chloride.

$$2C_6H_5O_7^{3-} + 3Ca^{2+} \rightarrow Ca_3(C_6H_5O_7)_2\downarrow$$

$Cd(CH_3COO)_2$
↓
white $Cd_3(C_6H_5O_7)_2$

4. *Cadmium acetate solution* white, gelatinous precipitate of cadmium citrate $Cd_3(C_6H_5O_7)_2$, practically insoluble in boiling water, but readily soluble in warm acetic acid (tartrate gives no precipitate).

$$2C_6H_5O_7^{3-} + 3Cd^{2+} \rightarrow Cd_3(C_6H_5O_7)_2\downarrow$$

$HgSO_4$ + $KMnO_4$
↓
white ppt
+ decolourization

5. *Denigés's test* add 0.5 ml acidified mercury(II) sulphate solution (**POISON**) to 3 ml citrate solution, heat to boiling and then add a few drops of 0.02M solution of potassium permanganate. Decolourization of the permanganate will take place rapidly and then, somewhat suddenly, a heavy white precipitate, consisting of the double salt of basic mercury(II) sulphate with mercury(II) acetone dicarboxylate

$$HgSO_4 \cdot 2HgO \cdot 2[CO(CH_2 \cdot CO_2)_2Hg]\downarrow$$

is formed. Salts of the halogen acids interfere and must therefore be removed before carrying out the test.

Heating
↓
decomposition

6. *Action of heat* citrates and citric acid char on heating; carbon monoxide, carbon dioxide, and acrid-smelling vapours are evolved.

4.40 Salicylates, $C_6H_4(OH)COO^-$ or $C_7H_5O_3^-$

Solubility salicylic acid, $C_6H_4(OH)COOH$ (*o*-hydroxybenzoic acid), forms colourless needles, which melt at 155°C. The acid is sparingly soluble in cold water, but more soluble in hot water, from which it can be recrystallized. It is readily soluble in alcohol and ether. With the exception of the lead, mercury, silver, and barium salts, the monobasic salts, $C_6H_4(OH)COOM$ – these are the most commonly occurring salts – are readily soluble in cold water.

$C_7H_5O_3^-$ reactions

To study these reactions use a 0.5M solution of sodium salicylate, $C_6H_4(OH)COONa$ or $Na \cdot C_7H_5O_3$.

cc H_2SO_4
↓
decomposition

1. *Concentrated sulphuric acid* solid sodium salicylate dissolves on warming; charring occurs slowly, accompanied by the evolution of carbon monoxide and sulphur dioxide.

cc $H_2SO_4 + CH_3OH$
↓
methyl salicylate,
pleasant odour

2. *Concentrated sulphuric acid and methanol ('oil of wintergreen' test)* when 0.5 g of a salicylate or of salicylic acid is treated with a mixture of 1.5 ml concentrated sulphuric acid and 3 ml methanol and the whole gently warmed, the characteristic, fragrant odour of the ester, methyl salicylate(I) ('oil of wintergreen'), is obtained. The odour is readily detected by pouring the mixture into dilute sodium carbonate solution contained in a porcelain dish.

$$C_6H_4(OH)COOH + CH_3OH \rightarrow C_6H_4(OH)COO \cdot CH_3 \uparrow (I) + H_2O$$

(The sulphuric acid acts as a dehydrating agent.)

HCl
↓
white salicylic acid

3. *Dilute hydrochloric acid* crystalline precipitate of salicylic acid from solutions of the salts. The precipitate is moderately soluble in hot water, from which it crystallizes on cooling.

$AgNO_3$
↓
white ppt

4. *Silver nitrate solution* heavy, crystalline precipitate of silver salicylate, $Ag \cdot C_7H_5O_3$, from neutral solutions; it is soluble in boiling water and crystallizes from the solution on cooling.

$$C_6H_4(OH)COO^- + Ag^+ \rightarrow C_6H_4(OH)COOAg \downarrow$$

$FeCl_3$
↓
violet colour

5. *Iron(III) chloride solution* intense violet-red colouration with neutral solutions of salicylates or with free salicylic acid; the colour disappears upon the addition of dilute mineral acids, but not of a little acetic acid. The presence of a large excess of many organic acids (acetic, tartaric, and citric) prevents the development of the colour; but the addition of a few drops of dilute ammonia solution will cause it to appear.

HNO_3
↓
white ppt

6. *Dilute nitric acid* when a salicylate or the free acid is boiled with dilute nitric acid (2M) and the mixture poured into 4 times its volume of cold water, a crystalline precipitate of 5-nitrosalicylic acid is obtained. The precipitate is filtered off and recrystallized from boiling water; the acid, after drying, has a melting point of 226°C.

$$C_6H_4(OH)COOH + HNO_3 \rightarrow C_6H_3(NO_2)(OH)COOH \downarrow + H_2O$$

Polynitrosalicylic acids are formed when the concentration of nitric acid exceeds 2–3M. **NITROSALICYLIC ACIDS MAY EXPLODE ON STRONG HEATING.** This must be borne in mind when removing salicylates by oxidation with nitric acid and evaporating to dryness prior to the precipitation of Group IIIA. If salicylate is suspected, it is probably best to remove it by acidifying the concentrated solution with dilute hydrochloric acid: most of the salicylic acid will separate out upon cooling.

Soda lime
↓
phenol

7. *Soda lime* when salicylic acid or one of its salts is heated with excess of soda lime in an ignition tube, phenol, C_6H_5OH, is evolved, which may be recognized by its characteristic odour.

$$C_6H_4(OH)COONa + NaOH \rightarrow C_6H_5OH \uparrow + Na_2CO_3$$

Heating
↓
sublimation

8. *Action of heat* salicylic acid, when gradually heated above its melting point, sublimes. If it is heated rapidly, it is decomposed into carbon dioxide and phenol. Salicylates char on heating and phenol is evolved.

$$C_6H_4(OH)COOH \rightarrow C_6H_5OH \uparrow + CO_2 \uparrow$$

4.41 Benzoates, $C_6H_5COO^-$ or $C_7H_5O_2^-$

Solubility benzoic acid, C_6H_5COOH or $H \cdot C_7H_5O_2$, is a white crystalline solid; it has a melting point of 121°C. The acid is sparingly soluble in cold water, but more soluble in hot water, from which it crystallizes out on cooling; it is soluble in alcohol and in ether. All the benzoates, with the exception of the silver and basic iron(III) salts, are readily soluble in cold water.

$C_6H_5COO^-$ reactions

To study these reactions use a 0.5M solution of potassium benzoate, C_6H_5COOK or $K \cdot C_7H_5O_2$.

1. *Concentrated sulphuric acid* no charring occurs on heating solid benzoic acid with this reagent; the acid forms a sublimate on the sides of the test-tube and irritating fumes are evolved.

H_2SO_4
↓
white C_6H_5COOH

2. *Dilute sulphuric acid* white, crystalline precipitate of benzoic acid from cold solutions of benzoates. The acid may be filtered off, dried between filter paper or upon a porous tile and identified by means of its melting point (121°C). If the latter is somewhat low, it may be recrystallized from hot water.

$$C_6H_5COO^- + H^+ \rightarrow C_6H_5COOH\downarrow$$

cc $H_2SO_4 + C_2H_5OH$
↓
ethyl benzoate,
pleasant odour

3. *Concentrated sulphuric acid and ethanol* heat 0.5 g of a benzoate or of benzoic acid with a mixture of 1.5 ml concentrated sulphuric acid and 3 ml ethanol for a few minutes. Allow the mixture to cool, and note the pleasant and characteristic aromatic odour of the ester, ethyl benzoate. The odour is more apparent if the mixture is poured into dilute sodium carbonate solution contained in an evaporating basin; oily drops of ethyl benzoate will separate out.

$$C_6H_5 \cdot COOH + C_2H_5OH \rightarrow C_6H_5 \cdot COOC_2H_5\uparrow + H_2O$$

AgNO$_3$
↓
white AgC_6H_5COO

4. *Silver nitrate solution* white precipitate of silver benzoate, $Ag \cdot C_7H_5O_2$, from neutral solutions. The precipitate is soluble in hot water and crystallizes from the solution on cooling; it is also soluble in dilute ammonia solution.

$$C_6H_5 \cdot COO^- + Ag^+ \rightarrow C_6H_5 \cdot COOAg\downarrow$$

$$C_6H_5 \cdot COOAg\downarrow + 2NH_3 \rightarrow [Ag(NH_3)_2]^+ + C_6H_5 \cdot COO^-$$

The solution obtained after dissolving in ammonia should be discarded quickly to avoid a serious **explosion!** (Cf. Section 3.6, reaction 1.)

FeCl$_3$
↓
buff ppt

5. *Iron(III) chloride solution* buff-coloured precipitate of basic iron(III) benzoate from neutral solutions. The precipitate is soluble in hydrochloric acid, with the simultaneous separation of benzoic acid (see reaction 2).

$$3C_6H_5COO^- + 2Fe^{3+} + 3H_2O \rightarrow (C_6H_5COO)_3Fe \cdot Fe(OH)_3\downarrow + 3H^+$$

Soda lime
↓
benzene vapours

6. *Soda lime* benzoic acid and benzoates, when heated in an ignition tube with excess soda lime, are decomposed into benzene, C_6H_6, which burns with a smoky flame, and carbon dioxide, the latter combining with the alkali present.

$$C_6H_5 \cdot COOH + 2NaOH \rightarrow C_6H_6\uparrow + Na_2CO_3 + H_2O\uparrow$$

Heating
↓
decomposition

7. *Action of heat* when benzoic acid is heated in an ignition tube, it melts, evaporates, and condenses in the cool parts of the tube. An irritating vapour is simultaneously evolved, but no charring occurs. If a little of the acid or of one of its salts is heated upon platinum foil or broken porcelain it burns with a blue, smoky flame (distinction from succinate).

4.42 Succinates, $C_4H_4O_4^{2-}$

Solubility succinic acid, $HOOC \cdot CH_2 \cdot CH_2 \cdot COOH$ or $H_2 \cdot C_4H_4O_4$ (a dibasic acid), is a white, crystalline solid with a melting point of 182°C; it boils at 235°C with the formation of succinic anhydride, $C_4H_4O_3$, by the loss of one molecule of water. The acid is fairly soluble in water ($68.4\,g\,l^{-1}$ at 20°C), but more soluble in hot water; it is moderately soluble in alcohol and in acetone, slightly soluble in ether, and sparingly soluble in chloroform. Most normal succinates are soluble in cold water; the silver, calcium, barium, and basic iron(III) salts are sparingly soluble.

$C_4H_4O_4^{2-}$ reactions

To study these reactions use a 0.5M solution of sodium succinate, $C_2H_4(COONa)_2 \cdot 6H_2O$ or $Na_2 \cdot C_4H_4O_4 \cdot 6H_2O$.

cc H_2SO_4 + heat
↓
decomposition

1. *Concentrated sulphuric acid* the acid or its salts dissolve in warm, concentrated sulphuric acid without charring; if the solution is strongly heated, slight charring takes place and sulphur dioxide is evolved.

Dilute sulphuric acid has no visible action.

AgNO$_3$
↓
white Ag$_2$C$_4$H$_4$O$_4$

2. *Silver nitrate solution* white precipitate of silver succinate, $Ag_2 \cdot C_4H_4O_4$, from neutral solutions, readily soluble in dilute ammonia solution.

$$C_2H_4(COO)_2^{2-} + 2Ag^+ \rightarrow C_2H_4(COOAg)_2\downarrow$$

$$C_2H_4(COOAg)_2\downarrow + 4NH_3 \rightarrow 2[Ag(NH_3)_2]^+ + C_2H_4(COO)_2^{2-}$$

The solution obtained after dissolving in ammonia should be discarded quickly to avoid a serious **explosion!** (Cf. Section 3.6, reaction 1.)

FeCl$_3$
↓
brown ppt

3. *Iron(III) chloride solution* light-brown precipitate of basic iron(III) succinate with a neutral solution; some free succinic acid is simultaneously produced, and the solution becomes slightly acidic.

$$3C_2H_4(COO)_2^{2-} + 2Fe^{3+} + 2H_2O \rightarrow$$
$$\rightarrow 2C_2H_4(COO)_2Fe(OH)\downarrow + C_2H_4(COOH)_2$$

BaCl$_2$
↓
white BaC$_4$H$_4$O$_4$

4. *Barium chloride solution* white precipitate of barium succinate from neutral or slightly ammoniacal solutions (distinction from benzoate). Precipitation is slow from dilute solutions, but may be accelerated by

vigorous shaking or by rubbing the inner wall of the test-tube with a glass rod.

$$C_2H_4(COO)_2^{2-} + Ba^{2+} \rightarrow C_2H_4(COO)_2Ba\downarrow$$

CaCl$_2$
↓
white CaC$_4$H$_4$O$_4$

5. *Calcium chloride solution* a precipitate of calcium succinate is very slowly produced from concentrated neutral solutions.

$$C_2H_4(COO)_2^{2-} + Ca^{2+} \rightarrow C_2H_4(COO)_2Ca\downarrow$$

Resorcinol + cc H$_2$SO$_4$
↓
green fluorescence

6. *Fluorescein test* about 0.5 g of the acid or one of its salts is mixed with 1 g resorcinol, a few drops of concentrated sulphuric acid added, and the mixture gently heated; a deep-red solution is formed. On pouring the latter into a large volume of water, an orange-yellow solution is obtained, which exhibits an intense green fluorescence. The addition of excess sodium hydroxide solution intensifies the fluorescence, and the colour of the solution changes to bright red.

Under the influence of concentrated sulphuric acid, succinic anhydride (1) is first formed, and this condenses with the resorcinol (2) to yield succinyl-fluorescein (3).

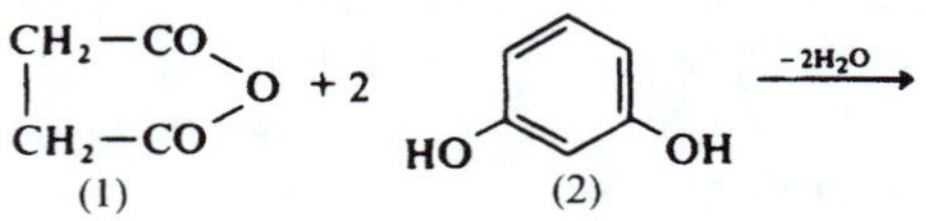

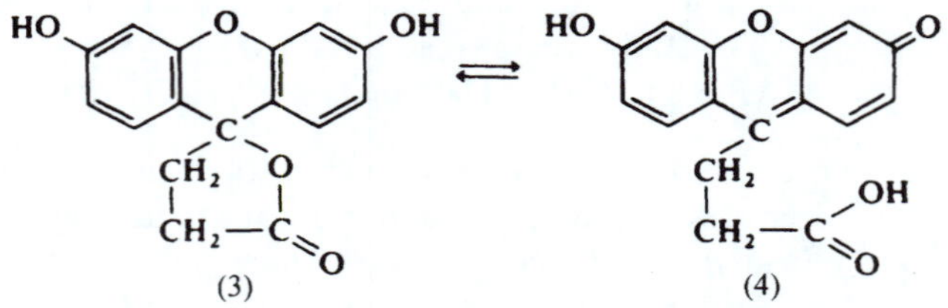

The quinoid structure (4) is predominant in alkaline solutions.

Heat
↓
decomposition

7. *Action of heat* when succinic acid or its salts are strongly heated in an ignition tube, a white sublimate of succinic anhydride, C$_4$H$_4$O$_3$, is formed and irritating vapours are evolved. If the ignition is carried out on platinum foil or upon broken porcelain, the vapour burns with a non-luminous, blue flame, leaving a residue of carbon (distinction from benzoate).

4.43 Hydrogen peroxide, H$_2$O$_2$ (the perhydroxyl ion, OOH$^-$)

Hydrogen peroxide is marketed as '20 volumes' (6%), '40 volumes' (12%) and '100 volumes' (30%) solutions. It decomposes slowly in cold, but rapidly on heating, to water and oxygen gas:

$$2H_2O_2 \rightarrow 2H_2O + O_2\uparrow$$

H$_2$O$_2$ and OOH$^-$
reactions

To study its reactions use a 3% (approximately M) solution of hydrogen peroxide. Alternatively, a molar solution of sodium peroxoborate NaBO$_3 \cdot 4H_2O$ can be used, which hydrolyses rapidly to form hydrogen peroxide:

$$BO_3^- + H_2O \rightarrow H_2O_2 + BO_2^-$$

The metaborate ion which is formed does not interfere with the reactions described below.

K$_2$Cr$_2$O$_7$ + ether
↓
blue CrO$_5$ solution in the organic layer

1. *Chromium pentoxide test*   if an acidified solution of hydrogen peroxide is mixed with a little diethyl ether (highly inflammable), amyl alcohol or amyl acetate and a few drops of potassium dichromate solution added and the mixture gently shaken, the upper organic layer is coloured a beautiful blue. For further details, see Section 4.33, reaction 4, and Section 3.24, reaction 6b. The test will detect 0.1 mg of hydrogen peroxide.

If ether is employed, a blank test must always be carried out with the ether alone because after standing it may contain organic peroxides which give the test. The peroxides may be removed by shaking 1 l of ether with a solution containing 60 g FeSO$_4 \cdot$7H$_2$O, 6 ml concentrated H$_2$SO$_4$, and 110 ml water, then with chromic acid solution (to oxidize any acetaldehyde produced), followed by washing with alkali and redistilling. The peroxides may also be removed by treatment with aqueous sodium sulphite. The use of amyl alcohol is, however, to be preferred.

KI + starch
↓
oxidation to blue complex

2. *Potassium iodide and starch*   if potassium iodide and starch are added to hydrogen peroxide, acidified previously with dilute sulphuric acid, iodine is formed slowly and the solution turns gradually to deeper and deeper blue:

$$H_2O_2 + 2H^+ + 3I^- \rightarrow I_3^- + 2H_2O$$

Molybdate ions accelerate this reaction. In the presence of 1 drop of 0.025M (0.4%) ammonium molybdate solution the reaction is instantaneous.

KMnO$_4$ + H$^+$
↓
reduction to colourless Mn^{2+}

3. *Potassium permanganate solution*   decolourized in acid solution and oxygen is evolved:

$$2MnO_4^- + 5H_2O_2 + 6H^+ \rightarrow 2Mn^{2+} + 5O_2\uparrow + 8H_2O$$

TiCl$_4$
↓
orange-red per-acid

4. *Titanium(IV) chloride solution*   orange-red colouration (see under titanium, Section 6.17, reaction 6). This is a very sensitive test.

K$_3$Fe(CN)$_6$ + FeCl$_3$
↓
slowly
Prussian blue

5. *Potassium hexacyanoferrate(III) and iron(III) chloride*   to a nearly neutral solution of iron(III) chloride add some potassium hexacyanoferrate(III) solution. A yellow solution is obtained. To this, add a nearly neutral solution of hydrogen peroxide. The solution turns to green and Prussian blue separates slowly.

Hydrogen peroxide reduces hexacyanoferrate(III) ions to hexacyanoferrate(II):

$$2[Fe(CN)_6]^{3-} + H_2O_2 \rightarrow 2[Fe(CN)_6]^{4-} + 2H^+ + O_2\uparrow$$

Prussian blue is then produced from iron(III) and hexacyanoferrate(II) ions:

$$4Fe^{3+} + 3[Fe(CN)_6]^{4-} \rightarrow Fe_4[Fe(CN)_6]_3\downarrow$$

(cf. Section 3.22, reaction 6 and Section 3.21, reaction 7).

Other reducing agents, e.g. tin(II) chloride, sodium sulphite, sodium thiosulphate, etc., will reduce hexacyanoferrate(III) ions to hexacyanoferrate(II), so this reaction is not always a reliable test.

The spot-test technique is as follows. In adjacent depressions of a spot plate place a drop of distilled water and a drop of the test solution. Add a drop of the reagent (equal volumes of 0.5M iron(III) chloride solution and 0.033M potassium hexacyanoferrate(III) solution) to each. A blue colouration or precipitate is formed.

Sensitivity 0.1 µg H_2O_2.

Concentration limit 1 in 600 000.

AuCl₃
↓
reduction to Au

6. *Gold chloride solution* reduced to finely divided metallic gold, which appears greenish-blue by transmitted light and brown by reflected light.

$$2Au^{3+} + 3H_2O_2 \rightarrow 2Au\downarrow + 3O_2\uparrow + 6H^+$$

The spot-test technique is as follows. Mix a drop of the neutral test solution with a drop of 0.33M gold chloride solution in a micro crucible and warm. After a short time, the solution is coloured red or blue with colloidal gold.

Sensitivity 0.07 µg H_2O_2.

Concentration limit 1 in 700 000.

Black PbS
↓
white PbSO₄

7. *Lead sulphate test* hydrogen peroxide converts black lead sulphide into white lead sulphate:

$$PbS\downarrow + 4H_2O_2 \rightarrow PbSO_4\downarrow + 4H_2O$$

Place a drop of the neutral or slightly acid test solution upon drop-reaction paper impregnated with lead sulphide. A white spot is formed on the brown paper.

Sensitivity 0.04 µg H_2O_2.

Concentration limit 1 in 1 250 000.

The lead sulphide paper is prepared by soaking drop-reaction paper in a 0.0025M solution of lead acetate, exposing it to a little hydrogen sulphide gas and then drying in a vacuum desiccator. The paper will keep in a stoppered bottle.

Lucigenin
↓
green
chemiluminescence

8. *Chemiluminescence test* this test has to be carried out in a dark room.

To 20 ml 2M sodium hydroxide add 5 ml 1% aqueous solution of lucigenin (bis-*N*-methylacridinium nitrate or *N,N'*-dimethyl-9,9'-biacridinium dinitrate). Switch off the light. Add 5 ml 3% hydrogen peroxide solution and mix. A steady greenish-yellow light is emitted; the phenomenon lasts for about 15 minutes.

To one-half of the solution add 1 drop of 1% osmium tetroxide solution and mix. The light flares up, turns to bluish-violet, then quickly fades away.

For the explanation of these phenomena consult the literature.[14]

---

[14]See L. Erdey in E. Bishop (ed.), *Indicators*, Pergamon Press 1972, p. 713 ff.

4.44 Dithionites, $S_2O_4^{2-}$

Dithionites are obtained by the action of reducing agents, such as zinc, upon hydrogen sulphites

$$4HSO_3^- + Zn \rightarrow S_2O_4^{2-} + 2SO_3^{2-} + Zn^{2+} + 2H_2O$$

Sulphur dioxide may also be passed into a cooled suspension of zinc dust in water:

$$Zn + 2SO_2 \rightarrow Zn^{2+} + S_2O_4^{2-}$$

Zinc ions, which are produced in both processes, can be removed from the solution with sodium carbonate, when zinc carbonate is precipitated. By saturating the solution with sodium chloride, sodium dithionite, $Na_2S_2O_4 \cdot 2H_2O$, is precipitated.

Sodium dithionite is a powerful reducing agent. A solution of the salt, containing an excess of sodium hydroxide, is used as an absorbent for oxygen in gas analysis.

$S_2O_4^{2-}$ reactions

To study these reactions use a freshly prepared 0.5M solution of sodium dithionite, $Na_2S_2O_4$.

H_2SO_4
↓
transitional orange colour

1. *Dilute sulphuric acid* orange colouration, which soon disappears, accompanied by the evolution of sulphur dioxide and the precipitation of sulphur:

$$2S_2O_4^{2-} + 4H^+ \rightarrow 3SO_2\uparrow + S\downarrow + 2H_2O$$

cc H_2SO_4
↓
white S

2. *Concentrated sulphuric acid* immediate evolution of sulphur dioxide and the precipitation of white or pale yellow sulphur:

$$2S_2O_4^{2-} + 2H_2SO_4 \rightarrow 3SO_2\uparrow + S\downarrow + 2SO_4^{2-} + 2H_2O$$

$AgNO_3$
↓
black Ag

3. *Silver nitrate solution* black precipitate of metallic silver:

$$S_2O_4^{2-} + 2Ag^+ + 2H_2O \rightarrow 2Ag\downarrow + 2SO_3^{2-} + 4H^+$$

If the reagent is added in excess, white silver sulphite is also precipitated (cf. Section 4.4, reaction 3).

$CuSO_4$
↓
red Cu

4. *Copper sulphate solution* red precipitate of metallic copper:

$$S_2O_4^{2-} + Cu^{2+} + 2H_2O \rightarrow Cu\downarrow + 2SO_3^{2-} + 4H^+$$

$HgCl_2$
↓
grey Hg

5. *Mercury(II) chloride solution* (**POISON**) grey precipitate of metallic mercury:

$$S_2O_4^{2-} + HgCl_2 + 2H_2O \rightarrow Hg\downarrow + 2SO_3^{2-} + 2Cl^- + 4H^+$$

Methylene blue
↓
decolourization

6. *Methylene blue solution* immediate decolourization in the cold. The methylene blue is reduced to the colourless leuco compound.

Indigo
↓
decolourization

7. *Indigo solution* reduced to the colourless leuco compound, known as indigo white.

$$S_2O_4^{2-} + C_{16}H_{10}O_2N_2 + 2H_2O \rightarrow C_{16}H_{12}O_2N_2 + 2SO_3^{2-} + 2H^+$$

indigo   indigo white

$K_4Fe(CN)_6 + FeSO_4$
↓
white ppt,
turns to
Prussian blue

8. *Potassium hexacyanoferrate(II) and iron(II) sulphate* if potassium hexacyanoferrate(II) and iron(II) sulphate are mixed, white precipitate of dipotassium iron(II) hexacyanoferrate(II) turns quite rapidly into Prussian blue (cf. Section 3.21, reaction 6), because of oxidation caused by dissolved and atmospheric oxygen. In the presence of dithionite the precipitate remains durably white.

Heating
↓
decomposition

9. *Action of heat* upon heating solid sodium dithionite, it suddenly evolves sulphur dioxide at about 190°C (exothermic reaction). The residue is a mixture of sodium sulphite and sodium thiosulphate:

$$2Na_2S_2O_4 \rightarrow Na_2SO_3 + Na_2S_2O_3 + SO_2\uparrow$$

On dissolving the residue, it can no longer reduce indigo or methylene blue solutions.

Chapter 5

Selected tests and separations

5.1 Introduction

Having become familiar with the reactions of cations and anions, the reader should improve his/her skills in qualitative analysis by carrying out special tests and separations.

First the analysis of a dissolved sample, containing *one unknown cation*, should be attempted and, if time allows, repeated with further cations belonging to different groups. Once the cation is found, its presence should be confirmed by other, characteristic tests.

Given enough time, a systematic separation and identification of several cations in mixtures should be tried. It would be futile to start with a mixture containing *all* cations mentioned so far, but perhaps a mixture of five or six ions, taken at random from different groups, should be attempted.

Finally, special tests for individual anions and their mixtures should be carried out, concentrating on combinations which are often encountered in actual samples.

Any analysis of a real sample must begin with a **preliminary examination** of the material, which could be (a) a liquid (usually a solution), (b) a solid, non-metallic substance, (c) a metal or an alloy, or (d) an insoluble material. The description of preliminary tests will be followed by hints on dissolution or fusion, as the main testing and separation has to be carried out in solution.

5.2 Preliminary tests on liquid samples (samples in solution)

1. *Appearance* observe the colour, odour and any special physical properties. The most frequently occurring colours are listed below:

    Blue solutions of copper(II) salts, complex nickel(II) salts
    Green simple nickel(II), iron(II) and chromium(III) salts, manganates
    Yellow chromate, hexacyanoferrate(II) and iron(III) ions
    Orange-red dichromate ions
    Purple permanganate ions
    Pink cobalt(II) and manganese(II) ions

2. *Reaction* test its reaction towards litmus paper or an appropriate, narrow-range indicator paper.

    (a) *The solution is neutral* free acids, free bases, acid salts or salts which give an acid or alkaline reaction, owing to hydrolysis, are absent.

(b) *The solution reacts alkaline* this may be due to hydroxides of the alkali and alkaline earth metals, to the presence of carbonates, borates, sulphides, cyanides, hypochlorites, silicates, peroxo-salts and peroxides of the alkali metals.

(c) *The solution reacts acid* this may be due to the presence of free acids, acid salts, and salts of weak bases, which by hydrolysis produce an acid reaction. It is best to neutralize such a solution with sodium carbonate before testing for anions.

3. *Evaporation* evaporate a known volume of the sample to dryness on a water bath; carefully smell vapours evolved from time to time. If a *solid residue* remains, it can be tested as described in Section 5.3 (if the original solution is acid, it should be neutralized with a few drops of 2M ammonia before evaporation, otherwise some volatile acids may be lost). If the residue is a *liquid*, try to evaporate it at higher temperatures; any solid residue should be examined as already stated. If charring occurs, organic matter is present, which has to be removed in the systematic analysis before attempting the separation of Group IIIA cations. If *no residue* remains, then the liquid contains some volatile substance which may be water, or an aqueous solution of certain gases or volatile substances (like CO_2, NH_3, SO_2, H_2S, HCl, HBr, HI, H_2O_2, $(NH_4)_2CO_3$, etc.) all of which can be detected by special tests.

4. *Tests on alkaline solutions* if the solution has an alkaline reaction, the following tests should be performed:

(i) Test for peroxides and peroxo salts (e.g. H_2O_2, $NaBO_3$).

(a) Heat a small portion of the solution with a few drops of 0.5M cobalt(II) nitrate. In the presence of peroxides and peroxo salts a black precipitate of cobalt(III) hydroxide is formed (sulphides and hypochlorites interfere and must be absent).

(b) Add a few drops of 10% titanium(IV) chloride solution: a yellow colouration occurs in the presence of hydrogen peroxide.

(ii) Test for hydroxides and carbonates.

Boil the solution to decompose hydrogen peroxide (if present). Add a slight excess of 0.25M barium chloride solution; filter off the precipitate. If the solution reacts alkaline, hydroxyl ions are present. Test the precipitate for carbonate (Section 4.2, reaction 1).

5.3 Preliminary tests on non-metallic solid samples

1. *Appearance* the appearance of the substance should be carefully noted: a lens or a microscope should be used if necessary. Observe whether the material is crystalline or amorphous, whether it is magnetic and whether it possesses any characteristic odour or colour.

Colour Some of the commonly occurring coloured compounds are listed below:

Red Pb_3O_4, HgO, HgI_2, HgS, Sb_2S_3, CrO_3, Cu_2O, $K_3[Fe(CN)_6]$
Orange-red dichromates
Purple permanganates, chrome-alum

Pink hydrated manganese(II) and cobalt(II) salts

Yellow S (if pure), CdS, As_3S_3, SnS_2, PbI_2, $K_4[Fe(CN)_6]$, chromates, $FeCl_3$ and $Fe(NO_3)_3$

Green Cr_2O_3, Hg_2I_2, $Cr(OH)_3$, iron(II), nickel(II) and chromium(III) salts, $CuCl_2 \cdot 2H_2O$, $CuCO_3$, K_2MnO_4

Blue Hydrated copper(II) salts, anhydrous cobalt(II) salts, $Fe_4[Fe(CN)_6]$ (Prussian blue)

Brown PbO_2, Ag_3AsO_4, SnS, Fe_2O_3 and $Fe(OH)_3$

Black PbS, CuS, CuO, HgS, FeS, MnO_2, Co_3O_4, CoS, NiS, Ni_2O_3, Ag_2S, C

2. *Dry tests* carry out the tests described in Section 2.2, observe and evaluate results.

Fig. 5.1

3. *Test for ammonium ions* take 1 ml test solution (or 0.1 g solid material), and boil it gently with an excess of 2M sodium hydroxide solution. The evolution of ammonia can be detected by its odour and its action upon red litmus paper. Great care must be taken when heating the mixture because of the destructive effects of hot alkalis upon the eyes and skin. It is best to use the apparatus shown in Fig. 5.1. An all-glass apparatus can be constructed from Quickfit test-tube MF 24/2 and joint CBB 19. The test paper is supported on the end of the wide glass tube. If ammonia is present, the litmus paper should show a gradual development of blue colour from the bottom upwards, and should eventually become uniform in colour; scattered blue spots indicate that droplets of the alkaline solution have come into contact with the paper. The spray may be trapped, if desired, by a loosely fitting plug of cotton wool inserted in the upper part of the test-tube.

Alternatively, the ammonia gas formed in this reaction can be absorbed in 3–4 ml of water or 2M HCl, using the apparatus shown in Fig. 4.1. Care must be taken that no 'sucking back' should occur during the process. In the cold solution ammonium ions can be identified through a precipitate caused by Nessler's reagent (Section 3.38, reaction 2).

Test for NO_2^- and NO_3^-

4. *Test for nitrate and/or nitrite* if ammonium ions were found in the previous test, boil the mixture until ammonia can no longer be detected. Add 0.1 g aluminium powder (or zinc dust, or finely powdered Devarda's alloy) to the cool solution, and warm gently. Test for ammonia as described above. A positive test indicates the presence of nitrate and/or nitrite. Nitrite alone can be detected by the test with dilute sulphuric acid (see in this section under 6). Cyanides, thiocyanates, hexacyanoferrate(II)s and hexacyanoferrate(III)s also yield ammonia, but the reaction is much slower.

Test for BO_3^{3-}

5. *Test for borate* mix the original dry solid sample with calcium fluoride, and by adding a few drops of concentrated sulphuric acid make a thick-flowing paste. Hold some of this in a platinum loop just outside the base of a Bunsen flame. A green colouration, due to boron trifluoride, indicates the presence of borate.

Test with H_2SO_4

6. *Test with dilute sulphuric acid* treat about 0.1 g of the substance in a small test-tube with 2 ml M sulphuric acid, and note whether any reaction

takes place in the cold (indicated in the list below by *C*). Warm gently and observe the effect. Most common effects are listed below:

(a) A colourless gas is evolved with effervescence; gas is odourless and produces turbidity when passed into lime water (*C*). The gas is CO_2 which originates from carbonates. Some carbonate rocks (magnesite, dolomite) react only when the acid is warmed.

(b) Reddish-brown nitrous fumes appear, especially when observed at the mouth of the tube, turning starch–potassium iodide paper bluish-black. The gas is NO_2 originating from nitrite.

(c) Yellowish-green chlorine gas evolves from hypochlorites. The gas can be identified by starch paper, treated with 1–2 drops of 0.1M potassium iodide. A bluish-black colour, caused by iodine, indicates the presence of chlorine in the gas.

(d) Acetylene is produced from carbides; it can be identified by its characteristic odour. It burns with a luminous, smoky flame.

(e) Colourless SO_2 is formed from sulphites. It has a suffocating odour, turning filter paper moistened with acidified potassium dichromate green. If a white precipitate is also formed, thiosulphates are present. Thiocyanates also release SO_2, but only when the solution is boiled for a while.

(f) Colourless H_2S is produced from sulphides; the gas has the distinct odour of rotten eggs. The gas turns filter paper moistened with cadmium acetate solution yellow. If polysulphides are present, a white precipitate is also formed. **THE GAS IS POISONOUS!**

(g) Acetic acid, with the smell of vinegar, is produced from acetates.

(h) Colourless HCN gas with the odour of bitter almond originates from cyanides. **THE GAS IS HIGHLY POISONOUS!**

(i) Colourless gas with a pungent odour can be a mixture of CO_2 and HCHO originating from cyanate.

(j) Colourless O_2 gas, which rekindles a glowing splint, is produced from peroxides, peroxo salts of alkali metals.

5.4 Preliminary tests on metal samples

The sample of the metal or alloy should be in the form of drillings, turnings or filings. Its analysis is simplified by the fact that no anions need to be looked for. Many alloys contain small amounts of C, P, S and Si. Of these, phosphorus is converted to phosphate on dissolution in dilute nitric acid, and can be identified with an appropriate test.

Treat about 0.5 g of the sample with 10 ml 8M nitric acid in a porcelain dish in a fume cupboard, first at room temperature, then by heating on a water bath. When the evolution of red fumes ceases, evaporate the mixture almost to dryness. Add 10 ml water, heat and stir the mixture for a few minutes, and filter if necessary. There are three possible outcomes:

1. *The metal or alloy dissolves completely* the sample does not contain noble metals such as gold or platinum. The solution can be used for the separation and detection of cations, as described under Section 5.7.

2. *The metal or alloy does not dissolve completely*   analyse the filtrate as described in Section 5.7. Examine the insoluble residue according to Section 5.6.

3. *The metal or alloy is unattacked*   if the alloy is not attacked by 8M nitric acid, treat a separate 0.5 g sample with 20 ml aqua regia in a porcelain dish, covered with a watch-glass. Heat the mixture gently (in a fume cupboard) until the alloy has disintegrated completely. Raise the watch-glass and evaporate on a water bath to dryness. Add 5 ml concentrated hydrochloric acid, heat gently, dilute with 15 ml water, stir and heat to boiling. Cool to room temperature and filter. The residue may consist of AgCl, $PbCl_2$ and SiO_2. The filtrate may contain the noble metals, together with arsenate, phosphate and sulphate. Metals can be identified after separation, or directly from the solution by atomic emission or atomic absorption spectroscopy.

If the alloy resists the action of aqua regia, fuse it with sodium hydroxide pellets in a silver dish or crucible **(CAUTION: USE A FUME CUPBOARD)**. When decomposition is complete, allow to cool, transfer the silver vessel to a beaker and extract the melt with water with gentle heating. Remove the silver vessel from the beaker. Strongly acidify the contents of the beaker with 8M nitric acid, evaporate to dryness on a water bath and proceed as above. Note that the silver vessel will be attacked during this procedure; thus do not test for silver in the melt.

The composition of some common alloys are given below; the chief constituents are listed in their order of mass per cent:

Brasses	Cu, Zn, Sn, Pb	Monel metal	Ni, Cu
Bronzes	Cu, Sn, Zn, Pb	Constantan	Cu, Ni
Phosphor bronzes	Cu, Sn, Pb, P	Nichrome	Ni, Fe, Cr
Solders	Sn, Pb, Bi	Manganin	Cu, Mn, Ni
Pewter	Sn, Sb, Pb, Cu	Wood's alloy	Bi, Pb, Sn, Cd
Type metals	Pb, Sb, Sn	Rose's alloy	Bi, Pb, Sn
German silver	Cu, Ni, Zn	Devarda's alloy	Cu, Al, Zn

5.5   Dissolution of the sample

The preliminary tests, described in the previous sections, have already revealed whether the substance is soluble in water or in acids. If such information is not available, the following procedure should be adopted.

Use small quantities (5–10 mg) of the powdered solid and examine the solubility in the following solvents in the order given: (1) water, (2) 6M hydrochloric acid, (3) concentrated hydrochloric acid, (4) 8M nitric acid, (5) concentrated nitric acid, (6) aqua regia. Assess the solubility first when cold, then on warming. If in doubt whether some of the substance has dissolved, evaporate a little of the clear solution on a watch-glass. If the substance dissolves in water, proceed immediately to test for metal ions. If the use of 6M hydrochloric acid results in the formation of a precipitate, this may consist of metals of Group I; the precipitate may either be filtered off and examined for this group, or the original substance should be dissolved in 8M nitric acid.

If concentrated hydrochloric acid is employed for dissolution, it will be necessary to evaporate most of the acid, since certain metals of Group II (e.g. cadmium and lead) are not completely precipitated with hydrogen sulphide in the presence of large concentrations of acid. When nitric acid has been used for dissolution, all the acid should be removed by evaporation, followed by the addition of small amounts of hydrochloric acid, evaporating again to a small bulk and then diluting with water. The same procedure should be applied with aqua regia.

Once a suitable solvent has been found, prepare a solution for analysis from 0.5–1 g of the solid; the volume of the final solution should be 20–50 ml.

If the substance is insoluble in any of the solvents listed above, follow the procedures described in Section 5.6.

If the sample is a metal, decide on the best solvent after the preliminary tests described in Section 5.4. The insoluble part, if any, should be treated according to Section 5.6.

Unless the sample is readily soluble in water, the solution is unsuitable for the detection of anions, as during dissolution in acids some of these might decompose. In such case a soda-extract must be prepared according to Section 5.9.

5.6 Examination of the insoluble residue

It may well happen that after treating the substance with the solvents as described above, some insoluble material remains or the whole sample may prove to be insoluble. The most common insoluble substances, together with hints for their identification, are described below.

Ag halides

1. *Silver halides* AgCl, AgBr, AgI and related compounds (AgCN). These are white or pale yellow substances and can be dissolved in potassium cyanide (**POISON**) or sodium thiosulphate. In the solution both silver (cf. Section 3.6) and the halides can be identified (see appropriate sections in Chapter 4, as well as Section 5.9, paragraphs 5, 6 and 18).

Sulphates

2. *Certain sulphates* like $SrSO_4$, $BaSO_4$, $PbSO_4$. These white substances can be dissolved in 5% sodium ethylenediamine tetraacetate in the presence of 2M ammonia. Alternatively, they can be partially transformed into carbonates by boiling with saturated sodium carbonate. The precipitate can then be partially decomposed with 2M HCl and the metals can be identified with the usual tests (cf. Section 3.31, reaction 4).

Metal oxides

3. *Refractory metal oxides* like Al_2O_3, Cr_2O_3, Fe_2O_3, SnO_2, the mixed antimony(III) antimony(V) oxide Sb_2O_4, TiO_2, ThO_2, WO_3. These should be finely powdered and mixed with an excess of potassium pyrosulphate $K_2S_2O_7$ and fused in a porcelain crucible. After cooling the melt can be dissolved in warm, 6M hydrochloric acid.

For the fusion it is best to start with placing 5 g analytical grade potassium hydrogen sulphate $KHSO_4$ into a clean porcelain crucible. Heat gently over a Bunsen flame, when the hydrogen sulphate slowly decomposes to the pyrosulphate and water. Once foaming has abated, increase the temperature to red heat. Using forceps, lift the crucible off the flame, and, with careful movements, spread the molten pyrosulphate along the inner walls of the crucible. Allow to cool. Add 50–100 mg of the insoluble material, put a lid on the crucible and heat the mixture again to a reddish-yellowish glow for a prolonged period, as the reactions in the molten phase are rather slow. From time to time examine the molten mixture; if no dark, unreacted particles are to be seen, fusion is complete. Allow to cool, place the crucible, lying on its side, into a small beaker. Pour in 6M hydrochloric acid until the crucible is covered, and heat on a water bath for a few hours. Decant the solution which should contain the metals.

Chromates

4. *Insoluble chromates (minerals)* like $PbCrO_4$ (yellow) and $FeCrO_4$ (green). The lead chromate ('chromium yellow') can be decomposed by gently boiling with 5M sodium hydroxide; both chromium and lead can be detected in the supernatant liquid (cf. Sections 3.4 and 3.24).

Iron chromate (as the mineral *chromite*) has to be powdered extremely finely and then fused in a nickel crucible with sodium peroxide. First melt about 1 g of solid Na_2O_2 in a nickel crucible by very gentle heating, allow to cool and add 10–50 mg of the mineral. Heat the mixture gently until the fusion is complete. Leach the cooled metal with hot water until the mixture dissolves. Remove the crucible and carefully acidify the solution with 6M hydrochloric acid. The metal hydroxides should dissolve completely; the solution is green because of the presence of nickel ions, originating from the walls of the crucible. Iron(III) and chromium can be tested in the solution (cf. Sections 3.22 and 3.24). Examine the crucible and discard it if the walls are worn thin.

Hexacyanoferrate(II)s

5. *Insoluble hexacyanoferrate(II)s* like $Cu_2[Fe(CN)_6]$, $Zn_2[Fe(CN)_6]$, $Fe_4[Fe(CN)_6]_3$ (Prussian blue). These can be best decomposed by prolonged heating with 5M sodium hydroxide. In all cases hexacyanoferrate(II) ions appear in the solution and can be detected (cf. Section 4.11). Zinc ions can also be detected in the solution (cf. Section 3.29). Copper and iron remain as oxide precipitates. After filtration or decantation and washing, these materials will dissolve in hot 6M HCl; from the solution the metals can be identified (cf. Sections 3.10 and 3.22).

CaF$_2$

6. *Calcium fluoride (fluorspar)* CaF_2 is a very stable compound. The easiest way to decompose this is to mix the finely powdered sample with silica (or broken glass), and heat the mixture with concentrated H_2SO_4 in a platinum crucible. Silicon tetrafluoride gas is formed and can be detected (cf. Section 4.17, reaction 1). The residue contains calcium sulphate. This can be extracted with 5% sodium ethylenediamine tetraacetate in the presence of 2M ammonia, and calcium ions can be tested for in the solution (cf. Section 3.33).

SiO$_2$ and silicates

7. *Silica (SiO$_2$) and insoluble silicates* can be fused in a platinum crucible with anhydrous sodium carbonate (Na_2CO_3). Place 5 g Na_2CO_3 in a platinum crucible, heat until the substance melts; spread the molten material on the walls of the crucible by holding it with tongs (preferably equipped with platinum tips) and gentle agitation. After cooling, add 10–50 mg of

the well-powdered sample, and heat again until the molten metal mixture appears to be homogeneous when red hot. Extract the cold metal with hot water, allow the solution to cool and cautiously acidify with 6M hydrochloric acid until the vigorous foaming ceases. All the metals present can be identified in the solution: silica precipitates on the addition of the acid. By filtering and washing the precipitate, it can be mixed with some solid calcium fluoride and decomposed with concentrated H_2SO_4, as described in the previous paragraph. If only silica is to be detected, the original sample can be mixed with calcium fluoride; silica can be detected after heating the mixture with concentrated H_2SO_4.

Carbon

8. *Carbon* sometimes remains after the dissolution of metals (e.g. iron or steel) as a black insoluble substance. When heated in a small porcelain crucible, it will start to glow and will disappear almost completely.

Sulphur

9. *Sulphur*, if present in pure form, is yellow, but even small amounts of contaminations will make it appear dark. When heated in a porcelain crucible, it first melts, then volatilizes and, if present in larger quantities, burns with a pungent odour, because of the formation of sulphur dioxide. On heating sulphur with concentrated nitric acid, it dissolves with the formation of sulphate ions, which can be identified in the solution (cf. Section 4.24).

5.7 Testing for a single cation in solution

The scheme follows the classification of cations into groups, as described in Section 3.1.

1. *Group I cations*   to the solution add an excess of dilute HCl. If there is no change, follow step 2. A white precipitate may contain Pb^{2+}, Hg_2^{2+} or Ag^+. Filter, wash and add NH_3 solution. If the precipitate

Does not change	Pb^{2+} present
Turns black	Hg_2^{2+} present
Dissolves	Ag^+ present

2. *Group IIA cations*   to the acidified solution add H_2S in excess. If there is no change, follow step 4. A precipitate may result if Pb^{2+}, Hg^{2+}, Bi^{3+}, Cu^{2+}, Cd^{2+}, As^{3+}, As^{5+}, Sb^{3+}, Sn^{2+} or Sn^{4+} were originally present. Filter the precipitate, wash with dilute HCl, and add an excess of $(NH_4)_2S_x$. If the precipitate dissolves, follow step 3. If the remaining precipitate is

Yellow	Cd^{2+} present

Otherwise take a fresh sample and add dilute H_2SO_4. If the precipitate is

White	Pb^{2+} present

If there is no precipitate, take a fresh sample and add dilute NaOH. If the precipitate is

Blue	Cu^{2+} present
Yellow	Hg^{2+} present
White	Bi^{3+} present

3. *Group IIB cations*   to the filtrate add dilute HCl in excess, when the precipitate reappears. Examine its colour:

Brown precipitate Sn^{2+} present

An orange precipitate indicates Sb. To identify its oxidation state, take a fresh sample, acidify with 6M HCl and add KI:

No colouration Sb^{3+} present
Brown colouration Sb^{5+} present

A yellow precipitate indicates As or Sn^{4+}. Add $(NH_4)_2CO_3$ in excess. If the precipitate

Remains undissolved Sn^{4+} present

To identify the oxidation state of As present in the solution, take a fresh sample, acidify with 6M HCl and add KI:

No colouration As^{3+} present
Brown colouration As^{5+} present

4. *Group III cations*   neutralize the solution with NH_3 and add $(NH_4)_2S$ in excess. If there is no change, follow step 5. Examine the precipitate.
A green precipitate indicates Cr^{3+}. To a fresh sample, add NaOH:

Green precipitate which dissolves in an excess of
the reagent Cr^{3+} present

A pink (flesh-like) precipitate indicates Mn^{2+}. To a fresh sample, add NaOH:

White precipitate, which turns darker on standing Mn^{2+} present

A white precipitate may be caused by Al^{3+} or Zn^{2+}. To a fresh sample add NH_3, first in moderate amounts, then in excess:

White precipitate, which dissolves in excess NH_3 Zn^{2+} present
White precipitate, which remains unchanged
if excess NH_3 is added Al^{3+} present

A black precipitate occurs if Co^{2+}, Ni^{2+}, Fe^{2+} or Fe^{3+} were present originally. Filter, wash and mix the precipitate with 6M HCl. The precipitate dissolves if Fe^{2+} or Fe^{3+} were present, otherwise it remains unchanged.
To a fresh sample add NaOH in excess:

Green precipitate, turning dark on standing Fe^{2+} present
Dark brown precipitate Fe^{3+} present
Blue precipitate, turning pink if excess NaOH
is added Co^{2+} present
Green precipitate, which remains unchanged
on standing Ni^{2+} present

5. *Group IV cations*   to the solution add $(NH_4)_2CO_3$ in excess. If there is no precipitation, follow step 6. A white precipitate indicates the presence of Ba^{2+}, Sr^{2+} or Ca^{2+}.

To a fresh sample add a fourfold (in volume) of saturated $CaSO_4$ solution:

Immediate white precipitate	Ba^{2+} present
A white precipitate is slowly formed	Sr^{2+} present
No precipitation occurs	Ca^{2+} present

These cations may be identified also by their flame colourations (see Table 2.1).

6. *Group V cations*   to the solution add Na_2HPO_4 in excess:

White precipitate	Mg^{2+} present

Heat a fresh sample gently with some dilute NaOH:

Characteristic odour of ammonia	NH_4^+ present

(See also Section 5.3 (3) for the best experimental arrangements.)
Carry out a flame test with the original sample (see Table 2.1):

Yellow colouration	Na^+ present
Pale violet colouration	K^+ present

5.8   Separation and identification of cations in mixtures

The separation scheme below was first outlined by R. Fresenius in 1841, though in the course of time modifications were introduced. Before attempting such a task, it is necessary to test for ammonium ions in the *original* sample (cf. Section 5.3 (3)). During the course of separations most of the reagents are added in the form of ammonium compounds; thus by the time Group V is reached, a considerable amount of ammonium ions have built up in the test solution.

Separations[1] are carried out according to the scheme outlined in Tables 5.1 to 5.10. Note that it is important to verify the presence of a given cation by further tests.

Table 5.1   **Separation of cations into groups**   (Anions of organic acids, borates, fluorides, silicates, and phosphates being present). Add a few drops of 2M HCl to the cold solution. If a precipitate forms, continue adding 2M HCl until no further precipitation takes place. Filter.

Separation into groups

Residue	Filtrate
White. Group I present. Examine by Group Separation Table 5.2.	Add 1 ml of 3% H_2O_2 solution. Adjust the HCl concentration to 0.3M. Heat nearly to boiling, and saturate with H_2S under 'pressure'. Filter.

Residue	Filtrate
Coloured. Group II present.	Boil down in a porcelain dish to about 10 ml and thus ensure that all H_2S has been removed (test with lead acetate paper). Add 3–4 ml concentrated HNO_3 to oxidize any Fe^{2+} to Fe^{3+} etc. and evaporate cautiously to dryness; moisten with 2–3 ml concentrated HNO_3 and heat gently: this will remove organic acids.

[1] A more detailed description of these separations can be found in the 5th edition of this book: *Vogel's Textbook of Macro and Semimicro Qualitative Inorganic Analysis*, revised by G. Svehla. Longman, 1979, p. 395 ff.

Table 5.1   **Separation of cations into groups**   (contd.)

Residue	*Filtrate*
Separation Tables 5.3 to 5.5	If borate and fluoride are present, evaporate the residue repeatedly with 5–10 ml concentrated HCl.
	If borate is present and fluoride is absent, treat the residue with 5 ml methanol and 10 ml concentrated HCl and evaporate on a water bath.
	Add about 15 ml 2M HCl, warm and filter from any residue originating from any silicate present.
	Test 0.5 ml of the filtrate (or solution, if silicate is absent) for phosphate with 3 ml ammonium molybdate reagent and a few drops of concentrated HNO_3, and warm to about 40°C: a yellow ppt. indicates phosphate is present. If phosphate is present, remove all phosphate ions as detailed in Table 5.6.
	If phosphate is absent, add 1–2 g solid NH_4Cl, heat to boiling, add 2M NH_3 solution until mixture is alkaline and then 1 ml in excess, boil for 1 minute and filter immediately.

Residue	*Filtrate*
Group IIIA present. Examine by Group Separation Table 5.7.	Add 2–3 ml of 2M NH_3, heat, pass H_2S (under 'pressure') for 0.5–1 minute, filter and wash.

Residue	*Filtrate*
Group IIIB present. Examine by Group Separation Table 5.8.	Transfer to a porcelain dish and acidify with dilute acetic acid. Evaporate to a pasty mass (FUME CUPBOARD), allow to cool, add 3–4 ml concentrated HNO_3 so as to wash the solid round the walls to the centre of the dish, and heat cautiously until the mixture is dry. Then heat more strongly until all ammonium salts are volatilized. Cool. Add 3 ml 2M HCl and 10 ml water: warm and stir to dissolve the salts. Filter, if necessary. Add 0.25 g solid NH_4Cl (or 2.5 ml of 20% NH_4Cl solution), render alkaline with concentrated NH_3 solution and then add, with stirring, M $(NH_4)_2CO_3$ solution in slight excess. Keep and stir the solution in a water bath at 50–60°C for 3–5 minutes. Filter and wash with a little hot water.

Residue	*Filtrate*
White. Group IV present. Examine by Group Separation Table 5.9	Evaporate to a pasty mass in a porcelain dish (FUME CUPBOARD), add 3 ml concentrated HNO_3 so as to wash solid from walls to centre of dish, evaporate cautiously to dryness and then heat until white fumes of ammonium salts cease to be evolved. White residue. Group V present. Examine by Table 5.10.

Table 5.2 **Separation and identification of Group I cations (silver group)** The precipitate may contain $PbCl_2$, $AgCl$, and Hg_2Cl_2. Wash the precipitate on the filter first with 2 ml of 2M HCl, then 2–3 times with 1 ml portions of cold water and reject the washings with water. Transfer the precipitate to a small beaker or to a boiling tube, and boil with 5–10 ml water. Filter hot.

Group I

Residue	Filtrate
May contain Hg_2Cl_2 and $AgCl$. Wash the ppt. several times with hot water until the washings give no ppt. with 0.1M K_2CrO_4 solution: this ensures the complete removal of the Pb. Pour 3–4 ml warm dilute NH_3 solution over the ppt. and collect the filtrate.	May contain $PbCl_2$. Cool a portion of the solution: a white crystalline ppt. of $PbCl_2$ is obtained if Pb is present in any quantity. Divide the filtrate into 3 parts: (1) Add 0.1M K_2CrO_4 solution. Yellow ppt. of $PbCrO_4$, insoluble in dilute acetic acid. (2) Add 0.1M KI solution. Yellow ppt. of PbI_2, soluble in boiling water to a colourless solution, which deposits brilliant yellow crystals upon cooling. (3) Add M H_2SO_4. White ppt. of $PbSO_4$, soluble in ammonium acetate solution. Pb present.

Residue	Filtrate
If black, consists of $Hg(NH_2)Cl + Hg$. Hg_2^{2+} present.	May contain[a] $[Ag(NH_3)_2]^+$. Divide into 2 parts: (1) Acidify with 2M HNO_3. White ppt. of $AgCl$. (2) Add a few drops of 0.1M KI solution. Pale-yellow ppt. of AgI. Ag present.

[a] The solution obtained after dissolving in ammonia should be discarded quickly to avoid a serious **explosion**! (Cf. Section 3.6, reaction 1.)

Table 5.3 **Separation of Group II into Group IIA and Group IIB** The precipitate may consist of the sulphides of the Group IIA metals (HgS, PbS, Bi_2S_3, CuS, CdS) and those of Group IIB (As_2S_3, Sb_2S_3, Sb_2S_5, SnS, SnS_2). Wash the precipitated sulphides with a little M NH_4Cl solution that has been saturated with H_2S (the latter to prevent conversion of CuS into $CuSO_4$ by atmospheric oxidation), transfer to a porcelain dish, add about 5 ml yellow ammonium polysulphide solution, heat to 50–60°C, and maintain at this temperature for 3–4 minutes with constant stirring. Filter.

Group II

Residue	Filtrate
May contain HgS, PbS, Bi_2S_3, CuS, and CdS. Wash once or twice with small volumes of dilute (1 + 100) ammonium sulphide solution, then with 2% NH_4NO_3 solution and reject all washings. Group IIA present. Follow Table 5.4.	May contain solutions of the thio salts $(NH_4)_3AsS_4$, $(NH_4)_2SbS_4$, and $(NH_4)_2SnS_3$. Just acidify by adding concentrated HCl dropwise (test with litmus paper), and warm gently. A yellow or orange ppt., which may contain As_2S_5, Sb_2S_5, and SnS_2, indicates Group IIB present. Follow Table 5.5.

Table 5.4 **Separation of Group IIA cations** The precipitate may contain HgS, PbS, Bi_2S_3, CuS and CdS. Transfer to a beaker or porcelain basin, add 5–10 ml 2M HNO_3, boil gently for 2–3 minutes, filter and wash with a little water.

Group IIA

Residue	Filtrate
Black. HgS. Dissolve in a mixture of M NaOCl solution and 0.5 ml of 2M HCl. Add 1 ml 2M HCl, boil off excess Cl_2 and cool. Add $SnCl_2$ solution. White ppt. turning grey or black. Hg^{2+} present.	May contain nitrates of Pb, Bi, Cu, and Cd. Test a small portion for Pb by adding dilute H_2SO_4 and alcohol. A white ppt. of $PbSO_4$ indicates Pb present. If Pb present, add M H_2SO_4 to the remainder of the solution, concentrate in the fume cupboard until white fumes (from the decomposition of the H_2SO_4) appear. Cool, add 10 ml of water, stir, allow to stand 2–3 minutes, filter and wash with a little water.

Residue	Filtrate
White: $PbSO_4$. Pour 2 ml of 3M ammonium acetate through the filter several times, add to the filtrate a few drops of 2M acetic acid and then 0.1M K_2CrO_4 solution. Yellow ppt. of $PbCrO_4$. Pb present.	May contain nitrates and sulphates of Bi, Cu and Cd. Add concentrated NH_3 solution until solution is distinctly alkaline. Filter.

Residue	Filtrate
White: may be $Bi(OH)_3$. Dissolve in the minimum volume of 2M HCl and pour into cold sodium tetrahydroxo-stannate(II). Black ppt. Bi present. Alternatively, dissolve a little of ppt. in 2–3 drops 2M HNO_3. Place 1 drop of this solution upon filter paper moistened with cinchonine–KI reagent. Orange-red spot. Bi present.	May contain $[Cu(NH_3)_4]^{2+}$ and $[Cd(NH_3)_4]^{2+}$. If deep blue in colour, Cu is present in quantity. Confirm Cu by acidifying a portion of the filtrate with 2M acetic acid and add $K_4[Fe(CN)_6]$ solution. Reddish-brown ppt. Cu present. To the remainder of the filtrate, add M KCN solution dropwise until colour is discharged, and add a further ml in excess. Pass H_2S for 20–30 seconds. Yellow ppt., sometimes discoloured, of CdS. Cd present. Filter off ppt. and dissolve a portion of it in 1 ml 2M HCl: boil to expel H_2S and most of the acid and apply the 'cadion-2B' test on 1 drop of the solution. A pink spot confirms Cd.

Table 5.5 **Separation of Group IIB cations** Treat the yellow ammonium polysulphide extract of the Group II precipitate (see Table 5.3) with 2M HCl, with constant stirring, until it is slightly acid (test with litmus paper), warm and shake or stir for 1–2 minutes. A fine white or yellow precipitate is sulphur only. A yellow or orange flocculant precipitate indicates As, Sb, and/or Sn present. Filter and wash the precipitate, which may contain As_2S_5, As_2S_3, Sb_2S_5 and S, with a little H_2S water; reject the washings.

Transfer the precipitate to a small conical flask, add 5–10 ml concentrated HCl and boil gently for 5 minutes (with funnel in mouth of flask). Dilute with 2–3 ml water, pass H_2S for 1 minute to reprecipitate small amounts of arsenic that may have dissolved, and filter.

Group IIB

Table 5.5 **Separation of Group IIB cations** (contd.)

Residue	Filtrate
May contain As_2S_5, As_2S_3, and S (yellow). Dissolve the ppt. in 3–4 ml warm 2M NH_3 solution, filter (if necessary), add 3–4 ml 3% H_2O_2 solution and warm for a few minutes to oxidize arsenite to arsenate. Add a few ml of the $Mg(NO_3)_2$ reagent. Stir and allow to stand. White, crystalline ppt. of $Mg(NH_4)AsO_4 \cdot 6H_2O$. As present. Filter off ppt., and pour 1 ml of $AgNO_3$ solution containing 6–7 drops of 2M acetic acid onto residue on filter. Brownish-red residue of Ag_3AsO_4.	May contain Sb^{3+} and Sn^{4+}. Boil to expel H_2S, and divide the cold solution into three parts. (1) Render just alkaline with 2M NH_3 solution, disregard any slight ppt., add 1–2 g solid oxalic acid, boil, and pass H_2S for *c.* 1 minute into the hot filtrate. Orange ppt. of Sb_2S_3. Sb present. (2) To 2 drops of the solution on a spot plate, add a minute crystal of $NaNO_2$ and then 2 drops of Rhodamine-B reagent. Violet solution or ppt. Sb present. (3) Partially neutralize the liquid with 2M NH_3, add 10 cm clean iron wire to 1 ml of the solution. (If much Sb is present, it is better to reduce with Mg powder.) Warm gently to reduce the tin to the divalent state, and filter into a solution of $HgCl_2$. White ppt. of Hg_2Cl_2 or grey ppt. of Hg. Sn present.

Table 5.6

Removal of PO_4^{3-}

Removal of phosphate If a phosphate has been found (cf. Table 5.1), proceed as follows. Dissolve the precipitate produced by the action of NH_4Cl and a slight excess of NH_3 solution in the minimum volume of 2M HCl. [The precipitate may contain $Fe(OH)_3$, $Al(OH)_3$, $Cr(OH)_3$, $MnO_2 \cdot xH_2O$, traces of CaF_2, and the phosphates of Mg and the Group IIIA, IIIB, and IV metals.] Test about 0.5 ml for Fe by the addition of $K_4[Fe(CN)_6]$ or 0.1M NH_4SCN solution. To the main volume of the cold solution, add 2M NH_3 solution dropwise, with stirring, until either a faint permanent precipitate is just obtained or until the solution is just alkaline (test with litmus paper). Then add 2–3 ml 9M acetic acid and 5 ml 6M ammonium acetate solution. Disregard any precipitate which may form at this stage. If the solution is red or brownish-red, sufficient iron(III) is present in the solution to combine with all the phosphate ions. If the solution is not red in colour, add almost neutral $FeCl_3$ solution drop by drop and with stirring, until the solution acquires a deep brownish-red colour. Dilute the solution to about 150 ml with hot water, boil gently for 1–2 minutes, filter hot and wash the residue with a little boiling water.

Residue	Filtrate
May contain the phosphates (and, possibly, the basic acetates) of Fe, Al, and Cr and also $Fe(OH)_3$. Group IIIA present. Rinse the ppt. into a porcelain dish by means of 10 ml cold water, add 1–1.5 g sodium peroxoborate, $NaBO_3 \cdot 4H_2O$ (or add 5 ml 2M NaOH solution, followed by 5 ml 3% H_2O_2 solution); and boil gently until the evolution of O_2 ceases (2–3 minutes). Filter and wash with a little hot water.	Boil down in an evaporating dish to 20–25 ml. Add 0.5 g NH_4Cl and then 2M NH_3 solution in slight excess. Filter, if necessary.

Residue	Filtrate	Residue	Filtrate
$FePO_4 + Fe(OH)_3$. Reject.	May contain $[Al(OH)_4]^-$ and CrO_4^{2-}. Examine for Al and Cr as described in Group Separation Table 5.7.	Examine for Al and Cr, if not previously tested for. In general, no ppt. will be obtained here.	Examine for Groups IIIB, IV and for Mg as detailed in Tables 5.8 and 5.9.

Table 5.7 **Separation of Group IIIA cations** The precipitate produced by adding NH_4Cl and NH_3 solution and boiling may contain $Fe(OH)_3$, $Cr(OH)_3$, $Al(OH)_3$, and a little $MnO(OH)_2$. Wash with a little hot 1% NH_4Cl solution. Transfer the precipitate with the aid of 5–10 ml water to a small evaporating basin or a small beaker, add 1–1.5 g sodium peroxoborate, $NaBO_3 \cdot 4H_2O$ (or add 5 ml 2M NaOH solution, followed by 5 ml 3% H_2O_2 solution). Boil gently until the evolution of O_2 ceases (2–3 minutes). Filter.

Group IIIA

Residue	*Filtrate*

May contain $Fe(OH)_3$ and $MnO(OH)_2$. Wash with a little hot water.
 	
Dissolve a small portion of the ppt. in 1 ml of 8M HNO_3 with the aid (if necessary) of 3–4 drops of 3% H_2O_2 solution or 1 drop of saturated H_2SO_3 solution. Boil (to decompose H_2O_2), cool thoroughly, add 0.05–0.1 g sodium bismuthate, shake, and allow the solid to settle.
    Violet solution of MnO_4^-.
      Mn present.
    Dissolve another portion of the ppt. in 2M HCl (filter, if necessary).
    Either – add a few drops of 0.1M NH_4SCN solution.
    Deep red colouration.
      Fe present.
    Or – add $K_4[Fe(CN)_6]$ solution.
    Blue ppt.
      Fe present.
    The original solution or substance should be tested with $K_4[Fe(CN)_6]$ and with 0.1M NH_4SCN to determine whether Fe^{2+} or Fe^{3+}.
	May contain CrO_4^{2-} (yellow) and $[Al(OH)_4]^-$ (colourless).
	
	If colourless, Cr is absent and need not be considered further.
	If the solution is yellow, Cr is indicated.
	Divide the liquid into three portions.
	(1) Acidify with 2M acetic acid and add 0.25M lead acetate solution.
	    Yellow ppt. of $PbCrO_4$.
	      Cr present.
	(2) Acidify 2 ml with 2M HNO_3, cool thoroughly, add 1 ml amyl alcohol, and 4 drops of 3% H_2O_2 solution. Shake well and allow the two layers to separate.
	    Blue upper layer (containing chromium pentoxide; it does not keep well).
	      Cr present.
	(3) Acidify with 2M HCl (test with litmus paper), then add 2M NH_3 solution until just alkaline. Heat to boiling. Filter.
	    White gelatinous ppt. of $Al(OH)_3$.
	      Al present.
	    Dissolve a small portion of the ppt. in 1 ml hot 2M HCl. Cool, add 1 ml 6M ammonium acetate solution and 0.5 ml of the M 'aluminon' reagent. Stir the solution and render basic with ammonium carbonate solution. A red ppt. confirms the presence of Al.

Table 5.8 **Separation of Group IIIB cations** The precipitate may contain CoS, NiS, MnS and ZnS. Wash well with 1% NH_4Cl solution to which 1% by volume of $(NH_4)_2S$ has been added; reject the washings. Transfer the precipitate to a small beaker. Add 5 ml water and 5 ml 2M HCl, stir well, allow to stand for 2–3 minutes and filter.

Group IIIB

Residue	*Filtrate*

If black, may contain CoS and NiS. Test residue with borax bead. If blue, Co is indicated.
    Dissolve the ppt. in a mixture of 1.5 ml M NaOCl solution and 0.5 ml 2M HCl. Add 1 ml 2M HCl, and boil until all Cl_2 is expelled. Cool and dilute to about 4 ml.
    Divide the solution into two equal parts.
	May contain Mn^{2+} and Zn^{2+} and, possibly, traces of Co^{2+} and Ni^{2+}. Boil until H_2S removed (test with lead acetate paper), cool, add excess 2M NaOH solution, followed by 1 ml 3% H_2O_2 solution. Boil for 3 minutes. Filter.

Table 5.8   **Separation of Group IIIB cations**   (contd.)

Group IIB

Residue	Residue	Filtrate
(1) Add 1 ml amyl alcohol, 2 g solid NH_4SCN and shake well. Amyl alcohol layer coloured blue. 　Co present. (2) Add 2 ml NH_4Cl M solution, 2M NH_3 solution until alkaline, and then excess of dimethylglyoxime reagent. 　Red ppt. 　　Ni present.	Largely $MnO(OH)_2$ and perhaps traces of $Ni(OH)_2$ and $Co(OH)_3$. Dissolve the ppt. in 5 ml of 8M HNO_3, with the addition of a few drops of 3% H_2O_2 solution, if necessary. Boil to decompose excess H_2O_2 and cool. Add 0.05 g $NaBiO_3$, stir and allow to settle. Purple solution of MnO_4^-. 　Mn present.	May contain $[Zn(OH)_4]^{2-}$. 　Divide into two parts. (1) Acidify with 2M acetic acid and pass H_2S. White ppt. of ZnS. 　Zn present. (2) Just acidify with M H_2SO_4, add 0.5 ml of 0.1M cobalt acetate solution and 0.5 ml of the ammonium tetrathiocyanato-mercurate(II) reagent: stir. 　Pale blue ppt. 　　Zn present.

Table 5.9   **Separation of Group IV cations**   The precipitate may contain $BaCO_3$, $SrCO_3$ and $CaCO_3$. Wash with a little hot water and reject the washings. Dissolve the precipitate in 5 ml hot 2M acetic acid by pouring the acid repeatedly through the filter paper. Test 1 ml for barium by adding 0.1M K_2CrO_4 solution dropwise to the nearly boiling solution. A yellow precipitate indicates Ba.

Group IV

Ba present. Heat the remainder of the solution almost to boiling and add a slight excess of 0.1M K_2CrO_4 solution (i.e. until the solution assumes a yellow colour and precipitation is complete). Filter and wash the precipitate (*C*) with a little hot water. Render the hot filtrate and washings basic with 2M NH_3 solution and add excess M $(NH_4)_2CO_3$ solution or, better, a little solid Na_2CO_3. A white precipitate indicates the presence of $SrCO_3$ and/or $CaCO_3$. Wash the precipitate with hot water, and dissolve it in 4 ml warm 2M acetic acid: boil to remove excess CO_2 (solution *A*).

Ba absent. Discard the portion used in testing for barium, and employ the remainder of the solution (*B*), after boiling for 1 minute to expel CO_2, to test for strontium and calcium.

Residue (C)	Solution A *or* Solution B
Yellow: $BaCrO_4$. 　Wash well with hot water. Dissolve the ppt. in a little concentrated HCl, evaporate almost to dryness and apply the flame test. 　Green (or yellowish-green) flame. 　Ba present. 　(Use spectroscope, if available.)	The volume should be about 4 ml. 　Either – To 2 ml of the cold solution, add 2 ml saturated $(NH_4)_2SO_4$ solution, followed by 0.2 g sodium thiosulphate, heat in a beaker of boiling water for 5 minutes and allow to stand for 1–2 minutes. Filter. 　Or – To 2 ml of the solution add 2 ml triethanolamine, 2 ml saturated $(NH_4)_2SO_4$ solution, heat on a boiling water bath with continuous stirring for 5 minutes and allow to stand for 1–2 minutes. Dilute with an equal volume of water and filter.

Residue	Filtrate
Largely $SrSO_4$. Wash with a little water. Transfer ppt. and filter paper to a small crucible, heat until ppt. has charred (or burn filter paper and ppt., held in a Pt wire, over a crucible), moisten ash with a few drops concentrated HCl and apply the flame test. 　Crimson flame. 　Sr present. 　(Use spectroscope, if available.)	May contain Ca complex. 　(If Sr is absent, use 2 ml of solution *A* or *B*.) Add a little 0.1M $(NH_4)_2C_2O_4$ solution, 2 ml 2M CH_3COOH and warm on a water bath. 　White ppt. of CaC_2O_4. 　Ca present. 　Confirm by flame test on ppt. – brick-red flame. 　(Use spectroscope, if available.)

Table 5.10 **Identification of Group V cations** Treat the dry residue from Group IV with 4 ml
Group V water, stir, warm for 1 minute and filter.

If the residue dissolves completely (or almost completely) in water, dilute the resulting solution (after filtration, if necessary) to about 6 ml and divide it into three
approximately equal parts: (1) use the major portion to test for Mg with the 2%
8-hydroxyquinoline solution: confirm Mg by applying the magneson test to 3–4
drops of the solution; (2) and (3) test for Na and K, respectively, as described below.

Residue	*Filtrate*
Dissolve in a few drops of 2M HCl and add 2–3 ml water. Divide the solution into two unequal parts. (1) Larger portion: Treat 1 ml 2% 8-hydroxyquinoline solution in 2M acetic acid with 5 ml 2M ammonia solution and, if necessary, warm to dissolve any precipitated oxine. Add a little M NH_4Cl solution to the test solution, followed by the ammoniacal oxine reagent, and heat to boiling point for 1–2 minutes (the odour of NH_3 should be discernible). Pale yellow ppt. of Mg 'oxinate'. Mg present. (2) Smaller portion: To 3–4 drops, add 2 drops of the magneson reagent, followed by several drops of 2M NaOH solution until alkaline. A blue ppt. confirms Mg.	Divide into two parts (*a*) and (*b*). (*a*) Add a little uranyl magnesium acetate reagent, shake and allow to stand for a few minutes. Yellow crystalline ppt. Na present. Confirm by flame test: persistent yellow flame. (*b*) Add a little sodium hexanitrito-cobaltate(III) solution (or *c*. 4 mg of the solid) and a few drops of 2M acetic acid. Stir and, if necessary, allow to stand for 1–2 minutes. Yellow ppt. of $K_3[Co(NO_2)_6]$. K present. Confirm by flame test and view through two thicknesses of cobalt glass: red colouration (usually transient).

5.9 Special tests for mixtures of anions

It is not possible to devise a comprehensive separation scheme for anions;
however, it is possible to detect them individually in most cases, after perhaps
a 1–2 stage separation. It is advantageous to remove all heavy metals from
the sample by extracting the anions through boiling with sodium carbonate
solution; heavy metal ions are precipitated out in the form of carbonates,
while the anions remain in solution accompanied by sodium ions. The
selected tests are listed in order of increasing complexity; it is advisable to
do them in the order they are described.

Soda extract ***Preparation of the soda extract*** Intimately mix the solid sample with 3–4
times (in volume) of anhydrous sodium carbonate, add just sufficient
water to dissolve the sodium carbonate. In the case of dissolved samples,
add 5 g anhydrous sodium carbonate to 10 ml solution. Heat the mixture
under reflux for 15 minutes and filter. Wash and discard the residue. Acidify
the combined filtrate and washings gently with 8M nitric acid, boil gently
to expel carbon dioxide and allow to cool. Add 2M ammonia solution
portionwise and with stirring, until the solution just turns alkaline. Boil
gently for 1–2 minutes to remove excess ammonia. Cool, and filter again if
necessary. The filtrate will be termed here the 'neutralized soda extract'.

$CO_3^{2-} + SO_3^{2-}$ 1. *Carbonate in the presence of sulphite* sulphites, on treatment with dilute
sulphuric acid, liberate sulphur dioxide which, like carbon dioxide, produces
a turbidity with lime or baryta water. The dichromate test for sulphites is,

however, not influenced by the presence of carbonates. To detect carbonates in the presence of sulphites, treat the solid mixture with M sulphuric acid and pass the evolved gases through a small wash bottle or boiling tube containing potassium dichromate solution and M sulphuric acid. The solution will be turned green and the sulphur dioxide will, at the same time, be completely removed; the residual gas is then tested with lime or baryta water in the usual manner.

An alternative procedure is to add a little sodium potassium dichromate or a small volume of 3% hydrogen peroxide solution to the mixture and then to warm with M sulphuric acid; the evolved gas is then passed through lime or baryta water.

$NO_3^- + NO_2^-$ 2. *Nitrate in the presence of nitrite*   the nitrite is readily identified in the presence of a nitrate by treatment with dilute mineral acid, potassium iodide, and starch paste (or potassium iodide–starch paper). The nitrate cannot, however, be detected in the presence of a nitrite since the latter gives the brown ring test with iron(II) sulphate solution and dilute sulphuric acid. The nitrite is therefore completely decomposed first by adding some sulphamic acid to the solution.

The nitrate can then be tested for by the brown ring test.

$NO_3^- + Br^- + I^-$ 3. *Nitrate in the presence of bromide and iodide*   the brown ring test for nitrates cannot be applied in the presence of bromides and iodides since the liberation of the free halogen with concentrated sulphuric acid will obscure the brown ring due to the nitrate. The solution is therefore boiled with 2M sodium hydroxide solution until ammonium salts, if present, are completely decomposed; powdered Devarda's alloy or aluminium powder (or wire) is then added and the solution gently warmed. The evolution of ammonia, detected by its smell and its action upon red litmus paper, indicates the presence of a nitrate.

$NO_3^- + ClO_3^-$ 4. *Nitrate in the presence of chlorate*   the chlorate obscures the brown ring test. The nitrate is reduced to ammonia as described under paragraph 3; the chlorate is at the same time reduced to chloride, which may be tested for with silver nitrate solution and dilute nitric acid.

If a chloride is originally present, it may be removed first by the addition of silver sulphate solution or ammoniacal silver sulphate solution. The ammoniacal solution should be discarded quickly to avoid a serious **explosion!** (Cf. Section 3.6, reaction 1.)

$Cl^- + Br^- + I^-$ 5. *Chloride in the presence of bromide and/or iodide*   this procedure involves the removal of the bromide and iodide (but not chloride) by oxidation with potassium or ammonium peroxodisulphate in the presence of dilute sulphuric acid. The free halogens are thus liberated, and may be removed either by simple evaporation (addition of water may be necessary to maintain the original volume) or by evaporation at about 80°C in a stream of air.

It must be emphasized that these reactions take place only in acid media. Add solid potassium peroxodisulphate and 2M sulphuric acid to the neutralized soda extract of the mixed halides contained in a conical flask; heat the

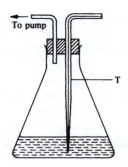

Fig. 5.2

flask to about 80°C, and aspirate a current of air through the solution with the aid of a filter pump until the solution is colourless (Fig. 5.2; T is a drawn-out capillary tube). An all-glass device can be assembled from Quickfit conical flask FE 50/2 and trap head MF 28/52. Add more potassium peroxodisulphate or water as may be found necessary. Test the residual colourless liquid for chloride with 0.1M silver nitrate solution and 2M nitric acid.

Do not mistake precipitated silver sulphate for silver chloride. In a modification of this method, lead dioxide is substituted for potassium peroxodisulphate. Acidify the solution with 2M acetic acid, add lead dioxide, and boil the mixture until bromine and iodine are no longer evolved. Filter, and test the filtrate, which should be colourless, with silver nitrate solution and dilute nitric acid.

$Cl^- + Br^-$

6. *Chloride in the presence of bromide*   warm the solid mixture with a little solid potassium dichromate and concentrated sulphuric acid in a small distilling flask (Fig. 5.1), and pass the vapours, which contain chromyl chloride, into sodium hydroxide solution. Test for chromate, which proves the presence of a chloride, with hydrogen peroxide and amyl alcohol or with the diphenylcarbazide reagent.

$Cl^- + I^-$

7. *Chloride in the presence of iodide*   add excess 0.1M silver nitrate solution to the neutralized soda extract and filter; reject the filtrate. Wash the precipitate with 2M ammonia solution and filter again. Add 2M nitric acid to the washings; a white precipitate of silver chloride indicates the presence of chloride.

$Br^- + I^- + Cl^-$

8. *Bromide and iodide in the presence of each other and of chloride*   the presence of a chloride does not interfere with the reactions described below. To the soda extract, strongly acidified with 6M hydrochloric acid, add 1–2 drops of chlorine water (the solution obtained by carefully acidifying a M solution of sodium hypochlorite with 2M hydrochloric acid may also be used) and 2–3 ml chloroform; shake; a violet colour indicates iodide. Continue the addition of chlorine water or of acidified sodium hypochlorite solution drop by drop to oxidize the iodine to iodate and shake after each addition. The violet colour will disappear, and a reddish-brown colouration of the chloroform, due to dissolved bromine (and/or bromine monochloride, BrCl), will be obtained if a bromide is present. If iodide alone is present, the solution will be colourless after the violet colour has disappeared.

$Cl^- + ClO_3^- + ClO_4^-$

9. *Chloride, chlorate, and perchlorate in the presence of each other*   each anion must be tested for separately. Divide the soda extract into 3 parts.

   (a) *Chloride*   acidify with 2M nitric acid and boil to expel carbon dioxide. Add 0.1M silver nitrate solution; a white precipitate of silver chloride indicates the presence of chloride. Silver chlorate and perchlorate are soluble in water.

   (b) *Chlorate*   acidify with 2M nitric acid and add 0.1M silver nitrate solution; filter off the precipitated silver chloride. Now introduce a little chloride-free sodium nitrite (which reduces the chlorate to chloride) and more silver nitrate solution into the filtrate; a white, curdy precipitate of silver chloride indicates the presence of chlorate.

(c) *Perchlorate*   pass excess sulphur dioxide into the solution to reduce chlorate to chloride, boil off the excess sulphur dioxide, and precipitate the chloride with saturated silver sulphate solution, afterwards removing the excess silver with 0.5M sodium carbonate. Evaporate the resulting solution to dryness and heat to dull redness (better in the presence of a little halide-free lime) to convert the perchlorate into chloride. Extract the residue with water, and test for chloride with 0.1M silver nitrate solution and 2M nitric acid.

$IO_3^- + I^-$

10. *Iodate and iodide in the presence of each other*   the addition of dilute acid to a mixture of iodate and iodide results in the separation of free iodine, due to the interaction between these ions:

$$IO_3^- + 8I^- + 6H^+ \rightarrow 3I_3^- + 3H_2O$$

Neither iodates nor iodides alone do this when acidified with dilute hydrochloric, sulphuric, or acetic acids in the cold.

Test for iodide in the neutralized soda extract or in a solution of the sodium salts by the addition of a few drops of chlorine water (or acidified sodium hypochlorite solution) and 2–3 ml chloroform; the latter is coloured violet. Add excess saturated silver sulphate solution to another portion of the neutral solution and filter off the silver iodide; remove the excess silver sulphate with 0.5M sodium carbonate solution. Pass sulphur dioxide into the filtrate to reduce iodate to iodide, boil off the excess sulphur dioxide, and add 0.1M silver nitrate solution and 2M nitric acid. A yellow precipitate of silver iodide confirms the presence of iodate in the original substance.

$PO_4^{3-} + AsO_4^{3-}$

11. *Phosphate in the presence of arsenate*   both arsenate and phosphate give a yellow precipitate on warming with ammonium molybdate solution and nitric acid, the latter on gently warming and the former on boiling.

Acidify the soda extract with 6M hydrochloric acid, bubble through sulphur dioxide to reduce the arsenate to arsenite, boil off the excess sulphur dioxide (test with potassium dichromate paper), and pass hydrogen sulphide into the solution to precipitate the arsenic as arsenic(III) sulphide; continue the passage of hydrogen sulphide until no more precipitate forms. Filter, boil off the hydrogen sulphide, and test the filtrate for phosphate by the ammonium molybdate test or with the magnesium nitrate reagent.

$PO_4^{3-} + AsO_3^{3-} +$
AsO_4^{3-}

12. *Phosphate, arsenate, and arsenite in the presence of each other*   treat the soda extract with M sulphuric acid until acidic, warm to expel carbon dioxide, render just alkaline with 2M ammonia solution and filter, if necessary. Add a few millilitres of magnesium nitrate reagent and allow to stand for 5–10 minutes; shake or stir from time to time. Filter off the white, crystalline precipitate of magnesium ammonium phosphate and/or magnesium ammonium arsenate (*A*), and keep the filtrate (*B*). Test for arsenite in the filtrate (*B*) by acidifying with 2M hydrochloric acid and passing hydrogen sulphide, when a

yellow precipitate of arsenic(III) sulphide is immediately produced if arsenite is present.

Wash the precipitate (A) with 2M ammonia solution, and remove half of it to a small beaker with the aid of a clean spatula. Treat the residue on the filter with a little 0.1M silver nitrate solution containing a few drops of 2M acetic acid: a brownish-red residue (due to Ag_3AsO_4) indicates arsenate present. If the residue is yellow (largely Ag_3PO_4), pour 6M hydrochloric acid through the filter a number of times, and add a little 0.1M potassium iodide solution and 1–2 ml chloroform to the extract and shake; if the organic layer acquires a purple colour, an arsenate is present.

Dissolve the white precipitate in the beaker in 2M hydrochloric acid, reduce the arsenate (if present) with sulphur dioxide, precipitate the arsenic as arsenic(III) sulphide with hydrogen sulphide and boil off the hydrogen sulphide in the filtrate. Render the filtrate (C) slightly ammoniacal and add a little magnesium nitrate reagent: white, crystalline magnesium ammonium phosphate will be precipitated if a phosphate is present. Filter off the precipitate, wash with a little water, and pour a little 0.1M silver nitrate solution (containing a few drops of 2M acetic acid) over it: yellow silver phosphate will be formed. Alternatively, the ammonium molybdate test may be applied to the filtrate (C) after evaporating to a small volume.

$S^{2-} + SO_3^{2-} + S_2O_3^{2-} + SO_4^{2-}$

13. *Sulphide, sulphite, thiosulphate, and sulphate in the presence of each other* upon the addition of dilute acid to the mixture, the hydrogen sulphide, liberated from the sulphide, and the sulphur dioxide, liberated from the sulphite and thiosulphate, react and sulphur is precipitated; this complication necessitates the use of a special procedure for their separation (see Table 5.11).

Table 5.11 **Separation of sulphide, sulphite, thiosulphate, and sulphate** Shake the solution with excess of freshly precipitated $CdCO_3$, and filter.

Residue	Filtrate				
CdS and excess of $CdCO_3$. Wash and reject washings. Digest residue with 2M acetic acid to remove excess carbonate. A yellow residue indicates sulphide. Confirm by warming with 2M HCl and test the evolved H_2S with lead acetate paper.	Add $Sr(NO_3)_2$ solution in slight excess, shake, allow to stand overnight and filter.				
	Residue			Filtrate	
	$SrSO_3$ and $SrSO_4$. Wash, treat with 2M HCl and filter.			Contains SrS_2O_3. Acidify with 2M HCl and boil; SO_2 is evolved and S is slowly precipitated. Thiosulphate present.	
	Residue	Filtrate			
	$SrSO_4$. White. Sulphate present. Confirm by fusion with Na_2CO_3 and apply sodium nitroprusside test.	Contains sulphurous acid. Add a few drops of a 0.005M solution of iodine; the latter is decolourized. Sulphite present.			

A mixture of the sodium salts or soda extract is employed for this separation.

$S^{2-} + SO_3^{2-} + S_2O_3^{2-}$

14. *Sulphide, sulphite, and thiosulphate in the presence of each other* it is assumed that the solution is slightly alkaline and contains the sodium salts of the anions, e.g. the soda extract is used.

(a) Test a portion of the solution for sulphide with sodium nitroprusside solution. The formation of a purple colouration indicates the presence of a sulphide.

(b) If sulphide is present, remove it by shaking with freshly precipitated cadmium carbonate and filter. Test a portion of the filtrate with sodium nitroprusside solution to see that all the sulphide has been removed, When no sulphide is present, treat the remainder of the filtrate with a drop of phenolphthalein solution and bubble through carbon dioxide until the solution is decolourized by it. Test 2–3 ml of the colourless solution with 0.5–1 ml of the 0.1% fuchsin reagent. If the reagent is decolourized, the presence of sulphite is indicated.

(c) Treat the remainder of the colourless solution with 2M hydrochloric acid and boil for a few minutes. If sulphur separates, a thiosulphate is indicated.

$BO_3^{3-} + Cu^{2+} + Ba^{2+}$

15. *Borate in the presence of copper and barium salts* when carrying out the ethyl borate test for borates (Section 4.23, reaction 2), it must be remembered that copper and barium salts may also impart a green colour to the alcohol flame. The test may be carried out in the following way when salts of either or both of these metals are present. The mixture of the borate, concentrated sulphuric acid, and ethanol is placed in a small, round-bottomed flask, fitted with a glass jet and surmounted by a wide glass tube, which acts as a 'chimney' (Fig. 5.3). An all-glass apparatus can be put together from Quickfit

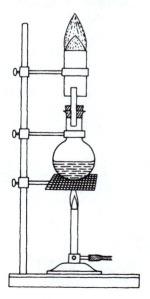

Fig. 5.3

round bottomed flask FR 50/1S and joint cone with stem MF 15/2. The mixture is gently warmed, and the vapours ignited at the top of the wide glass tube. A green flame confirms the presence of a borate.

$F^- + [SiF_6]^{2-} + SO_4^{2-}$

16. *Fluoride, hexafluorosilicate, and sulphate in the presence of each other* the following differences in solubilities of the lead salts are utilized in this separation: lead hexafluorosilicate is soluble in water; lead fluoride is insoluble in water but soluble in dilute acetic acid; lead sulphate is insoluble in water and in boiling dilute acetic acid, but soluble in concentrated ammonium acetate solution. (See Table 5.12.)

$(COO)_2^{2-} + F^-$

17. *Oxalate in the presence of fluoride* both calcium fluoride and calcium oxalate are precipitated by 0.5M calcium chloride solution in the presence of 2M acetic acid. The fluoride may be identified in the usual manner with concentrated sulphuric acid. The oxalate is most simply detected by dissolving a portion of the precipitate in hot M sulphuric acid and then adding a few drops of a 0.004M solution of potassium permanganate. The latter will be decolourized if an oxalate is present.

$Cl^- + CN^-$

18. *Chloride and cyanide in the presence of each other* both silver chloride and silver cyanide are insoluble in water, but soluble in dilute ammonia solution.

The concentrated solution, or preferably the solid mixture of sodium salts, is treated with about 5 times its weight of 30% hydrogen peroxide and the mixture gently warmed; ammonia is evolved, which is recognized by its action upon mercury(I) nitrate paper. The solution is then boiled to decompose all the hydrogen peroxide, and then tested for chloride with 0.1M silver nitrate solution and 2M nitric acid.

$$CN^- + H_2O_2 \rightarrow OCN^- + H_2O$$

$$OCN^- + 2H_2O \rightarrow HCO_3^- + NH_3$$

$[Fe(CN)_6]^{4-} +$
$[Fe(CN)_6]^{3-} + SCN^-$

19. *Hexacyanoferrate(II), hexacyanoferrate(III), and thiocyanate in the presence of each other* details are given in Table 5.13.

Table 5.12 **Detection of fluoride, hexafluorosilicate, and sulphate in the presence of each other**
Add excess 0.25M lead acetate solution to the solution of the alkali metal salts, and filter cold.

Residue	Filtrate
Wash well with cold water. Divide into two parts. (1) Smaller portion. Add excess 2M acetic acid and boil (this will dissolve any lead fluoride). White residue indicates sulphate. The residue is soluble in 6M ammonium acetate solution. (2) Larger portion. Treat cautiously with concentrated sulphuric acid and test with a moist glass rod. Milkiness of the water and etching of the tube indicate fluoride.	Add Ba(NO_3)_2 solution and warm gently. White crystalline precipitate indicates hexafluorosilicate.

Table 5.13 **Separation of hexacyanoferrate(II), hexacyanoferrate(III), and thiocyanate** Acidify the neutralized soda extract of the sodium salts with 2M acetic acid, add excess of a solution of thorium nitrate and some macerated filter paper (Whatman Ashless Floc) and shake well; filter.

Residue	Filtrate	
$Th[Fe(CN)_6]$. Wash with a little cold water. Treat the ppt. on the filter paper with 2M NaOH solution; acidify the alkaline extract with 0.5M HCl and add a few drops of 2M $FeCl_3$ solution.	Add $CdSO_4$ solution, shake, filter	
A precipitate of Prussian blue indicates hexacyanoferrate(II).	Residue	Filtrate
	Orange. $Cd_3[Fe(CN)_6]_2$. Extract with 2M NaOH solution, acidify extract with dilute HCl, add freshly prepared 0.5M $FeSO_4$ solution. Precipitate of Prussian blue indicates hexacyanoferrate(III).	Add 0.5M $FeCl_3$ solution and ether. Red colouration of ether proves presence of thiocyanate.

Organic acids

20. *Detection of organic salts (oxalate, tartrate, citrate, benzoate, succinate, acetate, formate and salicylate) in the presence of each other* this can be done after a partial separation, as described in Table 5.14.

Procedure A: separation of benzoate and succinate

Boil the precipitated iron(III) salts with a little 2M ammonia solution, filter off the precipitated iron(III) hydroxide and discard the precipitate; boil off the excess ammonia from the filtrate. Divide the filtrate into two parts.

Treat one portion with barium chloride solution and filter off the precipitate formed. The residue consists of barium succinate. Confirm the succinate by the fluorescein test. Acidify the second portion of the filtrate with 2M hydrochloric acid, when benzoic acid will separate out on cooling. It can be identified by its melting point.

Procedure B: formates and acetates in the presence of each other

Formate

(a) In the absence of tartrates, citrates, and reducing agents. Add 0.1M silver nitrate solution to the neutralized soda extract; a white precipitate, which is converted into a black deposit of silver on boiling, indicates formate. A little of the solid substance may be treated with concentrated sulphuric acid, when carbon monoxide (burns with blue flame) will be evolved in the cold.

(b) In the presence of tartrates, citrates, and reducing agents. Acidify the mixture with M sulphuric acid and distil. A solution of formic and acetic acids will pass over. Neutralize the distillate with 2M ammonia solution, boil off excess ammonia (if necessary) and test as in (a).

Acetate

If other organic acids are present, obtain the solution of mixed acids as in (b); otherwise, use the neutralized soda extract. Boil under

Table 5.14 **Separation and detection of organic acids** Add 0.5M $CaCl_2$ solution to the cold neutralized soda extract, rub sides of vessel with a glass rod, allow to stand for 10 minutes with occasional shaking; filter.

Residue	*Filtrate*					
May be Ca oxalate (formed immediately) or Ca tartrate (formed on standing). Wash. Boil with 2M acetic acid and filter.	Add more $CaCl_2$ solution, boil under reflux for at least 5 minutes. A ppt. may form gradually; filter					
	Residue	*Filtrate*				
	Ca citrate. White. Citrate present. Divide into 2 parts. (1) Confirm by Denigés's test. (2) Dissolve in 2M HCl, convert into neutral solution with 2M NH_3 solution, etc., and add Cd acetate solution. White gelatinous precipitate.	If citrate present, evaporate to dryness on water bath, add a little cold water, and filter from any residue of Ca citrate. If citrate absent, proceed with solution. Add 0.5M $FeCl_3$ solution; filter (if necessary).				

Residue	*Filtrate*
CaC_2O_4. Dissolve in a little 2M HCl, add NH_3 solution in slight excess. White ppt. insoluble in 2M acetic acid. Oxalate present. Confirm by decolouriz-ation of acidified 0.004M $KMnO_4$ at 60–70°C.	May contain Ca tartrate. Test neutral solution with Fenton's reagent, violet colouration, or by silver mirror test for tartrate. Confirm by copper hydroxide test.

Residue	*Filtrate*		
Iron(III) benzoate and/or iron(III) succinate. Separate by procedure A.	Coloured. Dilute, boil and filter.		
	Residue	*Filtrate*	
	Basic iron(III) formate and basic iron(III) acetate. Separate by procedure B.	Coloured. (a) Violet – salicylate. Confirm by 'oil of winter-green' test. (b) Deep blue or greenish black – gallate or tannate.	

reflux with an equal volume of potassium dichromate solution and M sulphuric acid; this will decompose the formic acid. Distil off the acetic acid, neutralize and test in the distillate with 0.5M iron(III) chloride solution.

Chapter 6

Reactions of some less common ions

6.1 Introduction

In the previous chapters the discussions were restricted to those cations and anions which occur most often in ordinary samples. Having studied the reactions, separation, and identification of those ions, the reader should now concentrate on the so-called 'rarer' elements. Many of these, like tungsten, molybdenum, titanium, vanadium, and beryllium, have important industrial applications.

No attempt has been made to give more than a short introduction to the subject; to economize on space, most of the simple equations have been omitted. The elements have been classified, in so far as is possible, in the simple groups with which the reader is already familiar, and methods of separation have been briefly indicated. Thus thallium and tungsten are in Group I; molybdenum, gold, platinum, selenium, tellurium, and vanadium in Group II; and beryllium, titanium, uranium, thorium, and cerium in Group III. It is hoped that the subject-matter of this chapter will suffice to enable the reader to detect the presence of one or two of the less common ions.

6.2 Thallium, Tl (A_r: 204.34) – thallium(I)

Thallium is a heavy metal with characteristics reminiscent of lead. It melts at 302.3°C. Thallium metal can be dissolved readily in nitric acid; it is insoluble in hydrochloric acid.

Thallium forms the monovalent thallium(I) and trivalent thallium(III) ions, the former being of greater analytical importance. Thallium(III) ions are less frequently encountered in solutions, as they tend to hydrolyse in aqueous media, forming thallium(III) hydroxide precipitate. Thallium(I) ions can be oxidized to thallium(III) ions in acid media with permanganate and hexacyanoferrate(III) ions as well as with lead dioxide, chlorine gas, bromine water or aqua regia (but not with concentrated nitric acid). The reduction of thallium(III) ions to thallium(I) is easily effected by tin(II) chloride, sulphurous acid, iron(II) ions, hydroxylamine or ascorbic acid.

Tl$^+$ reactions **_Reactions of thallium(I) ions_** To study these reactions a 0.025M solution of thallium(I) sulphate, Tl_2SO_4, or a 0.05M solution of thallium(I) nitrate, $TlNO_3$, should be used. All these compounds are **HIGHLY POISONOUS**.

HCl
↓
white TlCl

1. *Dilute hydrochloric acid* white precipitate of thallium(I) chloride, TlCl, sparingly soluble in cold, but more soluble in hot water (compare lead).

KI
↓
yellow TlI

2. *Potassium iodide solution* yellow precipitate of thallium(I) iodide, TlI, almost insoluble in water; it is also insoluble in cold sodium thiosulphate solution (difference, and method of separation, from lead).

K_2CrO_4
↓
yellow Tl_2CrO_4

3. *Potassium chromate solution* yellow precipitate of thallium(I) chromate, Tl_2CrO_4, insoluble in cold, dilute nitric or sulphuric acid.

H_2S or $(NH_4)_2S$
↓
black Tl_2S

4. *Hydrogen sulphide* no precipitate in the presence of dilute mineral acid. Incomplete precipitation of black thallium(I) sulphide, Tl_2S, occurs in neutral or acetic acid solution.

5. *Ammonium sulphide solution* black precipitate of thallium(I) sulphide, Tl_2S, soluble in mineral acids. The precipitate is oxidized to thallium(I) sulphate, Tl_2SO_4, upon exposure to air.

Owing to the slight solubility of thallium(I) chloride, some of the thallium is also precipitated with Group II ions.

$Na_3[Co(NO_2)_6]$
↓
red $Tl_3[Co(NO_2)_6]$

6. *Sodium hexanitritocobaltate(III) solution* light-red precipitate of thallium(I) hexanitritocobaltate(III), $Tl_3[Co(NO_2)_6]$.

$H_2[PtCl_6]$
↓
yellow $Tl_2[PtCl_6]$

7. *Hexachloroplatinic(IV) acid solution* pale-yellow precipitate of thallium(I) hexachloroplatinate(IV) $Tl_2[PtCl_6]$, almost insoluble in water (solubility: $0.06\,\mathrm{g\,l^{-1}}$ at 15°C).

$K_3[Fe(CN)_6]$
↓
brown $Tl(OH)_3$

8. *Potassium hexacyanoferrate(III) solution* brown precipitate of thallium(III) hydroxide in alkaline solution:

$$Tl^+ + 2[Fe(CN)_6]^{3-} + 3OH^- \rightarrow Tl(OH)_3\downarrow + 2[Fe(CN)_6]^{4-}$$

A similar result is obtained with sodium hypochlorite, sodium hypobromite or hydrogen peroxide in alkaline solution.

NH_4SCN
↓
white TlSCN

9. *Ammonium thiocyanate solution* white precipitate of thallium(I) thiocyanate, TlSCN; the precipitate is dissolved in hot water.

Atomic spectrometry

10. *Atomic spectrometric tests* in **emission** mode 10 ppm thallium can be identified at 535.05 nm with the air/acetylene flame. By **atomic absorption**, using a thallium hollow-cathode lamp and an air/acetylene flame the detection sensitivity is about 1 ppm at 276.79 nm. A 1000 ppm thallium **standard** solution can be prepared by dissolving 1.2350 g of analytical grade thallium(I) sulphate Tl_2SO_4 in water and diluting to 1 litre.

Flame colour: green

11. *Flame test* all thallium salts exhibit a characteristic green colouration when introduced into the colourless Bunsen flame. When examined through the spectroscope only one sharp line can be seen, at 535 nm, in contrast to barium, which exhibits several lines between 510 and 550 nm (cf. Fig. 2.5 in Section 2.2).

6.3 Thallium, Tl (A_r: 204.34) – thallium(III)

The general physical and chemical properties of thallium were discussed in Section 6.2.

Tl³⁺ reactions

Reactions of thallium(III) ions   To study these reactions use a 0.2M solution of thallium(III) chloride, $TlCl_3$. Note that all thallium salts are **HIGHLY POISONOUS!**

NaOH
↓
brown Tl(OH)₃

1. *Sodium hydroxide or ammonia solution*   brown precipitate of thallium(III) hydroxide, insoluble in excess of the reagent (difference from thallium(I) salts, which give no precipitate), but readily soluble in hydrochloric acid.

2. *Hydrochloric acid*   no precipitate [difference from thallium(I) salts].

3. *Potassium chromate solution*   no precipitate [difference from thallium(I) salts].

KI
↓
brown TlI + I₃⁻

4. *Potassium iodide solution*   brownish-black precipitate, probably a mixture of thallium(I) iodide and iodine.

H₂S
↓
Tl₂S + S ppt

5. *Hydrogen sulphide*   reduced to thallium(I) with the precipitation of sulphur. If the acid is neutralized, thallium(I) sulphide, Tl_2S, precipitates.

6. *Atomic spectrometric tests*   these were described in Section 6.2, reaction 10.

7. *Flame test*   see Section 6.2, reaction 11.

6.4 Tungsten, W (A_r: 183.85) – tungstate

Solid tungsten is a white-coloured metal; the powdered metal is grey. Its melting point is extremely high (3370°C). The metal is insoluble in acids, including aqua regia. To dissolve metallic tungsten, it should be ignited first in a stream of oxygen, and the tungsten trioxide, WO_3, which is formed, can then be fused with solid sodium hydroxide in an iron crucible. The solidified melt will dissolve in water when tungstate ions, WO_4^{2-}, are formed.

Tungstates form complex acids with phosphoric, boric, and silicic acids; tungstic acid cannot therefore be precipitated from these compounds by hydrochloric acid. The complexes may usually be decomposed by heating with concentrated sulphuric acid, tungstic acid being liberated.

WO₄²⁻ reactions

Reactions of tungstate ions   To study these reactions use a 0.2M aqueous solution of sodium tungstate $Na_2WO_4 \cdot 2H_2O$.

HCl
↓
white ppt

1. *Dilute hydrochloric acid*   white precipitate of hydrated tungstic acid, $H_2WO_4 \cdot H_2O$, in the cold; upon boiling the mixture, this is converted into yellow tungstic acid, H_2WO_4, insoluble in dilute acids. Similar results are

obtained with dilute nitric and sulphuric acids, but not with phosphoric acid. Tartrates, citrates, and oxalates inhibit the precipitation of tungstic acid. The precipitate is soluble in dilute ammonia solution (distinction from $SiO_2 \cdot xH_2O$).

H_3PO_4
↓
white ppt

2. *Phosphoric acid* white precipitate of phosphotungstic acid (dodeca-tungstatophosphoric acid) $H_3[PO_4(W_{12}O_{36})]$ or $H_3[P(W_3O_{10})_4]$, soluble in excess of the reagent.

3. *Hydrogen sulphide* no precipitate in acid solution.

$(NH_4)_2S + $ acid
↓
brown WS_3,
dissolves in excess

4. *Ammonium sulphide solution* no precipitate, but if the solution is afterwards acidified with dilute hydrochloric acid, a brown precipitate of tungsten trisulphide, WS_3, is produced. The precipitate dissolves in ammonium sulphide solution forming a thiotungstate ion WS_4^{2-}.

Reduction to tungsten
blue

5. *Zinc and hydrochloric acid* if a solution of tungstate is treated with hydrochloric acid and then a little zinc added, a blue colouration or precipitate is produced. The product is called 'tungsten blue' and has a composition approximate to the formula W_2O_5.

6. *Tin(II) chloride solution* yellow precipitate, which becomes blue upon warming with concentrated hydrochloric acid.

The spot-test technique is as follows. Mix 1–2 drops of the test solution with 3–5 drops of the tin(II) chloride reagent on a spot plate. A blue precipitate or colouration of tungsten blue W_2O_5.

Sensitivity 5 μg W.

Concentration limit 1 in 10 000.

Molybdenum gives a similar reaction. If, however, a thiocyanate is added, the red complex ion $[Mo(SCN)_6]^{3-}$ is formed, and upon the addition of concentrated hydrochloric acid the red colour disappears and the blue colour due to tungsten remains.

The spot test is conducted as follows in the presence of molybdenum.

Place a drop of concentrated hydrochloric acid upon filter or drop-reaction paper and a drop of the test solution in the centre of the spot. A tungstate produces a yellow stain. Add a drop of 10% potassium thiocyanate solution and a drop of saturated tin(II) chloride; a red spot, due to $[Mo(SCN)_6]^{3-}$, is produced, but this disappears when a drop of concentrated hydrochloric acid is added and a blue colour, due to tungsten blue, remains.

Sensitivity 4 μg W.

Concentration limit 1 in 12 000.

$FeSO_4$
↓
brown ppt,
difference from Mo

7. *Iron(II) sulphate solution* brown precipitate. This turns white upon adding dilute hydrochloric acid, and then yellow upon heating (difference from molybdates).

$AgNO_3$
↓
yellow Ag_2WO_4

8. *Silver nitrate solution* pale yellow precipitate of silver tungstate, soluble in ammonia solution, decomposed by nitric acid with the formation of white hydrated tungstic acid.

Phenol test

9. *KHSO₄–H₂SO₄–phenol test (Defacqz reaction)* a little of the solid (or the residue obtained by evaporating a little of the solution to dryness) is heated with 4–5 times its weight of potassium hydrogen sulphate slowly to fusion, and the temperature is maintained until the fluid melt is clear. The cold melt is stirred with concentrated sulphuric acid. Upon adding a few milligrams of phenol to a few drops of the sulphuric acid solution, an intense red colouration is produced (difference from molybdate). A reddish-violet colouration is obtained if hydroquinone replaces the phenol. The test is a highly sensitive one and will detect 2 µg of tungstate.

Atomic spectrometry

10. *Atomic spectrometric tests* for the detection of tungsten, high-temperature flames have to be used. In **emission** mode 1 ppm can be identified at 400.87 nm with the nitrous oxide/acetylene flame. By **atomic absorption**, using a tungsten hollow-cathode lamp and a nitrous oxide/acetylene flame the detection sensitivity is about 1 ppm at the same wavelength. More sensitive results can be achieved at the 255.13 nm resonance line. A 1000 ppm tungsten **standard** solution can be prepared by dissolving 1.7942 g of analytical grade sodium tungstate dihydrate $Na_2WO_4 \cdot 2H_2O$ in a mixture of 100 ml water and 20 ml 2M sodium hydroxide. After complete dissolution dilute with water to 1 litre.

Dry test

11. *Dry test* microcosmic salt bead: oxidizing flame – colourless or pale-yellow; reducing flame – blue, changing to blood-red upon the addition of a little iron(II) sulphate.

6.5 Separation and identification of Group I cations in the presence of thallium and tungsten

The separation and identification of Hg(I), Ag, W, Pb, and Tl(I) is described in Table 6.1.

6.6 Molybdenum, Mo (A_r: 95.94) – molybdate

Molybdenum is a silverish-white, hard, heavy metal. In powder form it is grey. It melts at 2622°C. The metal is resistant to alkalis and hydrochloric acid. Dilute nitric acid dissolves it slowly, concentrated nitric acid renders it passive. Molybdenum can readily be dissolved in aqua regia or in a mixture of concentrated nitric acid and hydrogen fluoride.

Molybdenum forms compounds with oxidation numbers +2, +3, +4, +5, and +6. Of these, molybdates are the most important (with oxidation number +6). Molybdates are the salts of molybdic acid, H_2MoO_4. This acid tends to polymerize with the splitting off of molecules of water. Thus, the commercial ammonium molybdate is in fact a *heptamolybdate* in which $[Mo_7O_{24}]^{6-}$ ions are present. For the sake of simplicity the formula MoO_4^{2-} will be used in this text whenever molybdates are discussed.

Table 6.1
Separations in Group I

Separation and identification of Group I cations in the presence of Tl and W   The precipitate may contain $PbCl_2$, $AgCl$, Hg_2Cl_2, $TlCl$, and tungstic acid ($WO_3 \cdot xH_2O$). Wash the precipitate on the filter with 2 ml portions of 2M HCl, then 2–3 times with 1 ml portions of cold water, and reject the washings. Transfer the precipitate[a] to a boiling tube or to a small beaker, and boil with 10–15 ml water. Filter hot.

Residue	*Filtrate*

May contain Hg_2Cl_2, $AgCl$, and tungstic acid. Wash the ppt. several times with hot water until the washings give no ppt. with 0.1M K_2CrO_4 solution; this ensures the complete removal of the Pb and Tl.
    Pour 5 ml warm, 2M NH_3 solution repeatedly through the filter.
	May contain Pb^{2+} and Tl^+; these may crystallize out on cooling.
	    Evaporate to fuming with 2–3 ml concentrated H_2SO_4, cool, dilute to 10–20 ml, cool and filter.

Residue	*Filtrate*	*Residue*	*Filtrate*

If black, consists of $Hg(NH_2)Cl$ + Hg.
    Hg(I) present.
	May contain $[Ag(NH_3)_2]^+$ and WO_4^{2-}. Nearly neutralize with 2M HCl, add just enough 2M NH_3 solution to redissolve any ppt. which forms. Add 0.1M KI solution and filter.[b]	If white – consists of $PbSO_4$. This is soluble in 6M ammonium acetate solution; K_2CrO_4 solution then precipitates yellow $PbCrO_4$, insoluble in 2M acetic acid.
		    Pb present.
			May contain Tl^+. Just neutralize with 2M NH_3 solution, and add 0.1M KI solution. Yellow ppt. of TlI, insoluble in cold 0.5M $Na_2S_2O_3$ solution.
			    Tl present.
			    Confirm by flame test; intense green flame. (Use spectroscope, if available.)

Residue	*Filtrate*

Pale yellow (AgI).
    Ag present.
	May contain WO_4^{2-}. Evaporate to a small volume, acidify with 2M HCl, add 3 ml 0.25M $SnCl_2$ solution, boil, add 3 ml concentrated HCl, and heat again to boiling.
	    Blue ppt. or colouration.
	    W present.
	    Confirm by the Defacqz reaction.

[a] A gelatinous precipitate may also be hydrated silica, which is partially precipitated here from silicates decomposed by acids.

[b] The solution obtained after dissolving in ammonia should be discarded quickly to avoid a serious **explosion!** (Cf. Section 3.6, reaction 1.)

MoO_4^{2-} reactions

Reactions of molybdate ions   To study these reactions use a 4.5% solution of ammonium molybdate, $(NH_4)_6Mo_7O_{24} \cdot 4H_2O$, which is approximately 0.25M for molybdate MoO_4^{2-}.

HCl
↓
white ppt

1. *Dilute hydrochloric acid*   white or yellow precipitate of molybdic acid, H_2MoO_4, from concentrated solutions, soluble in excess mineral acid.

H_2S
↓
blue colour
then brown MoS_3

2. *Hydrogen sulphide*   with a small quantity of the gas and an acidified molybdate solution, a blue colouration is produced; further passage of hydrogen sulphide yields a brown precipitate of the trisulphide, MoS_3; this is soluble in ammonium sulphide solution to a solution containing a

thiomolybdate, $[MoS_4]^{2-}$, from which MoS_3 is reprecipitated by the addition of acids. The precipitation in acid solution is incomplete in the cold; more extensive precipitation is obtained by the prolonged passage of the gas into the boiling solution and under pressure. Precipitation is quantitative with excess hydrogen sulphide at $0°C$ in the presence of formic acid.

Reduction to
molybdenum blue

3. *Reducing agents* e.g. zinc, tin(II) chloride solution, colour a molybdate solution acidified with dilute hydrochloric acid blue (due to 'molybdenum blue' Mo_2O_5), then green and finally brown (see also section 3.15, reaction 8).

NH_4SCN
↓
yellow colour

4. *Ammonium thiocyanate solution* yellow colouration in solution acidified with dilute hydrochloric acid, becoming blood-red upon the addition of zinc or of tin(II) chloride on account of the formation of hexathiocyanatomolybdate(III) $[Mo(SCN)_6]^{3-}$; the latter is soluble in ether. The red colouration is produced in the presence of phosphoric acid (difference from iron).

The spot-test technique is as follows. Place a drop of the test solution and a drop of 10% potassium thiocyanate solution upon quantitative filter paper or upon drop-reaction paper. Add a drop of saturated tin(II) chloride solution. A red spot is obtained.

Sensitivity 0.1 μg Mo.

Concentration limit 1 in 500 000.

If iron is present, a red spot will appear initially but this disappears upon the addition of tin(II) chloride solution (or of sodium thiosulphate solution). Tungstates reduce the sensitivity of the test (cf. Section 6.4, reaction 6).

Na_2HPO_4
↓
yellow ppt

5. *Disodium hydrogen phosphate solution* yellow, crystalline precipitate of ammonium phosphomolybdate in the presence of excess nitric acid (cf. phosphates, Section 4.28, reaction 4).

$K_4[Fe(CN)_6]$
↓
brown ppt

6. *Potassium hexacyanoferrate(II) solution* reddish-brown precipitate of molybdenum hexacyanoferrate(II), insoluble in dilute mineral acids, but readily soluble in solutions of caustic alkalis and ammonia [difference from uranyl and copper(II) hexacyanoferrate(II)s].

Cupron
↓
white ppt

7. *Benzoin α-oxime reagent (or 'cupron' reagent)*

$$\{C_6H_5 \cdot CHOH \cdot C(=NOH)C_6H_5\}$$

The molybdate solution is strongly acidified with dilute sulphuric acid and 0.5 ml of the reagent added. A white precipitate is produced.

K xanthate
↓
purple colour

8. *Ethyl potassium xanthate test* $\{SC(SK)OC_2H_5\}$ when a molybdate solution is treated with a little solid ethyl potassium xanthate and then acidified with dilute hydrochloric acid, a red-purple colouration is produced. With large amounts of molybdenum, the compound separates as dark, oily drops which are readily soluble in organic solvents such as benzene, chloroform, and carbon disulphide. The reaction product has been given the formula $MoO_2[SC(SH)(OC_2H_5)]_2$. The test is said to be specific for molybdates, although copper, cobalt, nickel, iron, chromium, and uranium under exceptional conditions interfere. Large quantities of oxalates, tartrates, and citrates decrease the sensitivity of the test.

The spot-test technique, for which the reaction is particularly well adapted, is as follows. Place a drop of the nearly neutral or faintly acid test solution on a spot plate, introduce a minute crystal of ethyl potassium xanthate, followed by 2 drops of 2M hydrochloric acid. An intense red-violet colouration is obtained.

Sensitivity 0.04 µg Mo.
Concentration limit 1 in 250 000.

Phenylhydrazine
↓
red colour

9. *Phenylhydrazine reagent* $(C_6H_5 \cdot NHNH_2)$ a red colouration or precipitate is produced when molybdates and an acid solution of phenylhydrazine react. The latter is oxidized by the molybdate to a diazonium salt, which then couples with the excess of base in the presence of the molybdate to yield a coloured compound.

Mix a drop of the test solution and a drop of the reagent on a spot plate. A red colouration appears.

Sensitivity 0.3 µg Mo.
Concentration limit 1 in 150 000.

$Fe_2(SO_4)_3$
↓
reduction, difference from W

10. *Iron(II) sulphate solution*   reddish-brown colour. Upon adding dilute mineral acid, the colour changes to blue; the colour becomes paler and more green upon warming but returns to blue on cooling (difference from tungstate).

Atomic spectrometry

11. *Atomic spectrometric tests*   to avoid interferences from certain metals (like Ca, Fe) it is best to use the nitrous oxide/acetylene flame. In **emission** mode 1 ppm molybdenum can be identified at 390.30 nm with either the air/acetylene or the nitrous oxide/acetylene flame. By **atomic absorption**, using a molybdenum hollow-cathode lamp and an air/acetylene flame the detection sensitivity is about 0.1 ppm at 313.26 nm. A 1000 ppm molybdenum **standard** solution can be prepared by dissolving 1.8403 g of analytical grade ammonium molybdate tetrahydrate $(NH_4)_6Mo_7O_{24} \cdot 4H_2O$ in water and diluting to 1 litre.

Dry tests

12. *Dry tests*

(a) Microcosmic salt bead: oxidizing flame – yellow to green while hot and colourless when cold; reducing flame – brown when hot, green when cold.

(b) Evaporation with concentrated sulphuric acid in a porcelain dish or crucible: a blue mass (containing 'molybdenum blue') is obtained. The blue colour is destroyed by dilution with water.

6.7   Gold, Au (A_r: 196.97) – gold(III)

Gold is a heavy metal with its characteristic yellow colour. In powderous form it is reddish-brown. It melts at 1064.8°C.

Gold is resistant to acids; only aqua regia dissolves it when tetrachloroaurate(III), $[AuCl_4]^-$, anions are formed. Gold dissolves slowly in potassium cyanide, when dicyanoaurate(I), $[Au(CN)_2]^-$, anions are formed. From both the monovalent and trivalent forms gold can easily be reduced to the metal. Gold(I) compounds are less stable than those of gold(III).

Au^{3+} reactions

Reactions of gold(III) [tetrachloroaurate(III)] ions To study these reactions use a 0.33M solution of commercial gold(III) chloride, which in fact is hydrogen tetrachloroaurate(III), H[AuCl$_4$]·3H$_2$O.

H$_2$S
↓
black Au$_2$S + S

1. *Hydrogen sulphide* black precipitate of gold(I) sulphide, Au$_2$S (usually mixed with a little free gold), in the cold. It is insoluble in dilute acids, but largely soluble in yellow ammonium sulphide solution, from which it is reprecipitated by dilute hydrochloric acid. A brown precipitate of metallic gold, together with gold(I) sulphide and sulphur, is obtained upon precipitation of a hot solution; this is also largely dissolved by yellow ammonium sulphide solution.

$$2[AuCl_4]^- + 3H_2S \rightarrow Au_2S\downarrow + 2S\downarrow + 6H^+ + 8Cl^-$$

NH$_3$
↓
yellow ppt

2. *Ammonia solution* yellow precipitate of 'fulminating gold'; this has been formulated as Au$_2$O$_3$·3NH$_3$ + NH(ClNH$_2$Au)$_2$ but the exact composition is not fully established. The dry substance **EXPLODES UPON HEATING** or upon percussion (**DANGER**).

(COOH)$_2$
↓
brown Au

3. *Oxalic acid solution* gold is precipitated as a fine brown powder (or sometimes as a mirror) from cold neutral solutions (difference from platinum and other Group II metals). Under suitable conditions, the gold is obtained in the colloidal state as a red, violet, or blue solution.

$$2[AuCl_4]^- + 3(COO)_2^{2-} \rightarrow 2Au\downarrow + 6CO_2\uparrow + 8Cl^-$$

Similar results are obtained with iron(II) sulphate solution. Reduction also occurs with hydroxylamine and hydrazine salts, and with ascorbic acid.

SnCl$_2$
↓
purple of Cassius

4. *Tin(II) chloride solution* purple precipitate, 'purple of Cassius', consisting of an adsorption compound of tin(II) hydroxide, Sn(OH)$_2$, and colloidal gold, in neutral or weakly acid solution. In extremely dilute solutions only a purple colouration is produced. If the solution is made strongly acid with hydrochloric acid, a dark-brown precipitate of pure gold is formed.

$$2[AuCl_4]^- + 4Sn^{2+} + 2H_2O \rightarrow$$
$$\rightarrow 2Au\downarrow + Sn(OH)_2\downarrow + 3Sn^{4+} + 2H^+ + 8Cl^-$$

H$_2$O$_2$ + NaOH
↓
Au

5. *Hydrogen peroxide* the finely divided metal is precipitated in the presence of sodium hydroxide solution (distinction from platinum). The precipitated metal appears brownish-black by reflected light and bluish-green by transmitted light.

$$2[AuCl_4]^- + 3H_2O_2 + 6OH^- \rightarrow 2Au\downarrow + 3O_2\uparrow + 8Cl^- + 6H_2O$$

NaOH
↓
brown Au(OH)$_3$

6. *Sodium hydroxide solution* reddish-brown precipitate of gold(III) hydroxide, Au(OH)$_3$, from concentrated solutions. The precipitate has amphoteric properties; it dissolves in excess alkali forming tetrahydroxoaurate(III), [Au(OH)$_4$]$^-$, ion.

Rhodanine
↓
violet ppt

7. *4-Dimethylaminobenzylidene-rhodanine (or rhodanine) reagent* (for formula see silver, Section 3.6, reaction 11): red-violet precipitate in neutral or faintly acid solution. Silver, mercury, and palladium salts give coloured compounds with the reagent and must therefore be absent.

Moisten a piece of drop-reaction paper with the reagent and dry it. Place a drop of the neutral or weakly acid test solution upon it. A violet spot or ring is obtained.

Sensitivity 0.1 µg Au.

Concentration limit 1 in 500 000.

Atomic spectrometry

8. *Atomic spectrometric tests* in **emission** mode 10 ppm gold can be identified at 268.60 nm with the air/acetylene flame. By **atomic absorption**, using a gold hollow-cathode lamp and an air/acetylene flame the detection sensitivity is about 1 ppm at 242.80 nm. A 1000 ppm gold **standard** solution can be prepared by dissolving 0.1000 g of analytical grade gold metal in a mixture of 15 ml concentrated hydrochloric acid and 5 ml nitric acid. The solution should be evaporated on a water bath to 5 ml, then diluted to 100 ml with water. Keep the solution in a dark bottle.

Dry test

9. *Dry test* all gold compounds when heated upon charcoal with sodium carbonate yield yellow, malleable, metallic particles, which are insoluble in nitric acid, but soluble in aqua regia. The aqua regia solution should be evaporated to dryness, dissolved in water, and tests 1, 3 or 4 applied.

6.8 Platinum, Pt (A_r: 195.09)

Platinum is a greyish-white, ductile and malleable heavy metal with a density of 21.45 g cm^{-3} and a melting point of 1773°C.

It is a noble metal, not attacked by dilute or concentrated acids, except aqua regia, which dissolves platinum forming hexachloroplatinate(IV) ions:

$$3Pt + 4HNO_3 + 18HCl \rightarrow 3[PtCl_6]^{2-} + 4NO\uparrow + 6H^+ + 8H_2O$$

Molten alkalis and alkali peroxides attack platinum, therefore these should never be melted in platinum crucibles. In its compounds, platinum can be mono-, di-, tri-, tetra- and hexavalent, tetravalent platinum being the most important in analytical practice.

$[PtCl_6]^{2-}$ reactions

Reactions of hexachloroplatinate(IV) ions To study these reactions use a 0.5M solution of hexachloroplatinic acid [hydrogen hexachloroplatinate(IV)].

H$_2$S
↓
black PtS$_2$

1. *Hydrogen sulphide* black (or dark brown) precipitate of the disulphide, PtS$_2$ (possibly containing a little platinum metal), is slowly formed in the cold, but rapidly on warming. The precipitate is insoluble in concentrated acids, but dissolves in aqua regia and also in ammonium polysulphide solution; it is reprecipitated from the latter by dilute acids.

$$[PtCl_6]^{2-} + 2H_2S \rightarrow PtS_2\downarrow + 4H^+ + 6Cl^-$$

KCl
↓
yellow K$_2$[PtCl$_6$]

2. *Potassium chloride solution*   yellow precipitate of potassium hexachloro-platinate(IV), K$_2$[PtCl$_6$], from concentrated solutions (difference from gold). A similar result is obtained with ammonium chloride solution.

3. *Oxalic acid solution*   no precipitate of platinum (difference from gold). Hydrogen peroxide and sodium hydroxide solution likewise do not precipitate metallic platinum.

Reducing agents
↓
black Pt

4. *Sodium formate*   black powder of metallic platinum from neutral boiling solutions.

$$[PtCl_6]^{2-} + 2HCOO^- \rightarrow Pt\downarrow + 2CO_2\uparrow + 2H^+ + 6Cl^-$$

5. *Zinc, cadmium, magnesium or aluminium*   all these metals precipitate finely divided platinum.

$$[PtCl_6]^{2-} + 2Zn \rightarrow Pt\downarrow + 2Zn^{2+} + 6Cl^-$$

6. *Hydrazine sulphate*   ready reduction in ammoniacal solution to metallic platinum, some of which is deposited as a mirror upon the sides of the tube.

$$[PtCl_6]^{2-} + N_2H_4 + 4NH_3 \rightarrow Pt\downarrow + N_2\uparrow + 4NH_4^+ + 6Cl^-$$

AgNO$_3$
↓
yellow Ag$_2$[PtCl$_6$]

7. *Silver nitrate solution*   yellow precipitate of silver hexachloroplati-nate(IV), Ag$_2$[PtCl$_6$], sparingly soluble in ammonia solution but soluble in solutions of alkali cyanides and of alkali thiosulphates.

KI
↓
red complex

8. *Potassium iodide solution*   intense brownish-red or red colouration, due to [PtI$_6$]$^{2-}$ ions. With excess of the reagent K$_2$[PtI$_6$] may be precipitated as an unstable brown solid. On warming, black PtI$_4$ may be precipitated.

SnCl$_2$
↓
yellow colloidal Pt

9. *Tin(II) chloride solution*   red or yellow colouration, due to tetrachloro-platinate(IV) ions, [PtCl$_4$]$^{2-}$, soluble in ethyl acetate or in ether.

   To employ this reaction as a spot test in the presence of other noble metals (gold, palladium, etc.), the noble metals are first precipitated with thallium(I) ions. From the mixture, gold, palladium, etc. can be dissolved with ammonia, but thallium(I) hexachloroplatinate remains.

   Place a drop of saturated thallium(I) nitrate solution upon drop-reaction paper, add a drop of the test solution and then another drop of the thallium(I) nitrate solution. Wash the precipitate with ammonia solution, and add a drop of strongly acid tin(II) chloride solution. A yellow or orange spot remains.

   Sensitivity 0.5 μg Pt.

   Concentration limit 1 in 80 000.

Rubeanic acid
↓
red ppt

10. *Dithio oxamide (rubeanic acid) reagent (0.02%)*

$$\begin{array}{c} H_2N-C-C-NH_2 \\ \parallel \ \parallel \\ S \ \ S \end{array}$$

a purplish-red precipitate of the complex

$$\left(Pt \begin{array}{c} NH_2-C=S \\ \quad\quad\quad | \\ S-\!\!\!-\!\!\!-C=NH \end{array} \right)_2$$

is formed. Palladium and a large proportion of gold interfere.

Place a drop of the test solution (acid with HCl) upon a spot plate and add a drop of the reagent. A purplish-red precipitate is produced.

Concentration limit 1 in 10 000.

Atomic spectrometry

11. *Atomic spectrometric tests*   these can be somewhat unreliable, as the analytical signal can be depressed by a large number of metals present in the mixture. In **emission** mode 10 ppm platinum can be identified at 265.95 nm with the air/acetylene flame. By **atomic absorption**, using a platinum hollow-cathode lamp and an air/acetylene flame the detection sensitivity is about 1 ppm at the same wavelength. A 1000 ppm platinum **standard** solution can be prepared by dissolving 0.2275 g of analytical grade ammonium hexachloroplatinate $(NH_4)_2PtCl_6$ in water and diluting to 100 ml.

Dry test

12. *Dry test*   all platinum compounds when fused with sodium carbonate upon charcoal are reduced to the grey, spongy metal (distinction from gold). The residue is insoluble in concentrated mineral acids, but dissolves in aqua regia. The solution is evaporated almost to dryness, dissolved in water, and reactions 1, 2, 5, 9 or 10 applied.

6.9   Palladium, Pd (A_r: 106.4)

Palladium is a light-grey metal which melts at 1555°C. Its most interesting physical characteristic is that it is able to dissolve (absorb) hydrogen gas in large quantities.

Unlike platinum, palladium is slowly dissolved by concentrated nitric acid, and by hot concentrated sulphuric acid, forming a brown solution of palladium(II) ions. Palladium can also be dissolved by fusing the metal first with potassium pyrosulphate and then leaching the frozen melt with water. The metal dissolves readily in aqua regia, when both Pd^{2+} and Pd^{4+} (more precisely, $[PdCl_4]^{2-}$ and $[PdCl_6]^{2-}$) ions are formed. If such a solution is evaporated to dryness, the latter loses chlorine so that on treating the residue with water a solution of palladium(II) ions is obtained. Palladium(II) ions are most stable; palladium(III) and (IV) compounds can easily be transformed into palladium(II).

Pd²⁺ reactions

Reactions of palladium(II) ions   To study these reactions use a 1% solution of palladium(II) chloride.

H₂S
↓
black PdS

1. *Hydrogen sulphide*   black precipitate of palladium(II) sulphide, PdS, from acid or neutral solutions. The precipitate is insoluble in ammonium sulphide solution.

NaOH
↓
brown ppt

2. *Sodium hydroxide solution*   reddish-brown, gelatinous precipitate of the hydrated oxide, $PdO \cdot nH_2O$ (this may be contaminated with a basic salt), soluble in excess of the precipitant.

NH_3
↓
red ppt,
dissolves in excess

3. *Ammonia solution*   red precipitate of $[Pd(NH_3)_4][PdCl_4]$, soluble in excess of the reagent to give a colourless solution of $[Pd(NH_3)_4]^{2+}$ ions. Upon acidifying the latter solution with hydrochloric acid, a yellow crystalline precipitate of $[Pd(NH_3)_2Cl_2]$ is obtained.

KI
↓
black PdI_2,
dissolves in excess

4. *Potassium iodide solution*   black precipitate of palladium(II) iodide, PdI_2, in neutral solution, soluble in excess of the reagent to give a brown solution of $[PdI_4]^{2-}$. In acid solution, black PdI_2 is precipitated.

$Hg(CN)_2$
↓
white $Pd(CN)_2$,
difference from Pt

5. *Mercury(II) cyanide solution*   white precipitate of palladium(II) cyanide, $Pd(CN)_2$ (difference from platinum), sparingly soluble in dilute hydrochloric acid, readily soluble in potassium cyanide solution and in ammonia solution.

1-Nitroso-2-naphthol
↓
brown ppt

6. *1-Nitroso-2-naphthol solution*   brown, voluminous precipitate of $Pd(C_{10}H_6O_2N)_2$ (difference from platinum).

Reducing agents
↓
black Pd

7. *Reducing agents (Cd, Zn or Fe in acid solution, formic acid, sulphurous acid, etc.)*   black, spongy precipitate of metallic palladium, 'palladium black'. Tin(II) chloride yields a brown suspension containing metallic palladium.

Dimethylglyoxime
↓
yellow ppt

8. *Dimethylglyoxime reagent*   yellow, crystalline precipitate of palladium dimethylglyoxime, $Pd(C_4H_7O_2N_2)_2$, insoluble in M hydrochloric acid (difference from nickel and from other platinum metals) but soluble in dilute ammonia solution and in potassium cyanide solution (cf. nickel, Section 3.27, reaction 8).

Salicylaldoxime reagent also precipitates palladium quantitatively as $Pd(C_7H_6O_2N)_2$ (cf. copper, Section 3.10, reaction 9; difference from platinum).

Place a drop of the slightly acid solution on a microscope slide and add a minute crystal of dimethylglyoxime. After some minutes, a yellow precipitate is formed. This is seen to consist of long, very characteristic needles when examined under a microscope ($\times 75$).

Platinum does not interfere, but gold and nickel give a similar reaction.
Concentration limit 1 in 10 000.

Hg_2Cl_2 suspension
↓
reduction to Pd

9. *Mercury(I) chloride (calomel) suspension*   reduces palladium(II) ions to the metal

$$Pd^{2+} + 2Cl^- + Hg_2Cl_2\downarrow \rightarrow Pd\downarrow + 2HgCl_2$$

Shake the slightly acid test solution with solid mercury(I) chloride in the cold. The solid acquires a grey colour.
Concentration limit 1 in 100 000.

Atomic spectrometry

10. *Atomic spectrometric tests*   in **emission** mode 10 ppm palladium can be identified at 363.47 nm with the air/acetylene flame. By **atomic absorption**, using a palladium hollow-cathode lamp and an air/acetylene flame the detection sensitivity is about 1 ppm at 247.64 nm. An approximately 1000 ppm

palladium **standard** solution can be prepared by dissolving 0.17 g of reagent grade palladium(II) chloride $PdCl_2$ in water and diluting to 100 ml.

6.10 Selenium, Se (A_r: 78.96) – selenites, SeO_3^{2-}

Selenium resembles sulphur in many of its properties. It dissolves readily in concentrated nitric acid or in aqua regia to form selenious acid, H_2SeO_3. In analytical work selenites, SeO_3^{2-}, and selenates, SeO_4^{2-}, are both encountered; selenites being more stable. In this section selenites are dealt with; the reactions of selenates together with dry tests for selenium will be dealt with in Section 6.11.

SeO_3^{2-} reactions

Reactions of selenite ions To study these reactions use a 0.1M solution of selenious acid, H_2SeO_3 (or SeO_2), or sodium selenite, Na_2SeO_3.

H_2S
↓
yellow Se + S

1. *Hydrogen sulphide* yellow precipitate, consisting of a mixture of selenium and sulphur, in the cold, becoming red on heating. The precipitate is readily soluble in yellow ammonium sulphide solution.

$$SeO_3^{2-} + 2H_2S + 2H^+ \rightarrow Se\downarrow + 2S\downarrow + 3H_2O$$

Reducing agents
↓
red Se

2. *Reducing agents (sulphur dioxide, solution of tin(II) chloride, iron(II) sulphate, hydroxylamine hydrochloride, hydrazine hydrochloride or hydriodic acid (KI + HCl), zinc or iron)* red precipitate of selenium in hydrochloric acid solution. The precipitate frequently turns greyish-black on warming. When solutions in concentrated hydrochloric acid are boiled or evaporated, serious losses of selenium as $SeCl_4$ occur.

$CuSO_4$
↓
green $CuSeO_3$

3. *Copper sulphate solution* bluish-green, crystalline precipitate of copper selenite, $CuSeO_3$, in neutral solution (difference from selenate). The precipitate is soluble in dilute acetic acid.

$BaCl_2$
↓
white $BaSeO_3$

4. *Barium chloride solution* white precipitate of barium selenite, $BaSeO_3$, in neutral solution, soluble in dilute mineral acids.

Thiourea test

5. *Thiourea test, $CS(NH_2)_2$* solid or dissolved thiourea precipitates selenium as a red powder from cold dilute solutions of selenites. Tellurium and bismuth give yellow precipitates, whilst large amounts of nitrite and of copper interfere.

Place a little powdered thiourea on quantitative filter paper and moisten it with a drop of the test solution. Orange-red selenium separates out.

Sensitivity 2 μg Se.

KI
↓
reduced to Se

6. *Iodides* selenites are reduced by potassium iodide and hydrochloric acid

$$SeO_3^{2-} + 6I^- + 6H^+ \rightarrow Se\downarrow + 2I_3^- + 3H_2O$$

The iodine is removed by adding a thiosulphate, and the selenium remains as a reddish-brown powder. Tellurites react under these conditions forming the complex anion, $[TeI_6]^{2-}$, which also has a reddish-brown colour; it is, however, decomposed and decolourized by thiosulphate, thus permitting the detection of selenium in the presence of not too large an excess of tellurium.

Place a drop of concentrated hydriodic acid (or a drop each of 6M potassium iodide solution and of concentrated hydrochloric acid) upon drop-reaction paper and introduce a drop of the acid test solution into the middle of the original drop. A brownish-black spot appears. Add a drop of 0.5M sodium thiosulphate solution to the spot; a reddish-brown stain of elemental selenium remains.

Sensitivity 1 µg Se (in 0.025 ml).

Concentration limit 1 in 25 000.

Pyrrole
↓
blue colour,
distinction
from SeO_4^{2-}

7. *Pyrrole reagent*

$$\begin{array}{c} CH{-\!\!-}CH \\ \parallel \qquad \parallel \\ CH \quad CH \\ \diagdown \diagup \\ NH \end{array}$$

Selenious acid oxidizes pyrrole to a blue dyestuff of unknown composition ('pyrrole blue'). Iron salts accelerate the reaction when it is carried out in phosphoric acid solution. Selenic, tellurous and telluric acids do not react under the conditions given below: the test therefore provides a method of distinguishing selenites and selenates.

Place a drop of 0.5M iron(III) chloride solution and 7 drops of concentrated phosphoric acid (density $1.75\,g\,cm^{-3}$) on a spot plate containing 1 drop of the test solution and stir well. Add a drop of the pyrrole reagent and stir again. A greenish-blue colouration is obtained.

Sensitivity 0.5 µg Se.

Concentration limit 1 in 100 000.

NH_4SCN
↓
reduction to Se

8. *Ammonium thiocyanate and hydrochloric acid* selenites are reduced in acid solution to elemental selenium:

$$2SeO_3^{2-} + SCN^- + 4H^+ \rightarrow 2Se\downarrow + CO_2\uparrow + NH_4^+ + SO_4^{2-}$$

Mix 0.5 ml of the test solution with 2 ml of 10% potassium thiocyanate solution and 5 ml of 6M hydrochloric acid, and boil for 30 seconds. A red colouration, due to selenium, is produced.

6.11 Selenium, Se (A_r: 78.96) – selenates, SeO_4^{2-}

The most important properties of selenium were described in Section 6.10.

SeO_4^{2-} reactions

Reactions of selenate ions To study these reactions use a 0.1M solution of potassium selenate, K_2SeO_4, or sodium selenate, $Na_2SeO_4 \cdot 10H_2O$.

1. *Hydrogen sulphide* no precipitation occurs. If the solution is boiled with concentrated hydrochloric acid, the selenic acid is reduced to selenious acid; hydrogen sulphide then precipitates a mixture of selenium and sulphur.

$$SeO_4^{2-} + 2HCl \rightarrow SeO_3^{2-} + Cl_2\uparrow + H_2O$$

2. *Sulphur dioxide*   no reducing action.

3. *Copper sulphate solution*   no precipitate (difference from selenite).

4. *Barium chloride solution*   white precipitate of barium selenate, BaSeO₄, insoluble in dilute mineral acids. The precipitate dissolves when boiled with concentrated hydrochloric acid, and chlorine is evolved (distinction and separation from sulphate).

$$BaSeO_4\downarrow + 2HCl \rightarrow SeO_3^{2-} + Ba^{2+} + Cl_2\uparrow + H_2O$$

5. *Dry tests*

   (a) Selenium compounds mixed with sodium carbonate and heated upon charcoal: odour of rotten horseradish. A foul odour, due to hydrogen selenide, H_2Se, is obtained upon moistening the residue with a few drops of dilute hydrochloric acid. A black stain (due to Ag_2Se) is produced when the moistened residue is placed in contact with silver foil.

   (b) Elemental selenium dissolves in concentrated sulphuric acid: green solution, due to the presence of the compound $SeSO_3$. Upon dilution with water, red selenium is precipitated.

6.12   Tellurium, Te (A_r: 127.60) – tellurites, TeO_3^{2-}

Tellurium is less widely distributed in nature than selenium; both elements belong to Group VI of the periodic table. When fused with potassium cyanide, elemental tellurium is converted into potassium telluride, K_2Te, which dissolves in water to yield a red solution. If air is passed through the solution, the tellurium is precipitated as a black powder (difference and method of separation from selenium). Selenium under similar conditions yields the stable potassium selenocyanide, KSeCN; the selenium may be precipitated by the addition of dilute hydrochloric acid to its aqueous solution.

$$2KCN + Te \rightarrow K_2Te + (CN)_2\uparrow$$

$$2Te^{2-} + 2H_2O + O_2 \rightarrow 2Te\downarrow + 4OH^-$$

$$KCN + Se \rightarrow KSeCN$$

$$SeCN^- + H^+ \rightarrow Se\downarrow + HCN\uparrow$$

Tellurium is converted into the dioxide, TeO_2, by nitric acid. Like sulphur and selenium, it forms two anions, the tellurite, TeO_3^{2-}, and the tellurate, TeO_4^{4-}.

Reactions of tellurite ions   To study these reactions use a 0.1M solution of potassium tellurite, K_2TeO_3, or sodium tellurite, Na_2TeO_3.

1. *Hydrogen sulphide*   brown precipitate of the disulphide, TeS_2, from acid solutions. The sulphide decomposes easily into tellurium and sulphur, and is readily soluble in ammonium sulphide solution but insoluble in concentrated hydrochloric acid.

SO$_2$
↓
reduction to Te

2. *Sulphur dioxide* complete precipitation of tellurium from dilute (1–5M) hydrochloric acid solutions as a black powder. In the presence of much concentrated hydrochloric acid, no precipitate is formed (difference and method of separation from selenium).

3. *Iron(II) sulphate solution* no precipitation of tellurium (difference from selenium). A similar result is obtained with hydriodic acid (KI + HCl).

SnCl$_2$
↓
reduction to Te

4. *Tin(II) chloride or hydrazine hydrochloride solution or zinc* black tellurium precipitated.

HCl
↓
white H$_2$TeO$_3$

5. *Dilute hydrochloric acid* white precipitate of tellurous acid, H$_2$TeO$_3$ (difference from selenite), soluble in excess of the precipitant.

BaCl$_2$
↓
white BaTeO$_3$

6. *Barium chloride solution* white precipitate of barium tellurite, BaTeO$_3$, soluble in dilute hydrochloric acid but insoluble in 30% acetic acid.

KI
↓
black TeI$_4$

7. *Potassium iodide solution* black precipitate of TeI$_4$ in faintly acid solution, dissolving in excess of the reagent to form the red hexaiodo-tellurate(IV) ion, $[TeI_6]^{2-}$ (difference from selenite).

H$_3$PO$_2$
↓
reduction to Te

8. *Hypophosphorous acid test* both tellurites and tellurates are reduced to tellurium upon evaporation with hypophosphorous acid:

$$TeO_3^{2-} + H_2PO_2^- \rightarrow Te\downarrow + PO_4^{3-} + H_2O$$

$$2TeO_4^{2-} + 3H_2PO_2^- \rightarrow 2Te\downarrow + 3PO_4^{3-} + 2H^+ + 2H_2O$$

Selenites are likewise reduced to selenium. If, however, the solution of the selenites in concentrated sulphuric acid is treated with sodium sulphite, selenium separates but the tellurite is unaffected; the latter can be detected in the solution after removing the excess sulphide. Salts of silver, copper, gold, and platinum must be absent for they are reduced to the metal by the reagent.

Mix a drop of the test solution in mineral acid and a drop of 50% hypophosphorous acid in a porcelain micro crucible, and evaporate almost to dryness. Black grains or a grey stain of tellurium are obtained.

Sensitivity 0.1 μg tellurous acid.
Concentration limit 1 in 500 000.

Sensitivity 0.5 μg telluric acid.
Concentration limit 1 in 100 000.

6.13 Tellurium, Te (A_r : 127.60) – tellurates, TeO$_4^{2-}$

The most important characteristics of tellurium were described in Section 6.12.

TeO$_4^{2-}$ reactions

Reactions of tellurate ions To study these reactions use a 0.1M solution of sodium tellurate, Na$_2$TeO$_4$.

1. *Hydrogen sulphide* no precipitate in hydrochloric acid solution in the cold. In hot acid solution, the tellurate is first reduced to tellurite, and precipitation of the tellurium then occurs (compare Section 6.12). Other reducing agents give similar results.

2. *Hydrochloric acid* no precipitate in the cold. Upon boiling the solution, chlorine is evolved; tellurous acid, H_2TeO_3, is precipitated upon dilution (distinction from selenium).

$$TeO_4^{2-} + 4HCl \rightarrow H_2TeO_3\downarrow + Cl_2\uparrow + 2Cl^- + H_2O$$

BaCl$_2$
↓
white BaTeO$_4$

3. *Barium chloride solution* white precipitate of barium tellurate, $BaTeO_4$, from neutral solutions; the precipitate is readily soluble in dilute hydrochloric acid and in dilute acetic acid (distinction from selenate).

KI
↓
red complex

4. *Potassium iodide solution* yellow to red colour, due to hexaiodotellurate(VI) ions, $[TeI_6]^{2-}$, from dilute acid solutions (difference from tellurites).

Reducing agents act
only in hot solutions

5. *Reducing agents* no precipitate is produced in cold solutions with hydrogen sulphide or sulphur dioxide; with hot solutions, or with solutions that have been boiled with hydrochloric acid, precipitates of brown TeS (or Te + S) and of black Te respectively are formed. Tin(II) chloride, hydrazine or zinc in acid solution give black tellurium upon warming.

Dry tests

6. *Dry tests*

   (a) Fusion of any tellurium compound with sodium carbonate upon charcoal: formation of sodium telluride, Na_2Te, which produces a black stain (due to Ag_2Te) when placed in contact with moist silver foil.

   (b) Elemental tellurium dissolves in concentrated sulphuric acid to produce a red solution due to the presence of $TeSO_3$. Upon dilution with water, grey tellurium is precipitated.

6.14 Separation and identification of Group II cations in the presence of molybdenum, gold, platinum, palladium, selenium, and tellurium

The first step in this separation process is to separate cations into Groups IIA and IIB.

Hydrogen sulphide in acid solution precipitates Mo,[1] Au, Pt, Pd, Se, and Te in addition to the 'common' elements of Group II. Extraction of the group precipitate with ammonium polysulphide solution brings the greater part of the 'rarer' elements (excluding PdS) into Group IIB (arsenic group) but not completely, for appreciable quantities of Mo, Au, and Pt, as well as all the Pd, remain in the Group IIA (copper group) precipitate. The final three elements and also Pd are therefore also tested for in Group IIA.

---

[1] For the almost complete precipitation of Mo in Group II, it has been recommended that the solution be first saturated in the cold with hydrogen sulphide, then transferred to a pressure bottle and heated on a water bath.

Table 6.2 **Separation of Group II cations into Groups IIA and IIB in the presence of Mo, Au, Pt, Pd, Se, and Te** Transfer the Group II precipitate, which has been well washed with M NH$_4$Cl solution that has been saturated with H$_2$S, to a porcelain dish, add 5–10 ml ammonium polysulphide solution, heat to 50–60°C and maintain at this temperature for 3–4 minutes with constant stirring. Filter. Wash the precipitate with dilute (1 + 100) ammonium polysulphide solution.

Residue	Filtrate

May contain HgS, PbS, Bi$_2$S$_3$, CuS, CdS, PdS together with Au, Pt, traces of MoS$_3$ and SnS.
  Group IIA present.
	May contain solutions of the thio salts of As, Sb, and Sn together with Mo, Au, Pt, Se, and Te. Just acidify by adding concentrated HCl drop by drop (test with litmus paper) and warm gently.
	  A coloured precipitate indicates Group IIB present.

For the separation of cations into Groups IIA and IIB the procedures of Table 6.2 should be followed.

For the separation and identification of Group IIA cations, follow Table 6.3.

The separation and identification of Group IIB cations can be carried out by following the procedures given in Table 6.4.

Table 6.3 **Separation and identification of Group IIA cations in the presence of Pt, Au, and Pd** The precipitate may contain HgS, PbS, Bi$_2$S$_3$, CuS, CdS, and PdS, together with Au and Pt and also traces of MoS$_3$ and SnS. Transfer to a beaker or porcelain dish, add 5–10 ml 2M HNO$_3$, boil for 2–5 minutes and filter.

Residue			Filtrate

May contain HgS, Pt, and Au.
  Boil with concentrated HCl and a little bromine water, and filter, if necessary, from traces of SnO$_2$ and PbSO$_4$ which may separate here. Add M KCl solution and 2M HCl, and concentrate the solution. Filter.
			May contain Pb, Bi, Cu, Cd, and Pd ions.
			  Examine for Pb, Bi, Cu, and Cd by Table 5.4, in Section 5.8. After separation of Cu and Cd, acidify the solution with 2M HCl, introduce a few zinc granules and after several minutes filter off any solid and wash with water. Dissolve the ppt. in 2 ml aqua regia, evaporate just to dryness, dissolve the residue in 2M HCl and add dimethylglyoxime reagent.
			  Yellow ppt.
			  Pd present.

Residue	**Filtrate**		

Yellow and crystalline K$_2$[PtCl$_6$].
  Pt present.
	May contain [AuCl$_4$] and HgCl$_2$.
	  Boil to remove excess acid, render alkaline with 2M NaOH solution, and boil with excess 0.5M oxalic acid. Filter.		

	Residue	**Filtrate**	
	Brownish-black or purplish-black. Au present.	May contain HgCl$_2$. Add a few drops of 0.25M SnCl$_2$ solution. White or grey ppt. Hg present.	

Table 6.4   **Separation and identification of Group IIB cations in the presence of Pt, Au, Se, Te, and Mo**   Transfer the precipitate to a small conical flask, add 5 ml concentrated HCl, and boil gently for 5 minutes (with funnel in mouth of flask). Dilute with 2–3 ml water and filter.

Residue	*Filtrate*
May contain As, Au, Pt, Mo, Se, and Te as sulphides. Dissolve in concentrated HCl + little solid $KClO_3$; concentrate the solution to the crystallization point (use a water bath to reduce loss of Se to a minimum). Filter.	May contain Sb and Sn as chlorides or complex chloro-acids. Examine by Table 5.5 in Section 5.8.

Residue	*Filtrate*
Yellow $K_2[PtCl_6]$. Pt present. Confirm by dissolving in a little hot water and adding 0.1M KI solution. Red or brownish-red colouration.	May contain As, Au, Mo, Se, and Te as chlorides or acids. Render alkaline with 2M NH_3 solution, add $Mg(NO_3)_2$ reagent or magnesia mixture, allow to stand for 5 minutes with frequent stirring or shaking. Filter.

Residue	*Filtrate*
White crystalline $Mg(NH_4)$-$AsO_4 \cdot 6H_2O$. As present.	May contain Au, Mo, Se, and Te as chlorides or acids. Concentrate to remove ammonia, boil with several millilitres saturated oxalic acid solution, dilute, boil and filter. Extract ppt. with 2M HCl to remove coprecipitated tellurous acid.

Residue	*Filtrate*
Brownish-black or purplish-black. Au present.	Concentrate with 6M HCl on a water bath and, after removing the precipitated KCl, treat with a slight excess of solid Na_2SO_3. Filter.

Residue	*Filtrate*
Red. Se present.	Dilute with an equal volume of water, and add successively a little 0.1M KI solution and excess of solid Na_2SO_3 whereby the $[TeI_6]^{2-}$ ions are reduced to Te. Filter.

Residue	*Filtrate*
Black. Te present.	Boil with 2M HCl to remove dissolved SO_2 and treat successively with 10% KSCN soln and a little 0.25M $SnCl_2$ soln. Red colouration, soluble in ether. Mo present.

6.15   Vanadium, V (A_r: 50.94) – vanadate

Vanadium is a hard, grey metal. It melts at 1900°C. Vanadium cannot be dissolved in hydrochloric, nitric, or sulphuric acids or in alkalis. It dissolves readily in aqua regia, or in a mixture of concentrated nitric acid and hydrogen fluoride. In its compounds vanadium may have the oxidation numbers +2, +3, +4, +5, and +7; among these +5 is the most common and +4 also occurs frequently. Vanadates contain pentavalent vanadium; these are analogous to phosphates. Vanadic acid, like phosphoric acids, exists in the form of meta, pyro and ortho compounds (HVO_3, $H_4V_2O_7$ and H_3VO_4). In aqueous solutions polyvanadates are present, the degree of polymerization depending on the *p*H. Thus, in neutral solution the dodecavanadate ion, $V_{10}O_{28}^{6-}$, is predominant; in strongly alkaline solutions the divanadate (pyrovanadate) ion, $V_2O_7^{4-}$, is stable. In strongly acid solutions dioxovanadium(V) cations, VO_2^+, are present. For simplicity all these will be treated as the metavanadate, VO_3^-, ion. In tetravalent form vanadium is usually present as the vanadyl cation, VO^{2+}.

VO$_3^-$ reactions

Reactions of vanadate ions To study these reactions use a 0.1M solution of ammonium metavanadate, NH_4VO_3, or sodium metavanadate, $NaVO_3$. The addition of some sulphuric acid keeps these solutions stable.

H$_2$S
↓
reduction to V(IV)

1. *Hydrogen sulphide* no precipitate is produced in acid solution, but a blue solution (due to the production of tetravalent vanadium ions) is formed and sulphur separates. Other reducing agents, such as sulphur dioxide, oxalic acid, iron(II) sulphate, hydrazine, formic acid, and ethanol, also yield blue vanadium(IV) (VO^{2+}) ions (cf. molybdates, Section 6.6). The reaction takes place slowly in the cold, but more rapidly on warming.

Metals
↓
reduction to V(II)

2. *Zinc, cadmium or aluminium in acid solution* these carry the reduction still further. The solutions turn at first blue (VO^{2+} ions), then green (V^{3+} ions) and finally violet (V^{2+} ions).

(NH$_4$)$_2$S
↓
brown V$_2$S$_5$

3. *Ammonium sulphide solution* the solution is coloured claret-red, due to the formation of thiovanadates (probably VS_4^{3-}). Upon acidification of the solution, brown vanadium sulphide, V_2S_5, is incompletely precipitated, and the filtrate usually has a blue colour. The precipitate is soluble in solutions of alkalis, alkali carbonates, and sulphides.

H$_2$O$_2$
↓
red complex, non-extractable

4. *Hydrogen peroxide* a red colouration is produced when a few drops of hydrogen peroxide solution are added dropwise to a solution of vanadate containing 15–20% free sulphuric acid; excess hydrogen peroxide should be avoided. The colour is not removed by shaking the solution with ether, nor is it affected by phosphates or fluorides (distinction from titanium).

If more hydrogen peroxide is added or the solution is made alkaline or both, the colour changes to yellow.

The red colour is due to the formation of the peroxovanadium(V) cation, VO_2^{3+}:

$$VO_3^- + 4H^+ + H_2O_2 \rightarrow VO_2^{3+} + 3H_2O$$

The yellow colour, on the other hand, originates from the diperoxoorthovanadate(V) ions, $[VO_2(O_2)_2]^{3-}$, which are formed from the peroxovanadium(V) ions if more hydrogen peroxide is added and the solution is made alkaline:

$$VO_2^{3+} + H_2O_2 + 6OH^- \rightleftharpoons [VO_2(O_2)_2]^{3-} + 4H_2O$$

This reaction is reversible; on acidification the solution again turns red.

NH$_4$Cl
↓
colourless NH$_4$VO$_3$

5. *Ammonium chloride* the addition of solid ammonium chloride to a solution of an alkali vanadate results in the separation of colourless, crystalline ammonium vanadate, NH_4VO_3, sparingly soluble in a concentrated solution of ammonium chloride.

Pb(CH₃COO)₂
↓
white Pb(VO₃)₂

BaCl₂
↓
yellow Ba(VO₃)₂

CuSO₄
↓
green Cu(VO₃)₂

Hg₂(NO₃)₂
↓
white Hg₂(VO₃)₂

6. *Lead acetate solution*   yellow precipitate of lead vanadate, turning white or pale yellow on standing; the precipitate is insoluble in dilute acetic acid but soluble in dilute nitric acid.

7. *Barium chloride solution*   yellow precipitate of barium vanadate (distinction from arsenate and phosphate), soluble in dilute hydrochloric acid.

8. *Copper sulphate solution*   green precipitate with metavanadates. Pyrovanadates give a yellow precipitate.

9. *Mercury(I) nitrate solution*   white precipitate of mercury(I) vanadate from neutral solutions.

10. *Ammonium molybdate solution*   no precipitate is produced in the presence of ammonium nitrate and nitric acid, a soluble molybdovanadate being formed (compare phosphate). If the vanadate is mixed with a phosphate, much of the vanadium is coprecipitated with the ammonium phosphomolybdate.

Catalytic test

11. *Potassium chlorate-p-phenetidine catalytic test*   vanadium catalyses the reaction between *p*-phenetidine ($H_2N \cdot C_6H_4 \cdot OC_2H_5$) and potassium chlorate; potassium hydrogen tartrate has an activating effect upon the reaction.

Treat 0.5 ml of the test solution in a semimicro test-tube with 0.05 g potassium hydrogen tartrate, 1 ml of the *p*-phenetidine reagent (0.1%), 1 ml saturated potassium chlorate solution and dilute to 5 ml with distilled water. Immerse the test-tube in a water bath: a violet colour appears within a few minutes.

Interference of lead can be eliminated by adding 100 mg of solid sodium sulphate. In the presence of iron(III) 50 mg solid sodium nitrite should be added to make the test selective for vanadium. A similar result is obtained by replacing KClO₃ with a saturated solution of KBrO₃ but iodide, as well as lead and iron(III), interfere.

Sensitivity 0.001 µg V.

Tannin test

12. *Tannin test*   when a neutral or acetic acid solution of a vanadate is treated with an excess of 5% tannic acid, a deep blue (or blue-black) colouration is obtained. If ammonium acetate is present, a dark-blue (or blue-black) precipitate separates. The precipitate or colouration is destroyed by mineral acids.

Indirect test through Fe(II)

13. *Iron(III) chloride–dimethylglyoxime test*   the reaction

$$VO_3^- + 4H^+ + Fe^{2+} \rightleftarrows VO^{2+} + Fe^{3+} + 2H_2O$$

proceeds from left to right in acid solution and in the reverse direction in alkaline solution. The test for vanadates utilizes the deep-red colouration with dimethylglyoxime given by iron(II) salts (cf. iron(II), Section 3.21, reaction 10) and the fact that vanadates are readily reduced to the tetravalent state by heating with concentrated hydrochloric acid:

$$2VO_3^- + 8HCl \rightarrow 2VO^{2+} + Cl_2\uparrow + 6Cl^- + 4H_2O$$

All oxidizing agents interfere and must be removed.

Evaporate 1 drop of the test solution and 2 drops concentrated hydrochloric acid in a micro crucible almost to dryness. When cold, add a drop of 0.5M iron(III) chloride solution, followed by 3 drops dimethylglyoxime reagent, and render the mixture alkaline with 2M ammonia solution. Dip a strip of quantitative filter paper or of drop-reaction paper into the solution. The precipitated iron(III) hydroxide remains behind and the red solution of iron(II) dimethylglyoxime diffuses up the capillaries of the paper.

Sensitivity 1 μg V.

Concentration limit 1 in 50 000.

Atomic spectrometry

14. *Atomic spectrometric tests*   To avoid interferences by metals, it is best to use nitrous oxide/acetylene flame. In **emission** mode 1 ppm vanadium can be identified at 437.92 nm with the nitrous oxide/acetylene flame. By **atomic absorption**, using a vanadium hollow-cathode lamp and a nitrous oxide/acetylene flame the detection sensitivity is about 0.1 ppm at 318.54 nm. A 1000 ppm vanadium **standard** solution can be prepared by dissolving 2.2964 g of analytical grade ammonium metavanadate NH_4VO_3 in water and diluting to 1 litre.

Dry test

15. *Dry test*   borax bead: oxidizing flame – colourless (yellow in the presence of much vanadium); reducing flame – green.

6.16   Beryllium, Be (A_r: 9.01)

Beryllium is a greyish-white, light but very hard, brittle metal. It dissolves readily in dilute acids. In its compounds beryllium is divalent, otherwise it resembles closely aluminium in chemical properties; it also exhibits similarities to the alkaline earth metals. The salts react acid in aqueous solution and possess a sweet taste (hence the name glucinium formerly given to the element). Beryllium compounds are highly **POISONOUS**.

Be^{2+} reactions

Reactions of beryllium(II) ions   To study these reactions use a 0.1M solution of beryllium sulphate, $BeSO_4 \cdot 4H_2O$.

NH_3
↓
white $Be(OH)_2$

1. *Ammonia or ammonium sulphide solution*   white precipitate of beryllium hydroxide, $Be(OH)_2$, similar in appearance to aluminium hydroxide, insoluble in excess of the reagent, but readily soluble in dilute hydrochloric acid, forming a colourless solution. Precipitation is prevented by tartrates and citrates.

NaOH
↓
white $Be(OH)_2$
dissolves in excess

2. *Sodium hydroxide solution*   white gelatinous precipitate of beryllium hydroxide, readily soluble in excess of the precipitant, forming tetrahydroxoberyllate ions $[Be(OH)_4]^{2-}$; on boiling the latter solution (best when largely diluted), beryllium hydroxide is reprecipitated (distinction from aluminium). The precipitate is also soluble in 10% sodium hydrogen carbonate solution (distinction from aluminium).

$$Be(OH)_2\downarrow + 2OH^- \rightleftarrows [Be(OH)_4]^{2-}$$

On the other hand, the precipitate is insoluble in aqueous M ethylamine solution whereas aluminium hydroxide dissolves in a moderate excess of the reagent. Precipitation is prevented by tartrates and citrates.

$(NH_4)_2CO_3$
↓
white ppt

3. *Ammonium carbonate solution* white precipitate of basic beryllium carbonate, soluble in excess of the reagent (difference from aluminium). On boiling the solution, the white basic carbonate is reprecipitated.

4. *Oxalic acid or ammonium oxalate solution* no precipitate (difference from thorium, zirconium, and cerium).

5. *Sodium thiosulphate solution* no precipitate (difference from aluminium).

Extraction of acetate

6. *Basic acetate–chloroform test* upon dissolving beryllium hydroxide (reaction 1) in glacial acetic acid and evaporating to dryness with a little water, basic beryllium acetate, $BeO \cdot 3Be(CH_3COO)_2$, is produced, which dissolves readily upon extraction with chloroform. This forms the basis of a method for separating beryllium from aluminium, since basic aluminium acetate is insoluble in chloroform.

Quinalizarin
↓
blue complex

7. *Quinalizarin reagent*[2] cornflower-blue colouration with faintly alkaline solutions of beryllium salts. The reagent alone gives a characteristic violet colour with dilute alkali; but this is quite distinct from the blue of the beryllium complex; a blank test will render the difference clearly apparent.

Antimony, zinc, and aluminium salts do not interfere; aluminium should, however, be kept in solution by the addition of sufficient sodium hydroxide; the influence of copper, nickel, and cobalt salts can be eliminated by the addition of potassium cyanide solution; iron salts are 'masked' by the addition of a tartrate but if aluminium salts are also present a red colour is produced. Magnesium salts give a similar blue colour, but beryllium can be detected in the presence of this element by utilizing the fact that in ammoniacal solution the magnesium colour alone is completely destroyed by bromine water.

In adjacent depressions of a spot plate place a drop of the test solution and a drop of distilled water, and add a drop of the freshly prepared 0.05% quinalizarin reagent (in NaOH) to each. A blue colouration or precipitate, quite distinct from the violet colour of the reagent, is obtained.

Sensitivity 0.15 µg Be.

Concentration limit 1 in 350 000.

If magnesium is present, treat a drop of the solution on a spot plate with 2 drops of the reagent and 1 ml saturated bromine water. The original deep-blue colour becomes paler when the bromine is added, but remains more or less permanently blue.

4-(4-Nitrophenylazo)
resorcinol
↓
orange-red lake

8. *4-(4-Nitrophenylazo) resorcinol (magneson I) reagent (0.025%)*

orange-red lake with beryllium salts in alkaline solution. Magnesium salts yield a brownish-yellow precipitate; salts and hydroxides of the rare earths,

---

[2] See under aluminium, Section 3.23, reaction 9.

aluminium, and alkaline earths are without influence; the interfering effect of silver, copper, cadmium, nickel, cobalt, and zinc is eliminated by the addition of potassium cyanide solution.

Place a drop of the 0.025% reagent on drop-reaction paper and into the middle of the resulting yellow area introduce the tip of a capillary containing the test solution so that the latter runs slowly onto the paper. Treat the stain with a further drop of the reagent. The stain is coloured deep orange-red.

Sensitivity 0.2 µg Be.

Concentration limit 1 in 200 000.

Acetylacetone
↓
characteristic crystals

9. *Acetylacetone test*   acetylacetone reacts with beryllium salts to yield the complex $Be(C_5H_7O_2)_2$, which possesses a highly characteristic appearance under the microscope.

$$2CH_3 \cdot CO \cdot CH_2 \cdot CO \cdot CH_3 + Be^{2+} \rightarrow Be(C_5H_7O_2)_2 + 2H^+$$

Place a drop of the test solution on a microscope slide and add a drop of acetylacetone. Crystals separate immediately: these will be found to possess rhombic and hexagonal forms when observed under the microscope (use linear magnification of ×75).

Concentration limit 1 in 10 000.

Atomic spectrometry

10. *Atomic spectrometric tests*   to avoid interferences from a number of cations and anions, it is best to use the hot nitrous oxide/acetylene flame. In **emission** mode 1 ppm beryllium can be identified at 234.86 nm with the nitrous oxide/acetylene flame. By **atomic absorption**, using a beryllium hollow-cathode lamp and a nitrous oxide/acetylene flame the detection sensitivity is about 0.1 ppm at the same wavelength. A 1000 ppm beryllium **standard** solution can be prepared by dissolving 19.66 g of analytical grade beryllium sulphate tetrahydrate $BeSO_4 \cdot 4H_2O$ in water and diluting to 1 litre.

Dry test

11. *Dry test*   upon heating beryllium salts with a few drops of cobalt nitrate solution upon charcoal a grey mass is obtained (difference from aluminium).

6.17   Titanium, Ti (A_r: 47.90) – titanium(IV)

Titanium is a greyish, hard metal. It can be dissolved in hot, concentrated sulphuric acid and in hydrogen fluoride. Like tin, it is insoluble in concentrated nitric acid, because of the formation of titanic acid on the surface of the metal, which protects the rest of the metal from the acid.

Titanium forms the violet titanium(III), Ti^{3+}, and the colourless titanium(IV), Ti^{4+}, ions. Titanium(III) ions are rather unstable and are readily oxidized to titanium(IV) in aqueous solutions. Titanium(IV) ions exist only in strongly acid solutions; they tend to hydrolyse forming the titanyl ion, TiO^{2+}, first if the acidity of the solution is lowered; later titanium(IV) hydroxide is precipitated. In analytical work titanium(IV) ions are normally encountered.

Ti⁴⁺ reactions

Reactions of titanium(IV) [titanyl] ions   To study these reactions use a 0.2M solution of titanium(IV) sulphate.

Alkaline agents
↓
white H_4TiO_4

1. *Solutions of sodium hydroxide, ammonia or ammonium sulphide* all these reagents give a white gelatinous precipitate of orthotitanic acid, H_4TiO_4, or titanium(IV) hydroxide, $Ti(OH)_4$, in the cold; this is almost insoluble in excess reagent, but soluble in mineral acids. If precipitation takes place from hot solution, metatitanic acid, H_2TiO_3 or $TiO\cdot(OH)_2$, is said to be formed, which is sparingly soluble in dilute acids. Tartrates and citrates inhibit precipitation.

Hydrolysis to H_2TiO_3

2. *Water* a white precipitate of metatitanic acid is obtained on boiling a solution of a titanic salt with excess water.

Na_2HPO_4
↓
white $Ti(HPO_4)_2$

3. *Disodium hydrogen phosphate solution* white precipitate of titanium hydrogen phosphate, $Ti(HPO_4)_2$, in dilute sulphuric acid solution.

Reducing metals
↓
violet colour

4. *Zinc, cadmium or tin* when any of these metals is added to a solution of a titanium(IV) salt, containing excess hydrochloric acid, a violet colouration is produced, due to reduction to titanium(III) ions. No reduction occurs with sulphur dioxide or with hydrogen sulphide.

Cupferron
↓
yellow ppt

5. *Cupferron reagent* flocculent yellow precipitate of the titanium salt, $Ti(C_6H_5O_2N_2)_4$, in acid solution (distinction from aluminium and beryllium). If iron is present, it can be removed by precipitation with ammonia and ammonium sulphide solutions in the presence of a tartrate; the titanium may then be precipitated from the acidified solution by cupferron.

H_2O_2
↓
yellow peroxo compound

6. *Hydrogen peroxide* an orange-red colouration is produced in slightly acid solution. The colour is yellow with very dilute solutions. The colouration has been variously attributed to peroxotitanic acid, $HOO\text{-}Ti(OH)_3$, or to peroxodisulphatotitanium(IV) ions, $[TiO_2(SO_4)_2]^{2-}$.

Chromates, vanadates, and cerium salts give colour reactions with the reagent and should therefore be absent. Iron salts give a yellow colour with hydrogen peroxide, but this is eliminated by the addition of concentrated phosphoric acid. Fluorides bleach the colour (stable $[TiF_6]^{2-}$ ions are formed), and large amounts of nitrates, chlorides, bromides, and acetates as well as coloured ions reduce the sensitivity of the test. A decrease in the intensity of the yellow colouration upon the addition of ammonium fluoride indicates the presence of titanium.

Chromotropic acid
↓
brown colour

7. *Chromotropic acid (1,8-dihydroxynaphthalene-3,6-disulphonic acid) reagent*

reddish-brown colouration with titanium salts in the presence of hydrochloric or sulphuric acid. Appreciable concentrations of nitric acid inhibit the reaction.

Mix a drop of the test solution and a drop of the reagent on drop-reaction paper or upon a spot plate. A reddish-brown (or purplish-pink) spot or colouration results.

Sensitivity 5 µg TiO_2.
Concentration limit 1 in 10 000.

The reagent does not keep well, hence it is preferable to impregnate filter paper with the reagent solution and allow the paper to dry in the air. The impregnated papers are stable for several months. In use, a drop of the test solution and a drop of M H_2SO_4 are placed upon the impregnated paper; a purplish-pink colour results.

Catechol
↓
yellow colour

8. *Catechol (pyrocatechol)*

yellow colouration with neutral or weakly acid (sulphuric acid) solutions of titanium salts. Iron(III), chromium, cobalt, and nickel salts interfere as do large amounts of free mineral acids; alkali hydroxides and carbonates reduce the sensitivity of the test.

Place a drop of the sulphuric acid test solution on drop-reaction paper impregnated with the reagent. A yellow or yellowish-red spot is obtained.

Sensitivity 3 µg Ti.
Concentration limit 1 in 20 000.

Atomic spectrometry

9. *Atomic spectrometric tests* to avoid interferences it is best to use a hot nitrous oxide/acetylene flame. In **emission** mode 5 ppm titanium can be identified at 398.98 nm with the nitrous oxide/acetylene flame. By **atomic absorption**, using a titanium hollow-cathode lamp and a nitrous oxide/acetylene flame, the detection sensitivity is about 0.5 ppm at 365.35 nm. A 1000 ppm titanium **standard** solution can be prepared by dissolving 7.394 g of analytical grade potassium titanium oxalate dihydrate $K_2TiO(C_2O_4)_2 \cdot 2H_2O$ in water and diluting to 1 litre.

Dry test

10. *Dry test* microcosmic salt bead: oxidizing flame – colourless; reducing flame – yellow whilst hot and violet when cold (this result is obtained more rapidly if a little tin(II) chloride is added). If the bead is heated in the reducing flame with a trace of iron(II) sulphate, it acquires a blood-red colour.

6.18 Zirconium, Zr (A_r: 91.22)

Zirconium is a grey, very hard metal. It melts at 1860°C. Zirconium metal can be dissolved in aqua regia and in hydrogen fluoride.

Zirconium forms only one important oxide, zirconia ZrO_2, which is amphoteric in character. The normal zirconium salts, like $ZrCl_4$, are readily hydrolysed in solution giving rise chiefly to zirconyl salts, containing the divalent radical ZrO^{2+}. The zirconates, e.g. Na_2ZrO_3, are best produced from ZrO_2 by fusion methods. Zirconium also readily forms complex ions like hexafluorozirconate(IV) $[ZrF_6]^{2-}$, produced by heating zirconia with potassium hydrogen fluoride.

Ignited zirconium dioxide, or the mineral, is insoluble in all acids except hydrofluoric acid. It is soluble in fused caustic alkalis and in sodium carbonate: the resulting alkali zirconate is practically insoluble in water, being converted into zirconium hydroxide by this solvent. It is therefore best dissolved in hydrochloric acid, and the zirconium hydroxides precipitated by ammonia solution if necessary.

Zr^{4+} reactions

Reactions of zirconium(IV) [zirconyl] ions To study these reactions use a 0.1M solution of zirconyl nitrate, $ZrO(NO_3)_2 \cdot 2H_2O$, or zirconyl chloride, $ZrOCl_2 \cdot 8H_2O$. These solutions must contain some free acid.

NaOH
↓
white $Zr(OH)_4$

1. *Sodium hydroxide solution* white, gelatinous precipitate of the hydroxide, $Zr(OH)_4$ (or $ZrO_2 \cdot xH_2O$), in the cold, practically insoluble in excess of the reagent (difference from aluminium and beryllium), but soluble in dilute mineral acids (avoid sulphuric acid). With a hot solution of a zirconium salt, a white precipitate of $ZrO(OH)_2$ is obtained; it is sparingly soluble in dilute but soluble in concentrated mineral acids. Tartrates and citrates inhibit the precipitation of the hydroxide.

NH_3
↓
white $Zr(OH)_4$

2. *Ammonia or ammonium sulphide solution* white, gelatinous precipitate of the hydroxide, $Zr(OH)_4$ (or $ZrO_2 \cdot xH_2O$), insoluble in excess of the reagent.

Na_2HPO_4
↓
white $Zr(HPO_4)_2$

3. *Disodium hydrogen phosphate solution* white precipitate of zirconium hydrogen phosphate, $Zr(HPO_4)_2$ or $ZrO(H_2PO_4)_2$, even in solutions containing 10% sulphuric acid by weight and also tartrates and citrates. No other element forms an insoluble phosphate under these conditions except titanium. The latter element can be kept in solution as peroxotitanic acid by the addition of sufficient hydrogen peroxide solution (30%) before the sodium phosphate is introduced.

H_2O_2
↓
white ppt

4. *Hydrogen peroxide* white precipitate of peroxozirconic acid, HOO-$Zr(OH)_3$, from slightly acid solutions; this liberates chlorine when warmed with concentrated hydrochloric acid. When both hydrogen peroxide and sodium phosphate are added to a solution containing zirconium, the precipitate is zirconium hydrogen phosphate (see reaction 3).

$(NH_4)_2CO_3$
↓
white ppt

5. *Ammonium carbonate solution* white precipitate of basic zirconium carbonate, readily soluble in excess of the reagent, but reprecipitated on boiling.

Oxalate
↓
white $Zr(C_2O_4)_2$, soluble in excess

6. *Oxalic acid solution* white precipitate of zirconium oxalate, readily soluble in excess of the reagent and also in ammonium oxalate solution (difference from thorium).

7. *Ammonium oxalate solution* white precipitate of zirconium oxalate, soluble in excess of the reagent (distinction from aluminium and beryllium); the solution gives no precipitate with hydrochloric acid (difference from thorium).

Note: A solution of zirconium sulphate or a zirconium salt solution containing excess sulphate ions does not give a precipitate with either ammonium oxalate or oxalic acid. This is due to the fact that the zirconium is present as the anion $[ZrO(SO_4)_2]^{2-}$, hence sulphuric acid should be avoided in preparing solutions of zirconium salts.

K$_2$SO$_4$
↓
white ppt

8. *Saturated potassium sulphate solution* white precipitate of $K_2[ZrO(SO_4)_2]$, insoluble in excess of the reagent. When precipitation takes place in boiling solution, the resulting basic zirconium sulphate is insoluble in dilute hydrochloric acid (difference from thorium and cerium). No precipitate is obtained with sodium sulphate solution.

Phenylarsonic acid
↓
white ppt

9. *Phenylarsonic acid reagent* $\{C_6H_5 \cdot AsO(OH)_2\}$ white precipitate of zirconium phenylarsonate in the presence of 0.5–1M hydrochloric acid; it is best to boil the solution. Tin and thorium salts must be absent.

n-Propylarsonic acid
↓
white ppt

10. *n-Propylarsonic acid reagent*

$$\{CH_3 \cdot CH_2 \cdot CH_2 \cdot AsO(OH)_2\}$$

white precipitate of zirconium n-propylarsonate in dilute sulphuric acid solution (separation from most other metals including titanium but not tin).

The reagent consists of a 5% aqueous solution of n-propylarsonic acid.

KIO$_3$
↓
white Zr(IO$_3$)$_4$

11. *Potassium iodate solution* in faintly acid solution, a voluminous, white precipitate of basic zirconium iodate is obtained. The precipitate is soluble in warm hydrochloric acid (difference from aluminium).

Alizarin red S
↓
red ppt

12. *Alizarin red S*[3] *(2%)* red precipitate in a strongly acid medium. Fluorides discharge the colour because of the formation of the stable hexafluorozirconate(IV) ion $[ZrF_6]^{2-}$.

Place a drop of the test solution (which has been acidified with hydrochloric acid) on a spot plate, add a drop of the reagent and a drop of concentrated hydrochloric acid. A red precipitate results.

4-Dimethylamino-
benzene-azo-
phenylarsonic acid
↓
brown ppt

13. *4-Dimethylaminobenzene-azo-phenylarsonic acid reagent*

$(CH_3)_2N$—⟨⟩—$N=N$—⟨⟩—$AsO(OH)_2$

Acid solutions of zirconium salts give a brown precipitate with the reagent. If the test is conducted on filter paper, the brown precipitate remains in the pores of the paper and the excess of the coloured reagent may be washed out with dilute acid.

Impregnate some drop-reaction paper with the reagent and dry the paper. Place a drop of the acid test solution on the paper. Dip the paper into 2M hydrochloric acid at 50–60°C. A brown spot or ring remains.

Sensitivity 0.1 µg Zr (in M HCl).
Concentration limit 1 in 500 000.

Atomic spectrometry

14. *Atomic spectrometric tests* because of the low volatility of zirconium compounds, it is advisable to use the hot nitrous oxide/acetylene flame. In **emission** mode 5 ppm zirconium can be identified at 360.12 nm. By **atomic absorption**, using a zirconium hollow-cathode lamp the detection sensitivity

---

[3] For formula, see under aluminium, Section 3.23, reaction 8.

is about 1 ppm at the same wavelength. An approximately 1000 ppm zirconium **standard** solution can be prepared by dissolving 3.8 g of dry, technical grade zirconium nitrate $Zr(NO_3)_4$ in water and diluting to 1 litre.

15. *Dry test*   no characteristic results are obtained with the borax or microcosmic beads nor does zirconium yield a distinguishing flame test.

6.19   Uranium, U (A_r: 238.03)

Uranium is a grey, heavy metal. It is quite soft, and melts at 1133°C.

Uranium can be dissolved in dilute acids, when uranium(IV) ions, U^{4+}, and hydrogen gas are formed. Uranium(IV) ions can easily be oxidized to the hexavalent state, which is the most stable oxidation state of uranium. At this oxidation state, depending on the pH of the solution, two ions can be formed: the uranyl cation, UO_2^{2+}, is stable in acid solutions, while the diuranate anion, $U_2O_7^{2-}$, in alkaline media. The two ions are in equilibrium with each other:

$$2UO_2^{2+} + 3H_2O \rightleftharpoons U_2O_7^{2-} + 6H^+$$

Alkalizing the solution (removing H^+) will shift the equilibrium towards the formation of $U_2O_7^{2-}$, while acidifying it (adding H^+) will shift the equilibrium in the other direction.

In this text only the reactions of uranyl ions will be described.

Uranium compounds are all radioactive. In certain countries, the sale, storage, use and disposal of uranium compounds are regulated by law. Uranium compounds are **KIDNEY POISONS!**

UO_2^{2+} reactions

Reactions of uranyl ions   To study these reactions use a 0.1M solution of uranyl nitrate, $UO_2(NO_3)_2 \cdot 6H_2O$, or of uranyl acetate, $UO_2(CH_3COO)_2 \cdot 2H_2O$.

NH_3 or NaOH
↓
yellow ppt

1. *Ammonia solution*   yellow precipitate of ammonium diuranate, insoluble in excess of the reagent, but soluble in ammonium carbonate or sodium carbonate, forming the tricarbonatouranylate(VI) ion:

$$2UO_2^{2+} + 6NH_3 + 3H_2O \rightarrow (NH_4)_2U_2O_7\downarrow + 4NH_4^+$$

$$(NH_4)_2U_2O_7\downarrow + 6CO_3^{2-} + 3H_2O \rightarrow 2[UO_2(CO_3)_3]^{4-} + 2NH_4^+ + 6OH^-$$

No precipitation occurs in the presence of certain organic acids, such as oxalic, tartaric, and citric acids.

2. *Sodium hydroxide solution*   yellow amorphous precipitate of sodium diuranate, $Na_2U_2O_7$, soluble in ammonium carbonate solution.

$(NH_4)_2S$
↓
brown UO_2S

3. *Ammonium sulphide solution*   brown precipitate of uranyl sulphide, UO_2S, soluble in dilute acids and in ammonium carbonate solution.

H_2O_2
↓
yellow ppt

4. *Hydrogen peroxide*   pale-yellow precipitate of uranium tetroxide $UO_4 \cdot 2H_2O$ (sometimes called uranium peroxide), soluble in ammonium carbonate solution with the formation of a deep-yellow solution. Chromium, titanium, and vanadium interfere with this otherwise sensitive test.

5. *Cupferron reagent*   no precipitate (distinction from titanium).

Na$_2$HPO$_4$
↓
white UO$_2$HPO$_4$

6. *Disodium hydrogen phosphate solution*   white precipitate of uranyl hydrogen phosphate UO$_2$HPO$_4$, soluble in mineral acids but insoluble in dilute acetic acid. In the presence of ammonium sulphate or of ammonium acetate, uranyl ammonium phosphate UO$_2$(NH$_4$)PO$_4$ is precipitated.

(NH$_4$)$_2$CO$_3$
↓
white UO$_2$CO$_3$,
dissolves in excess

7. *Ammonium (or sodium) carbonate solution*   white precipitate of uranyl carbonate, UO$_2$CO$_3$, soluble in excess of the reagent forming a clear, yellow solution containing the tricarbonatouranylate(VI) ion (cf. reaction 1).

K$_4$[Fe(CN)$_6$]
↓
brown ppt

8. *Potassium hexacyanoferrate(II) solution*   brown precipitate of uranyl hexacyanoferrate(II), (UO$_2$)$_2$[Fe(CN)$_6$], in neutral or acetic acid solutions, soluble in dilute hydrochloric acid (difference from copper). The precipitate becomes yellow upon the addition of sodium hydroxide solution, due to its conversion into sodium diuranate (distinction from copper and from molybdenum).

$$(UO_2)_2[Fe(CN)_6]\downarrow + 2Na^+ + 6OH^- \rightarrow$$

$$\rightarrow Na_2U_2O_7\downarrow + [Fe(CN)_6]^{4-} + 3H_2O$$

Fluorescence

9. *Fluorescence*   uranium salts, when irradiated with a u.v. lamp, exhibit a characteristic green fluorescence. The phenomenon is especially interesting if in a dark room a bottle containing solid uranyl nitrate or uranyl acetate is irradiated.

In solution the intensity of fluorescence depends on the *p*H. In acid solutions the fluorescence is strong, but becomes gradually weaker as the *p*H of the solution is raised.

Atomic spectrometry

10. *Atomic spectrometric tests*   it is advisable to use the hot nitrous oxide/acetylene flame. In **emission** mode 5 ppm uranium can be identified at 544.8 nm. By **atomic absorption**, using a uranium hollow-cathode lamp the detection sensitivity is about 1 ppm at 358.49 nm. A 1000 ppm uranium **standard** solution can be prepared by dissolving 2.1095 g of analytical grade uranyl nitrate hexahydrate UO$_2$(NO$_3$)$_2$·6H$_2$O in water and diluting to 1 litre.

Owing to the toxic and radioactive nature of uranium compounds, there must be excellent ventilation facilities for FAES or FAAS.

Dry test

11. *Dry test*   borax or microcosmic salt bead: oxidizing flame – yellow; reducing flame – green.

6.20   Thorium, Th (*A*$_r$: 232.04)

Thorium is a greyish-white, moderately hard, heavy metal, which melts at 1750°C. It can be dissolved in concentrated hydrochloric acid or in aqua regia. The tetravalent Th^{4+} cations are stable in acid solutions.

In certain countries, the sale, storage, use and disposal of thorium compounds are regulated by law.

Th^{4+} reactions

Reactions of thorium(IV) ions To study these reactions use a 0.1M solution of thorium nitrate Th(NO$_3$)$_4 \cdot$4H$_2$O which contains some free nitric acid.

Alkaline reagents
↓
white Th(OH)$_4$

1. *Ammonia, ammonium sulphide or sodium hydroxide solution* white precipitate of thorium hydroxide, Th(OH)$_4$ or ThO$_2 \cdot x$H$_2$O, insoluble in excess of the reagent, but readily soluble in dilute acids when freshly precipitated. Tartrates and also citrates prevent the precipitation of the hydroxide.

Carbonates
↓
white ppt,
dissolves in excess

2. *Ammonium or sodium carbonate solution* white precipitate of basic carbonate, readily soluble in excess of the concentrated reagent forming the pentacarbonatothorate(IV) anion, [Th(CO$_3$)$_5$]$^{6-}$.

Oxalate
↓
white Th(C$_2$O$_4$)$_2$

3. *Oxalic acid solution* white, crystalline precipitate of thorium oxalate, Th(C$_2$O$_4$)$_2$ (distinction from aluminium and beryllium), insoluble in excess of the reagent and in 0.5M hydrochloric acid.

4. *Ammonium oxalate solution* white precipitate of thorium oxalate, which dissolves on boiling with a large excess of the reagent, forming the trioxalatothorate(IV) anion, [Th{(COO)$_2$}$_3$]$^{2-}$, but is reprecipitated upon the addition of hydrochloric acid (difference from zirconium).

K$_2$SO$_4$
↓
white ppt

5. *Saturated potassium sulphate solution* white precipitate of the complex salt potassium tetrasulphatothorate(IV), K$_4$[Th(SO$_4$)$_4$], insoluble in excess of the precipitant, but soluble in dilute hydrochloric acid.

H$_2$O$_2$
↓
white ppt

6. *Hydrogen peroxide* white precipitate of hydrated thorium heptoxide (thorium peroxide), Th$_2$O$_7 \cdot$4H$_2$O, in neutral or faintly acid solution.
   The compound is not a true peroxide, but an associate compound of thorium dioxide and hydrogen peroxide: 2ThO$_2 \cdot$3H$_2$O$_2 \cdot$H$_2$O.

Na$_2$S$_2$O$_3$
↓
white ppt

7. *Sodium thiosulphate solution* precipitate of thorium hydroxide and sulphur on boiling (distinction from cerium).

$$\text{Th}^{4+} + 2\text{S}_2\text{O}_3^{2-} + 2\text{H}_2\text{O} \rightarrow \text{Th(OH)}_4\downarrow + 2\text{S}\downarrow + 2\text{SO}_2\uparrow$$

KIO$_3$
↓
white Th(IO$_3$)$_4$

8. *Potassium iodate solution* white, bulky precipitate of thorium iodate, Th(IO$_3$)$_4$. Precipitation occurs even in the presence of 50% free nitric acid [difference from cerium(III)].

K$_4$[Fe(CN)$_6$]
↓
white ppt

9. *Potassium hexacyanoferrate(II) solution* white precipitate of thorium hexacyanoferrate(II), Th[Fe(CN)$_6$], in neutral or slightly acid solution.

KF
↓
white ThF$_4$

10. *Potassium fluoride solution* bulky, white precipitate of thorium fluoride, ThF$_4$, insoluble in excess of the reagent (distinction and method of separation for aluminium, beryllium, zirconium, and titanium).

Sebacic acid
↓
white ppt

11. *Saturated sebacic acid solution,* {*COOH·(CH$_2$)$_8$COOH*} white voluminous precipitate of thorium sebacate, Th(C$_{10}$H$_{16}$O$_4$)$_2$ (difference from cerium).

3-Nitrobenzoic acid
↓
white ppt

12. *3-Nitrobenzoic acid reagent, ($NO_2 \cdot C_6H_4COOH$)* upon addition of excess reagent to a neutral solution of a thorium salt at about 80°C, a white precipitate of the salt $Th(NO_2 \cdot C_6H_4 \cdot COO)_4$ is obtained (distinction from cerium).

6.21 Cerium, Ce (A_r: 140.12) – cerium(III)

Cerium is a white-grey, soft metal, which melts at 794°C. In its compounds cerium can be trivalent and tetravalent, forming cerium(III), Ce^{3+}, and cerium (IV), Ce^{4+}, ions respectively.

Ce^{3+} reactions

Reactions of cerium(III) ions To study these reactions use a 0.1M solution of cerium(III) nitrate $Ce(NO_3)_3 \cdot 6H_2O$.

NH_3 or $(NH_4)_2S$
↓
white $Ce(OH)_3$

1. *Ammonia or ammonium sulphide solution* white precipitate of cerium(III) hydroxide, $Ce(OH)_3$ (or $Ce_2O_3 \cdot xH_2O$), insoluble in excess of the precipitant, but readily soluble in acids. The precipitate slowly oxidizes in the air, finally becoming converted into yellow cerium(IV) hydroxide, $Ce(OH)_4$ (or $CeO_2 \cdot xH_2O$). Sodium hydroxide solution gives a similar result. Precipitation is prevented by tartrates and citrates.

Oxalate
↓
white $Ce_2(C_2O_4)_3$

2. *Oxalic acid or ammonium oxalate solution* white precipitate of cerium(III) oxalate, insoluble in excess reagent (compare thorium and zirconium), and in very dilute mineral acids.

3. *Sodium thiosulphate solution* no precipitate (distinction from thorium and from cerium(IV) ions).

K_2SO_4
↓
white ppt

4. *Saturated potassium sulphate solution* white, crystalline precipitate, having the composition $Ce_2(SO_4)_3 \cdot 3K_2SO_4$ in neutral solution and $Ce_2(SO_4)_3 \cdot 2K_2SO_4 \cdot 2H_2O$ from slightly acid solution (difference from aluminium and beryllium).

$NaBiO_3$
↓
oxidizes to Ce^{4+}

5. *Sodium bismuthate* the solid reagent, in the presence of dilute nitric acid, converts cerium(III) ions into cerium(IV) in the cold. A similar result is obtained by heating with ammonium peroxodisulphate or with lead dioxide and 8M nitric acid. In all cases, the solutions become yellow or orange in colour.

$(NH_4)_2CO_3$
↓
white $Ce_2(CO_3)_3$

6. *Ammonium carbonate solution* white precipitate of cerium(III) carbonate, $Ce_2(CO_3)_3$, nearly insoluble in excess of the precipitant (difference from beryllium, thorium, and zirconium) and insoluble in sodium carbonate solution.

H_2O_2
↓
oxidizes to Ce^{4+}.
brown ppt

7. *Hydrogen peroxide* when a cerium(III) salt is treated with ammonia solution and excess hydrogen peroxide is added, a yellowish-brown or reddish-brown precipitate or colouration occurs, due to cerium peroxide gel ($CeO_2 \cdot H_2O_2 \cdot H_2O$). This is not very stable. Upon boiling the mixture, yellow cerium(IV) hydroxide, $Ce(OH)_4$, is obtained. The test cannot be applied directly in the presence of iron since the colour of iron(III) hydroxide is similar to that of cerium dioxide. The precipitation of iron(III) hydroxide may be prevented by the addition of an alkali tartrate; this, however, reduces the sensitivity of the test for cerium.

NH$_4$F
↓
white CeF$_3$

8. *Ammonium fluoride solution* white, gelatinous precipitate of cerium(III) fluoride, CeF$_3$, in neutral or slightly acid solution. The precipitate becomes powdery upon standing.

KIO$_3$
↓
white Ce(IO$_3$)$_3$

9. *Potassium iodate solution* white precipitate of cerium(III) iodate, Ce(IO$_3$)$_3$, in neutral solution, soluble in nitric acid (difference from cerium(IV) iodate; cf. thorium, Section 6.20, reaction 8).

AgNO$_3$ + NH$_3$
↓
Ag$^+$ reduced,
Ce^{3-} oxidized

10. *Diammineargentato nitrate (ammoniacal silver nitrate) reagent* this reagent reacts with neutral solutions of cerium(III) salts to form cerium(IV) hydroxide and metallic silver [difference from cerium(IV)]; the former is coloured black by the finely divided silver:

$$Ce(OH)_3\downarrow + [Ag(NH_3)_2]^+ + OH^- \rightarrow Ce(OH)_4\downarrow + Ag\downarrow + 2NH_3$$

Iron(II), manganese(II), and cobalt(II) ions also give the higher metallic hydroxides and silver, and must therefore be absent.

Mix a drop of the neutral test solution and a drop of the reagent on a watch glass or in a porcelain micro crucible, and warm gently. A black precipitate or brown colouration appears.

Sensitivity 1 μg Ce.

Concentration limit 1 in 50 000.

The reagent should be freshly prepared and should be discarded quickly to avoid a serious **explosion!** (Cf. Section 3.6, reaction 1.)

6.22 Cerium, Ce (A_r: 140.12) – cerium(IV)

The most important physical and chemical properties of cerium have been described in Section 6.21.

Ce^{4+} reactions

Reactions of cerium(IV) ions To study these reactions use a 0.1M solution of cerium(IV) sulphate, Ce(SO$_4$)$_2$·4H$_2$O, or cerium(IV) ammonium sulphate, Ce(SO$_4$)$_2$·2(NH$_4$)$_2$SO$_4$·2H$_2$O. These solutions should contain a few per cent of free sulphuric acid. Both solutions are orange-yellow. On standing a fine precipitate is sometimes formed; it cannot be filtered easily, and the clear solution should preferably be decanted or syphoned before use.

Alkaline reagents
↓
yellow Ce(OH)$_4$

1. *Ammonia or sodium hydroxide solution* yellow precipitate of cerium(IV) hydroxide, Ce(OH)$_4$. If the precipitate is warmed with hydrochloric acid, chlorine is evolved and cerium(III) ions are formed.

Oxalate
↓
reduction to Ce^{3+}

2. *Oxalic acid or ammonium oxalate solution* reduction occurs, more rapidly on warming, to cerium(III) ions, and ultimately white cerium(III) oxalate is precipitated.

3. *Saturated potassium sulphate solution* no precipitate (distinction from cerium(III) salts).

Reducing agents
↓
reduction to Ce^{3+}

4. *Reducing agents (e.g. hydrogen sulphide, sulphur dioxide, hydrogen peroxide, and hydriodic acid)* these convert cerium(IV) ions into cerium(III).

5. *Sodium thiosulphate solution* yellow colour of solution disappears and sulphur precipitates, owing to reduction.

$$2Ce^{4+} + S_2O_3^{2-} + H_2O \rightarrow 2Ce^{3+} + S\downarrow + SO_4^{2-} + 2H^+$$

NH$_4$F
↓
yellow complex

6. *Ammonium fluoride solution* yellow colour of solution disappears but no precipitate is produced, due to the formation of hexafluorocerate(IV) anions, $[CeF_6]^{2-}$.

KIO$_3$
↓
white Ce(IO$_3$)$_4$

7. *Potassium iodate solution* white precipitate of cerium(IV) iodate, $Ce(IO_3)_4$, from concentrated nitric acid solution (difference from cerium(III); thorium and zirconium give a similar reaction).

Anthranilic acid
↓
brown colour

8. *Anthranilic acid reagent ($NH_2 \cdot C_6H_4 \cdot COOH$)* cerium(IV) oxidizes anthranilic acid to a brown compound. Cerium(III) does not react and must be oxidized first with lead dioxide and concentrated nitric acid to cerium(IV); other oxidizing agents cannot be used, since they react with the anthranilic acid. Iron(III) ions inhibit the test and must be masked by the addition of phosphoric acid. The ions of gold and vanadium, as well as chromate ion, react similarly and therefore interfere. Reducing agents must be absent.

Place a drop of the test solution (slightly acid with nitric acid) on a spot plate and add a drop of the reagent. A blackish-blue precipitate appears, which rapidly passes into a soluble product and colours the solution brown. *Concentration limit* 1 in 10 000.

Dry test

9. *Dry test* Borax bead: oxidizing flame – dark brown whilst hot and light yellow to colourless when cold: reducing flame – colourless.

Separation of Th and Ce

10. *Separation of thorium and cerium* cerium and thorium salts are precipitated in Group IIIA. They may be separated from the other metals of the group by dissolving the precipitate in 2M hydrochloric acid and adding oxalic acid solution, when the oxalates of both metals are precipitated. The thorium and cerium may be separated: (*a*) by dissolving the thorium oxalate in a mixture of ammonium acetate solution and acetic acid, cerium oxalate being insoluble under these conditions; (*b*) by treatment with a large excess of hot concentrated ammonium oxalate solution; only the thorium oxalate dissolves (a complex ion being formed), and may be reprecipitated from the resultant solution as oxalate by the addition of hydrochloric acid.

6.23 **Separation of Group III cations in the presence of titanium, zirconium, thorium, uranium, cerium, vanadium, thallium, and molybdenum**

When in the course of systematic analysis the separation of Group III cations is to be carried out, the solution should be tested for phosphate.

Table 6.5 **Separation and identification of Group IIIA cations in the presence of Ti, Zr, Ce, Th, U, and V**   Dissolve the precipitate in the minimum volume of 2M HCl. Pour the weakly acid solution into an equal volume of a solution which contains 2 ml 30% H_2O_2 and is 2.5M with respect to NaOH. (The latter solution should be freshly prepared.) Boil for 5 minutes, but no longer. Filter and wash the precipitate with hot 2% NH_4NO_3 solution.

Residue	Filtrate
May contain $Fe(OH)_3$; $TiO_2 \cdot xH_2O$; $ZrO_2 \cdot xH_2O$; $ThO_2 \cdot xH_2O$; $CeO_3 \cdot xH_2O$ (and some $MnO(OH)_2$).	May contain CrO_4^{2-}, $[Al(OH)_4]^-$, VO_3^-, and $U_2O_7^{2-}$.

Residue side:
Dissolve in 2M HCl, boil to expel Cl_2 and divide the solution into five parts.
(1) Add 0.1M KSCN solution. Red colouration. Fe present.
(2) If Fe present, add just sufficient H_3PO_4 to mask iron(III) ions, and then 3% H_2O_2. Orange-red colouration, discharged by the addition of solid NH_4F. Ti present. White ppt. Zr present.
(3) Add excess saturated oxalic acid solution. White ppt. Th and/or Ce present. For separation of Th and Ce, see Section 6.22, 10.
(4) Evaporate to fuming with 8M H_2SO_4 to expel HCl. Cool, dilute, add 2M HNO_3 and a little $NaBiO_3$. Stir and allow to stand. Purple colouration. Mn present.

Filtrate side:
Acidify with 2M HNO_3; add 3–4 ml 0.25M $Pb(NO_3)_2$ solution, followed by 2 g solid ammonium acetate. Stir well, filter and wash with hot water.

Residue	Filtrate
May contain $PbCrO_4$ and $Pb(VO_3)_2$.	May contain Al^{3+}, UO_2^{2+} and excess of Pb^{2+}. Pass H_2S to remove all the Pb as PbS. Filter, wash and boil the filtrate to expel H_2S. Almost neutralize with 2M NH_3 solution, cool and pour into an excess of concentrated $(NH_4)_2CO_3$ solution. Warm for 5 minutes. Allow to stand, filter and wash.

Residue (PbCrO4): Dissolve in the minimum volume of hot 2M HNO_3 (~5–6 ml), thoroughly cool the resulting solution, and transfer to a small separating funnel. Add an equal volume of amyl alcohol, and a little 3% H_2O_2. Shake well and allow the two layers to separate. A blue colouration in the upper layer indicates Cr present, and a red to brownish-red colouration in the lower layer indicates V present. Confirm V by the $KClO_3$-*p*-phenetidine test.

Residue	Filtrate
White: $Al(OH)_3$. Al present. Confirm by Thenard's blue test.	May contain U, probably as $[UO_2(CO_3)_3]^{4-}$. Evaporate to a small volume, acidify with 2M HCl and add $K_4[Fe(CN)_6]$ solution. Brown ppt. of $(UO_2)_2[Fe(CN)_6]$ becoming yellow upon the addition of 2M NaOH solution. U present.

If this test is positive, remove phosphate (see Table 5.6 in Section 5.8). Among the less common ions, titanium, zirconium, cerium, thorium, and uranium will be completely precipitated by ammonia solution, and will therefore appear together with the other Group IIIA cations. Vanadium will only partly be precipitated here; some of the vanadium present will be found in the filtrate of the sulphide precipitates of Group IIIB cations. Some of the thallium(I) partly removed with Group I will also appear in this Group; some molybdenum will appear in the filtrate of Group IIIB sulphides.

The precipitation of Group IIIA cations with ammonia solution is carried out according to the procedure given in Table 5.7. The separation and

Table 6.6   **Separation of Tl from the rest of Group IIIB cations**   Dissolve the precipitate in 2M HNO_3. Expel H_2S by boiling, add saturated H_2SO_3 and expel excess SO_2 again by boiling. Pour the solution into an excess of saturated Na_2CO_3 solution. Filter.

Residue	*Filtrate*
May contain $CoCO_3$, $NiCO_3$, $MnCO_3$ and $ZnCO_3$. Dissolve in 2M HCl, remove CO_2 by boiling, neutralize with NH_3, then add $(NH_4)_2S$. Treat the precipitate according to Table 5.8.	May contain Tl^+. Identify by reaction 6 in Section 6.2 or by flame test (Section 6.2, reaction 11). Tl present.

identification of Group IIIA cations in the presence of some less common ions, as described in Table 6.5, commences with the precipitate obtained with ammonia.

If *thallium* has been found in Group I, some of it may pass into Group IIIB because of the solubility of thallium(I) chloride in water and be precipitated as Tl_2S. It may be readily detected by the green flame colouration, preferably viewed through a hand spectroscope. For the separation of thallium from the rest of Group IIIB cations, commencing with the sulphide precipitate, follow the procedure given in Table 6.6.

6.24   Lithium, Li (A_r: 6.94)

Lithium is a silver-white metal; it is the lightest metal known (density $0.534\,\mathrm{g\,cm^{-3}}$ at 20°C) and floats upon petroleum. It melts at 186°C. It oxidizes on exposure to air, and reacts with water forming lithium hydroxide and liberating hydrogen, but the reaction is not so vigorous as with sodium and potassium. The metal dissolves in acids with the formation of salts. The salts may be regarded as derived from the oxide, Li_2O.

Some of the salts, notably the chloride, LiCl, and the chlorate, $LiClO_3$, are very deliquescent. The solubilities of the hydroxide, LiOH ($113\,\mathrm{g\,l^{-1}}$ at 10°C), the carbonate, Li_2CO_3 ($13.1\,\mathrm{g\,l^{-1}}$ at 13°C), the phosphate, Li_3PO_4 ($0.30\,\mathrm{g\,l^{-1}}$ at 25°C), and the fluoride, LiF ($2.7\,\mathrm{g\,l^{-1}}$ at 18°C), are less than the corresponding sodium and potassium salts, and in this respect lithium resembles the alkaline earth metals.

Li^+ reactions

Reactions of lithium(I) ions   To study these reactions use a M solution of lithium chloride, LiCl; alternatively dissolve lithium carbonate, Li_2CO_3, in the minimum volume of 2M hydrochloric acid.

Na_2HPO_4
↓
white Li_3PO_4

1. *Disodium hydrogen phosphate solution*   partial precipitation of lithium phosphate, Li_3PO_4, in neutral solutions; the precipitate is more readily obtained from dilute solutions on boiling. Precipitation is almost complete in the presence of sodium hydroxide solution. The precipitate is more soluble in ammonium chloride solution than in water (distinction from magnesium).

Upon boiling the precipitate with barium hydroxide solution, it passes into solution as lithium hydroxide (difference from magnesium).

Carbonate
↓
white Li₂CO₃

2. *Sodium or ammonium carbonate solution* white precipitate of li
carbonate, Li_2CO_3, from concentrated solutions and in the presence of
ammonia solution. No precipitation occurs in the presence of high concentrations of ammonium chloride since the carbonate-ion concentration is reduced to such an extent that the solubility product of Li_2CO_3 is not exceeded:

$$NH_4^+ + CO_3^{2-} \rightleftharpoons NH_3 + HCO_3^-$$

NH₄F
↓
white LiF

3. *Ammonium fluoride solution* a white, gelatinous precipitate of lithium fluoride, LiF, is slowly formed in ammoniacal solution (distinction from sodium and potassium).

4. *Tartaric acid, sodium hexanitritocobaltate(III) or hexachloroplatinic acid solution* no precipitate (distinction from potassium). A precipitate of lithium hexanitritocobaltate(III) is, however, produced in very concentrated (almost saturated) solutions of lithium salts; interference with the sodium hexanitritocobaltate(III) test for K^+ is therefore unlikely.

Fe(IO₄)₃
↓
white ppt

5. *Iron(III) periodate test* iron(III) salts react with periodates to yield a precipitate of iron(III) periodate: this precipitate is soluble in excess of the periodate solution and also in excess potassium hydroxide solution. The resulting alkaline solution is a selective reagent for lithium, since it gives a white precipitate, $KLiFe[IO_6]$, even from dilute solutions and in the cold. Neither sodium nor potassium gives a precipitate; ammonium salts, all metals of Groups I to IV, and magnesium should be absent.

Place a drop of the neutral or alkaline test solution in a micro test-tube, and add 1 drop of M sodium chloride solution and 2 drops of the iron(III) periodate reagent. Carry out a blank test with a drop of distilled water simultaneously. Immerse both tubes for 15–20 seconds in water at 40–50°C. A white (or yellowish-white) precipitate indicates the presence of lithium; the blank remains clear.

Sensitivity 0.1 μg Li.
Concentration limit 1 in 100 000.

Microscope test

6. *Ammonium carbonate solution (microscope test)* lithium carbonate, when freshly formed, has a characteristic appearance under the microscope.

Place a drop of the concentrated test solution on a microscope slide. Introduce a few minute specks of sodium or ammonium carbonate. Some lithium carbonate crystals are formed immediately. Examine under the microscope (magnification: ×200): the crystals are in the form of either hexagonal stars or plates (compare $CaSO_4 \cdot 2H_2O$, Section 3.33, reaction 10).

The cations of the alkaline earth metals and of magnesium must be absent.
Concentration limit 1 in 10 000.

Flame test
↓
characteristic red
colour

7. *Dry test* flame colouration. Lithium compounds impart a carmine-red colour to the non-luminous Bunsen flame. The colour is masked by the presence of considerable amounts of sodium salts, but becomes visible when observed through two thicknesses of cobalt glass.

The most distinctive test utilizes the spectroscope; the spectrum consists of a beautiful red line at 671 nm.

8. *Atomic spectrometric tests* in **emission** mode 0.5 ppm lithium can be identified at 670.78 nm with the air/acetylene flame; with the nitrous oxide/acetylene flame the sensitivity is better. By **atomic absorption**, using a lithium hollow-cathode lamp or an electrodeless discharge tube and an air/acetylene flame the detection sensitivity is about 0.1 ppm at the same wavelength. A 1000 ppm lithium **standard** solution can be prepared by dissolving 6.1086 g of analytical grade, dry lithium chloride (LiCl) in water and diluting to 1 litre.

9. *Separation of lithium from other alkali metals* in order to separate lithium from the other alkali metals, they are all converted into the chlorides (by evaporation with concentrated hydrochloric acid, if necessary), evaporated to dryness, and the residue extracted with absolute alcohol which dissolves the lithium chloride only. Better solvents are dry dioxan (diethylene dioxide, $C_4H_8O_2$) and dry acetone. Upon evaporation of the extract, the residue of lithium chloride is (*a*) subjected to the flame test, and (*b*) precipitated as the phosphate after dissolution in water and adding sodium hydroxide solution.

Appendix

Reagent solutions and gases

The reagents are listed alphabetically. One asterisk * indicates that a reagent has a limited stability and should not be kept for longer than 1 month. Two asterisks ** indicate that the reagent should be prepared freshly and discarded after use. Reagents with no asterisk can be kept for at least one year after preparation.

Acetic acid (concentrated, 'glacial'). The commercial concentrated or glacial acetic acid is a water-like solution with a characteristic smell, having a density of $1.06\,g\,cm^{-3}$. It contains 99.5% (w/w) of CH_3COOH ($1.06\,g\,CH_3COOH$ per ml) and is approximately 17.6 molar. When cooled to 0°C, the reagent freezes forming ice-like crystals, hence the name 'glacial'. It has to be handled with care.

Acetic acid (9M, 1 + 1). To 50 ml water add 50 ml glacial acetic acid and mix.

Acetic acid (30% v/v). Dilute 30 ml glacial acetic acid, CH_3COOH, with water to 100 ml.

Acetic acid (2M). Dilute 114 ml glacial acetic acid with water to 1 litre.

Acetone. The pure solvent is a clear colourless liquid with a characteristic odour. Its density is $0.79\,g\,cm^{-3}$ and it boils at 56.2°C.

Acetylacetone. $CH_3 \cdot CO \cdot CH_2 \cdot CO \cdot CH_3$. The commercial reagent is a colourless, sometimes yellowish liquid with a density of $0.97\,g\,cm^{-3}$. It boils at 137°C.

Albumin solution.* Dissolve 0.1 g albumin in 20 ml water to obtain a colloidal solution.

Alizarin (saturated solution in ethanol). To 2 g alizarin, $C_{14}H_8O_4$, add 10 ml 96% ethanol and shake. Use the clear solution for the tests.

Alizarin–hydrochloric acid solution. Mix 19 ml saturated alcoholic solution of alizarin and 1 ml of 2M hydrochloric acid.

Alizarin red S (2%). Dissolve 2 g alizarin red S (sodium alizarinsulphonate), $C_{14}H_7O_4 \cdot SO_3Na \cdot H_2O$, in 100 ml water.

Alizarin red S (0.1%). Dissolve 0.1 g alizarin red S (sodium alizarinsulphonate), $C_{14}H_7O_4 \cdot SO_3Na \cdot H_2O$, in 100 ml water.

Aluminium chloride (0.33M). Dissolve 80.5 g aluminium chloride hexahydrate, $AlCl_3 \cdot 6H_2O$, in water and dilute to 1 litre.

Aluminium sulphate (0.17M). Dissolve 107.2 g aluminium sulphate hexadecahydrate, $Al_2(SO_4)_3 \cdot 16H_2O$, in water and dilute to 1 litre. Alternatively, dissolve 158.1 g aluminium potassium sulphate icositetrahydrate (potash alum), $K_2SO_4 \cdot Al_2(SO_4)_3 \cdot 24H_2O$, in water and dilute to 1 litre.

Aluminon reagent (0.1%). Dissolve 0.1 g Aluminon (tri-ammonium arurine-tricarboxylate, $O:(COONH_4)C_6H_3{=}C[C_6H_3(OH)COONH_4]_2)$ in 100 ml water.

Ammonia solution (concentrated). The commercial concentrated ammonia solution is a water-like liquid with a characteristic smell, owing to the evaporation of ammonia gas. It has a density of $0.90\,g\,cm^{-3}$, contains 58.6% (w/w) NH_3 (or $0.53\,g\,NH_3$ per ml), and is approximately 15.1 molar. It should be handled with care, wearing eye protection. Direct smelling of the solution should be avoided. The solution should be kept far apart from concentrated hydrochloric acid to avoid the formation of ammonium chloride fumes.

Ammonia solution (7.5M, 1 + 1). To 500 ml water add 500 ml concentrated ammonia solution and mix.

Ammonia solution (2M).   Dilute 134 ml concentrated ammonia solution with water to 1 litre.

Ammonia solution (1.5M).   Dilute 100 ml concentrated ammonia solution to 1 litre.

Ammonia solution (2.5%).   Dilute 5 ml concentrated ammonia with water to 100 ml.

Ammonia solution (0.1M).   Dilute 5 ml 2M ammonia with water to 100 ml.

Ammonia solution (0.02M).   Dilute 1 ml 2M ammonia with water to 100 ml.

Ammonium acetate (6M).   Dissolve 46.3 g ammonium acetate, CH_3COONH_4, in water and dilute the solution to 100 ml.

Ammonium carbonate (M).*   Dissolve 96.1 g ammonium carbonate, $(NH_4)_2CO_3$, in water and dilute to 1 litre. Aged reagents must be boiled up and cooled before use to remove ammonium carbamate from the solution.

Ammonium carbonate (concentrated).*   Shake 20 g ammonium carbonate, $(NH_4)_2CO_3$, with 80 ml water. Allow to stand overnight and filter. Prepare the reagent freshly.

Ammonium chloride (saturated).   To 4 g ammonium chloride add 10 ml water. Warm the mixture on a water bath, and use the clear supernatant liquid of the cooled mixture as a reagent.

Ammonium chloride (20%).   Dissolve 20 g ammonium chloride, NH_4Cl, in water and dilute to 100 ml.

Ammonium chloride (M).   Dissolve 53.5 g ammonium chloride in water and dilute the solution to 1 litre.

Ammonium chloride (1%).   Dissolve 1 g ammonium chloride, NH_4Cl, in water and dilute to 100 ml.

Ammonium chloride–ammonium sulphide wash solution.* Dissolve 1 g ammonium chloride in 80 ml water, add 1 ml of M ammonium sulphide and dilute the solution to 100 ml.

Ammonium fluoride (0.1M).   Dissolve 3.7 g ammonium fluoride, NH_4F, in water and dilute to 1 litre.

Ammonium iodide (10%).   Dissolve 10 g ammonium iodide, NH_4I, in water and dilute to 100 ml.

Ammonium metavanadate (0.1M).   Dissolve 11.7 g ammonium metavanadate, NH_4VO_3, in 100 ml of M sulphuric acid and dilute with water to 1 litre.

Ammonium molybdate (0.25M for molybdenum).*   Dissolve 44.2 g ammonium molybdate, $(NH_4)_6Mo_7O_{24} \cdot 4H_2O$, in a mixture of 60 ml concentrated ammonia and 40 ml water. Add 120 g ammonium nitrate and after complete dissolution dilute the solution to 1 litre. Before use, the solution, to which the reagent is added, must be made acid by adding nitric acid.

Ammonium molybdate (0.025M).*   Dilute 1 ml of 0.25M ammonium molybdate with water to 10 ml.

Ammonium molybdate–quinine sulphate reagent.* Dissolve 4 g finely powdered ammonium molybdate, $(NH_4)_6Mo_7O_{24} \cdot 4H_2O$, in 20 ml water. Add to this solution, with stirring, a solution of 0.1 g quinine sulphate, $C_{20}H_{24}N_2O_2 \cdot H_2SO_4 \cdot 2H_2O$, in 80 ml concentrated nitric acid.

Ammonium nitrate (M).   Dissolve 80 g ammonium nitrate, NH_4NO_3, in water and dilute to 1 litre.

Ammonium nitrate (2%).   Dissolve 20 g ammonium nitrate, NH_4NO_3, in water and dilute the solution to 1 litre.

Ammonium nitrate wash solution.** Dissolve 0.5 g ammonium nitrate, NH_4NO_3, in 10 ml water and saturate the solution with hydrogen sulphide gas.

Ammonium oxalate (0.17M or 2.5%).   Dissolve 25 g ammonium oxalate mono-hydrate, $(COONH_4)_2 \cdot H_2O$, in 900 ml water and after complete dissolution dilute to 1 litre.

Ammonium oxalate (0.1M).   Dissolve 14.2 g ammonium oxalate monohydrate, $(COONH_4)_2 \cdot H_2O$, in water and dilute to 1 litre.

Ammonium peroxodisulphate (M).*   Dissolve 22.8 g ammonium peroxodisulphate, $(NH_4)_2S_2O_8$, in water and dilute to 100 ml.

Ammonium peroxodisulphate (0.1M).* Dissolve 22.8 g ammonium peroxodisulphate, $(NH_4)_2S_2O_8$, in water and dilute to 1 litre.

Ammonium polysulphide (M). To 1 litre ammonium sulphide solution (M) add 32 g sulphur, and heat gently until the latter dissolves completely and a yellow solution is formed. The formula of the reagent is $(NH_4)_2S_x$, where x is approx. 2.

Ammonium sulphate (saturated). Mix 45 g ammonium sulphate, $(NH_4)_2SO_4$, and 50 ml water. Heat the mixture on a water bath for 3 hours and allow to cool. Use the clear supernatant liquid for the tests.

Ammonium sulphate (M). Dissolve 132 g ammonium sulphate, $(NH_4)_2SO_4$, in water and dilute to 1 litre.

Ammonium sulphide (M).* Saturate 500 ml ammonia solution (2M) with hydrogen sulphide, until a small sample (1 ml) of the solution does not cause any precipitation in magnesium sulphate (M) solution. Then add 500 ml ammonia solution (2M). Store the solution in a well-stoppered bottle. The solution must be colourless. Yellow or orange colour indicates that considerable amounts of polysulphide are present in the solution.

Ammonium sulphide wash solution* (1 + 100). Dilute 1 ml M ammonium sulphide with water to 100 ml.

Ammonium tartrate (6M).* Dissolve 50 g tartaric acid, $C_4H_6O_6$, in 50 ml water and add 50 ml concentrated ammonia solution. The solution contains an excess of ammonia.

Ammonium tetrathiocyanatomercurate(II) (0.3M).** Dissolve 9 g ammonium thiocyanate, NH_4SCN, and 8 g mercury(II) chloride, $HgCl_2$, in water and dilute to 100 ml.

Ammonium thiocyanate (0.1M). Dissolve 7.61 g ammonium thiocyanate, NH_4SCN, in water and dilute to 1 litre.

Ammonium thiocyanate (saturated solution in acetone).* Shake 1 g ammonium thiocyanate, NH_4SCN, with 5 ml acetone, and use the clear supernatant liquid for the tests.

Amyl acetate (*iso*-amyl acetate, 3-methyl-butyl acetate). A colourless liquid with a pleasant, characteristic odour. It has a density of $0.87\,\mathrm{g\,cm}^{-3}$.

Amyl alcohol (*iso*-amyl alcohol, 3-methylbutan-1-ol or 2-methyl-butan-4-ol). A colourless liquid with a characteristic odour. It has a density of $0.81\,\mathrm{g\,cm}^{-3}$. It should be kept in a metal box or cupboard.

Aniline. The commercial product is an almost colourless oily liquid, which darkens on longer standing. Its density is $1.02\,\mathrm{g\,cm}^{-3}$.

Aniline sulphate (1%).* Dissolve 1 g aniline sulphate $(C_6H_5NH_2)_2 \cdot H_2SO_4$ in 100 ml of water.

Anthranilic acid (5%).* Dissolve 0.5 g anthranilic acid $NH_2 \cdot C_6H_4 \cdot COOH$ in 10 ml of 96% ethanol.

Antimony(III) chloride (0.2M).* Dissolve 45.6 g antimony(III) chloride $SbCl_3$ in 250 ml of 6M hydrochloric acid and dilute the solution to 1 litre. Alternatively, take 29.2 g antimony(III) oxide Sb_2O_3 and dissolve it in 500 ml of 6M hydrochloric acid by heating the mixture on a water bath. After cooling dilute the solution to 1 litre.

Aqua regia. To 3 volumes of concentrated hydrochloric acid add 1 volume of concentrated nitric acid. Mix and use immediately, the solution does not keep at all.

Arsenic trioxide (arsenious acid) (0.1M for arsenic; **POISON**). Boil 9.89 g arsenic trioxide with 500 ml water until complete dissolution. Cool the solution and dilute with water to 1 litre.

Ascorbic acid (0.05M).** Dissolve 0.9 g ascorbic acid in 100 ml water.

Barium chloride (0.25M). Dissolve 61.1 g barium chloride dihydrate, $BaCl_2 \cdot 2H_2O$, in water and dilute to 1 litre.

Barium nitrate (0.25M). Dissolve 65.3 g barium nitrate, $Ba(NO_3)_2$, in water and dilute to 1 litre.

Baryta water (saturated). Shake 5 g barium hydroxide octahydrate, $Ba(OH)_2 \cdot 8H_2O$, with 100 ml water. Allow to stand for 24 hours. Use the clear supernatant liquid for the tests.

Benzoin α-oxime (cupron) (5% in alcohol).** Dissolve 5 g benzoin α-oxime, $C_6H_5 \cdot$ $CH(OH) \cdot C(NOH) \cdot C_6H_5$, in 96% ethanol and dilute with the solvent to 100 ml.

Beryllium sulphate (0.1M). Dissolve 17.7 g beryllium sulphate tetrahydrate, $BeSO_4 \cdot 4H_2O$, in water and dilute to 1 litre.

1,1-Bi-2-naphtol (0.05%).* Dissolve 0.05 g 1,1-bi-2-naphtol (or di-β-naphtol, $(C_{10}H_6OH)_2$, in 100 ml concentrated sulphuric acid.

2,2'-Bipyridyl reagent.* Dissolve 0.01 g of the solid reagent in 0.5 ml of 0.1M hydrochloric acid. Alternatively, dissolve 0.01 g reagent in 0.5 ml 96% ethanol. The reagent is very expensive and it is worthwhile to use the minimum amount required.

Bis-(p-nitrophenyl) carbazide (0.1%).** Dissolve 0.1 g bis-(p-nitrophenyl) carbazide (p,p'-dinitro-1,5-diphenyl carbazide) $CO(NH \cdot NH \cdot C_6H_4 \cdot NO_2)_2$ in 100 ml 96% ethanol.

Bismuth nitrate (0.2M). To 500 ml water add cautiously 50 ml concentrated nitric acid, HNO_3. Dissolve 97.0 g bismuth nitrate pentahydrate, $Bi(NO_3)_3 \cdot 5H_2O$, in this mixture, and dilute with water to 1 litre.

Bromine water (saturated).** Shake 4 g (or 1 ml) liquid bromine with 100 ml water. Ensure that a slight excess of undissolved bromine is left at the bottom of the mixture. The solution keeps for 1 week. When handling bromine, exercise utmost care; use rubber gloves and eye protection always.

Bromothymol blue (0.04%). Dissolve 40 mg bromothymol blue in 100 ml 96% ethanol.

Cacotheline (0.25%).* Dissolve 0.25 g cacotheline (nitrobruciquinone hydrate, $C_{21}H_{21}O_7N_3$) in 100 ml water.

Cadmium acetate (0.5M). Dissolve 13.7 g cadmium acetate dihydrate, $Cd(CH_3COO)_2 \cdot 2H_2O$, in water and dilute to 100 ml.

Cadmium carbonate suspension (freshly precipitated).** To 20 ml of 0.5M cadmium acetate add 20 ml of 0.5M sodium carbonate solution. Allow the precipitate to settle and wash 4–5 times with water by decantation.

Cadmium sulphate (0.25M). Dissolve 64.13 g tricadmium sulphate octahydrate $3CdSO_4 \cdot 8H_2O$ in water and dilute to 1 litre.

Cadmium sulphate in ammonia (saturated) (*for the tetramminecadmium perchlorate test*). Equilibrate 10 g tricadmium sulphate octahydrate $3CdSO_4 \cdot 8H_2O$ overnight with 10 ml concentrated ammonia. Use the clear supernatant liquid for the test.

Calcium chloride (0.5M). Dissolve 109.5 g calcium chloride hexahydrate, $CaCl_2 \cdot 6H_2O$, in water and dilute to 1 litre. If the solid reagent is not pure enough, weigh 50 g calcium carbonate into a large porcelain dish, add slowly 100 ml of 6M hydrochloric acid, waiting after the addition of a new portion until effervescence ceases, then evaporate the mixture to dryness and dissolve the residue in water. Dilute finally to 1 litre.

Calcium nitrate (saturated). Mix 80 g calcium nitrate tetrahydrate, $Ca(NO_3)_2 \cdot 4H_2O$, and 20 ml water. Heat the mixture on a water bath until complete dissolution. Allow to cool.

Calcium sulphate (saturated). Shake 0.35 g calcium sulphate monohydrate, $CaSO_4 \cdot H_2O$, with 100 ml water. Allow the mixture to stand for 24 hours and use the clear supernatant liquid for the tests.

Carbon dioxide gas. This gas can be obtained from a Kipp apparatus using calcium carbonate pieces (broken marble) and 6M hydrochloric acid. The gas should be washed with concentrated sulphuric acid.

Catechol (pyrocatechol) (10%).* Dissolve 10 g catechol (o-dihydroxybenzene), $C_6H_4(OH)_2$, in water and dilute to 100 ml.

Cerium(III) nitrate (0.1M). Dissolve 43.5 g cerium(III) nitrate hexahydrate, $Ce(NO_3)_3 \cdot 6H_2O$, in 100 ml 2M nitric acid and dilute with water to 1 litre.

Cerium(IV) sulphate (0.1M). Dissolve 40.4 g cerium(IV) sulphate tetrahydrate, $Ce(SO_4)_2 \cdot 4H_2O$, or 63.6 g cerium(IV) diammonium sulphate dihydrate, $Ce(SO_4)_2 \cdot 2(NH_4)_2SO_4 \cdot 2H_2O$, in a cold mixture of 500 ml water and 50 ml concentrated sulphuric acid, and dilute the solution with water to 1 litre.

Chloramine-T (0.05M).** Dissolve 14.1 g chloramine-T, $CH_3 \cdot C_6H_4 \cdot SO_2 \cdot N \cdot NaCl \cdot 3H_2O$, in water and dilute to 1 litre.

Chlorine gas. This gas is available commercially in steel cylinders and should be taken from these. Alternatively, it can be produced in a Kipp apparatus (with glass joints only) from calcium hypochlorite (bleaching powder) and 6M hydrochloric acid. It is advisable to mix the bleaching powder with some gypsum powder, making the mixture wet, to produce lumps of the reagent. After drying, these lumps can be placed into the Kipp apparatus.

Chlorine water (about 0.1M).** Saturate 200 ml of water with chlorine gas.

Chloroform. The commercial reagent is a colourless liquid with a characteristic smell. Chloroform vapour should not be inhaled; it causes drowsiness. It has a high density ($1.47 \, g \, cm^{-3}$) and a high refractive index (1.44). Although it is not inflammable, it should be kept together with other organic solvents in a metal box or cupboard.

Chromic acid (25%). Pour 25 ml chromosulphuric acid cautiously, with constant stirring, into 70 ml water. After cooling, dilute the solution with water to 100 ml.

Chromium(III) chloride (0.33M). Dissolve 88.8 g chromium(III) chloride hexahydrate, $CrCl_3 \cdot 6H_2O$, in water and dilute to 1 litre.

Chromium(III) sulphate (0.167M). Dissolve 110.4 g chromium(III) sulphate pentadecahydrate, $Cr_2(SO_4)_3 \cdot 15H_2O$, in water and dilute to 1 litre.

Chromosulphuric acid (concentrated). To 100 g potassium or sodium dichromate ($K_2Cr_2O_7$ or $Na_2Cr_2O_7$) add 1 litre concentrated sulphuric acid. Stir the mixture occasionally and keep in a stoppered vessel. Because of its strong oxidizing and dehydrating properties, **handle the solution with the greatest care** (eye glasses, rubber gloves). Its cleansing action is slow but effective; best results are obtained if the mixture is left overnight in the vessel to be cleaned. **Used portions should be poured back into the stock of the mixture.** As the brownish colour changes slowly to green, discard the solution.

Chromotropic acid. To 0.1 g chromotropic acid sodium salt (1,8-dihydroxy-naphthalene-3,6-disulphonic acid; sodium salt) $C_{10}H_6O_8S_2Na_2$ add 5 ml water, mix thoroughly, and use the pure supernatant liquid for the tests.

Cinchonine–potassium iodide reagent (1%).* Dissolve 1 g cinchonine in a mixture of 99 ml water and 1 ml dilute nitric acid (2M) by boiling, let the mixture cool and dissolve 1 g potassium iodide, KI, in the solution. The reagent is stable for 2 weeks.

Cobalt(II) acetate (0.1M). Dissolve 2.5 g cobalt(II) acetate tetrahydrate, $(CH_3COO)_2Co \cdot 4H_2O$, in 100 ml water.

Cobalt(II) chloride (0.5M). Dissolve 119 g cobalt(II) chloride hexahydrate, $CoCl_2 \cdot 6H_2O$, in water and dilute to 1 litre.

Cobalt(II) nitrate (0.5M). Dissolve 146 g cobalt(II) nitrate hexahydrate, $Co(NO_3)_2 \cdot 6H_2O$, in water and dilute to 1 litre.

Cobalt(II) thiocyanate (10%).** Dissolve 1 g cobalt acetate tetrahydrate $(CH_3COO)_2Co \cdot 4H_2O$ in 5 ml water. In a separate vessel dissolve 1 g ammonium thiocyanate, NH_4SCN, in 5 ml water, and mix the two solutions.

Copper(I) chloride reagent (M).* Dissolve 9.9 g copper(I) chloride, CuCl, in a mixture of 60 ml water and 40 ml concentrated hydrochloric acid. Heating accelerates dissolution. Add to the solution strips of bright copper, which should be left there.

Copper(II) sulphate (0.25M). Dissolve 62.42 g copper sulphate pentahydrate, $CuSO_4 \cdot 5H_2O$, in water and dilute to 1 litre.

Copper(II) sulphate (0.1%). Dissolve 0.1 g copper sulphate pentahydrate, $CuSO_4 \cdot 5H_2O$, in 100 ml water.

Copper sulphide suspension. Dissolve 0.12 g copper sulphate pentahydrate $CuSO_4 \cdot 5H_2O$ in 100 ml water, add 5 drops of 2M ammonia solution and introduce hydrogen sulphide gas until the solution becomes cloudy. The suspension must be freshly prepared.

Cupferron (2%).** Dissolve 2 g cupferron, $C_6H_5 \cdot N(NO)ONH_4$, in 100 ml water. The solution does not keep well. Addition of 1 g ammonium carbonate, $(NH_4)_2CO_3$, enhances the stability.

Curcumin (saturated solution in ethanol).* Shake 0.5 g curcumin $[CH_3O \cdot C_6H_3(OH) \cdot CH:CH \cdot CO]_2CH_2$ with 10 ml 96% ethanol. Use the clear liquid to prepare turmeric paper.

Diammineargentatonitrate (0.1M).** To 10 ml 0.1M silver nitrate add 2M ammonia solution until the precipitate first formed is just redissolved. Discard the unused reagent because on standing it forms silver azide, which may cause **SERIOUS EXPLOSIONS**.

Diammonium hydrogen phosphate (0.5M).* Dissolve 66 g diammonium hydrogen orthophosphate, $(NH_4)_2HPO_4$, in water and dilute to 1 litre.

Diazine green (0.01%). Dissolve 0.01 g diazine green (Janus green, Colour Index 11050) in 100 ml water.

Di-(4-dimethylaminodiphenyl) methane (tetrabase) (0.5%). Dissolve 0.5 g tetrabase $[(CH_3)_2N \cdot C_6H_4]_2CH_2$ in a mixture of 20 ml glacial acetic acid and 80 ml 96% ethanol.

Di-(4-dimethylaminodiphenyl) methane (tetrabase) (1%). Dissolve 0.1 g tetrabase $[(CH_3)_2N \cdot C_6H_4]_2CH_2$ in 10 ml chloroform.

Diethyl ether. The commercial product is a clear, colourless, mobile liquid, which is highly inflammable. It has a density of 0.71 g cm^{-3}. It should be kept in a metal box or cupboard.

4-Dimethylaminobenzene-azo-phenyl-arsonic acid (0.1%).** Dissolve 0.1 g 4'-dimethylaminoazobenzene-4-phenylarsonic acid, $(CH_3)_2N \cdot C_6H_4 \cdot N:N \cdot C_6H_4AsO-(OH)_2$, in 5 ml concentrated hydrochloric acid, and dilute with 96% ethanol to 100 ml.

4-Dimethylaminobenzylidene–rhodanine (0.3% in acetone).** Dissolve 30 mg commercial 4-dimethylaminobenzylidene–rhodanine reagent in 10 ml acetone.

Dimethyl glyoxime (1%). Dissolve 1 g dimethyl glyoxime, $CH_3 \cdot C(NOH) \cdot C(NOH) \cdot CH_3$, in 100 ml 96% ethanol.

NN-Dimethyl-p-phenylenediamine dihydrochloride (1%).* Dissolve 1 g NN-dimethyl-p-phenylenediamine hydrochloride $NH_2 \cdot C_6H_4 \cdot N(CH_3)_2 \cdot 2HCl$ in water and dilute to 100 ml. Instead of the dihydrochloride, NN-dimethyl-p-phenylenediamine sulphate $NH_2 \cdot C_6H_4 \cdot N(CH_3)_2 \cdot H_2SO_4$ may be used.

Dioxan (1,4-dioxan, diethylene dioxide), $C_4H_8O_2$. The commercial solvent is a clear, colourless liquid with a characteristic odour. It has a density of 1.42 g cm^{-3}. The liquid freezes at 11.5°C and boils at 101°C.

Diphenylamine (0.5%).* Dissolve 0.5 g diphenylamine, $(C_6H_5)_2NH$, in 85 ml concentrated sulphuric acid, and dilute the solution with the greatest care with water to 100 ml.

1,5-Diphenylcarbazide (1% in ethanol).** Dissolve 0.10 g 1,5-diphenylcarbazide, $(C_6H_5 \cdot NH \cdot NH)_2CO$, in 10 ml of 96% ethanol. The solution decomposes rapidly if exposed to air.

1,5-Diphenylcarbazide (0.2% solution with acetic acid).** Dissolve 0.2 g 1,5-diphenylcarbazide, $(C_6H_5 \cdot NH \cdot NH)_2CO$, in 10 ml glacial acetic acid and dilute the mixture to 100 ml with 96% ethanol.

Diphenylthiocarbazone (dithizone) (0.005%).* Dissolve 5 mg diphenylthiocarbazone (dithizone), $C_6H_5 \cdot N:N \cdot CS \cdot NH \cdot NH \cdot C_6H_5$, in 100 ml chloroform.

p-Dipicrylamine (1%).** Boil a mixture of 0.2 g p-dipicrylamine (hexanitrodiphenylamine), $[C_6H_2(NO_2)_3]_2NH$, 4 ml of 0.5M sodium carbonate and 16 ml water. Allow to cool and filter. Use the filtrate for the tests.

Disodium hydrogen arsenate (0.25M). Dissolve 76 g disodium hydrogen arsenate, $Na_2HAsO_4 \cdot 7H_2O$, in water and dilute the solution to 1 litre.

Disodium hydrogen phosphate (0.033M). Dissolve 12 g disodium hydrogen orthophosphate dodecahydrate, $Na_2HPO_4 \cdot 12H_2O$, or 6 g disodium hydrogen orthophosphate dihydrate, $Na_2HPO_4 \cdot 2H_2O$, or 4.2 g anhydrous disodium hydrogen orthophosphate, Na_2HPO_4, in water and dilute to 1 litre.

Dithio-oxamide (rubeanic acid) (0.5%).** Dissolve 0.5 g dithio-oxamide $H_2N \cdot CS \cdot CS \cdot NH_2$ in 100 ml of 96% ethanol. The solution decomposes rapidly and should be prepared freshly each time.

Dithio-oxamide (rubeanic acid) (0.02%).** Dissolve 20 mg dithio-oxamide $H_2N \cdot CS \cdot CS \cdot NH_2$ in 100 ml glacial acetic acid. The solution decomposes rapidly and should be prepared freshly each time.

Ethanol (anhydrous or 'absolute'). The commercial product contains more than 99.7% C_2H_5OH.

Ethanol (ethyl alcohol) (96%). The commercial product contains about 95–96% C_2H_5OH and has a density of 0.81 g cm^{-3}. It is inflammable and should be kept in a metal box or cupboard.

Ethanol (80%). To 80 ml 96% ethanol add 16 ml water.

Ether. See Diethyl ether.

Ethyl acetate. A colourless liquid with a pleasant, characteristic odour. It has a density of 0.9 g cm^{-3}.

Ethylamine (M). Dissolve 4.5 g ethylamine, $C_2H_5NH_2$, in water and dilute to 100 ml. Alternatively, dilute 7 ml of 70% ethylamine solution with water to 100 ml.

Ethylenediamine (diamino-ethane). The commercial product is a liquid with a density of 0.90 g cm^{-3}.

Ferron (0.2%).* Dissolve 0.2 g ferron (8-hydroxy-7-iodoquinoline-5-sulphonic acid), $C_9H_6O_4NSI$, in 100 ml water.

Fluorescein reagent. To 1 g fluorescein, $C_{20}H_{12}O_5$ (Colour Index 45350), add a mixture of 50 ml 96% ethanol and 50 ml water. Shake, allow to stand for 24 hours, and filter the solution.

Formaldehyde solution (40%). Commercial formaldehyde solution (formalin) contains about 40 g HCHO per 100 ml. It normally contains some methanol as stabilizer. It has a characteristic, pungent odour. If kept cold for long periods, solid paraldehyde separates from the solution.

Formaldehyde (10%).* Dilute 10 ml of 40% formaldehyde solution (formalin) with 30 ml water.

Formaldehyde (4%)–sodium carbonate (0.05M) reagent. To 50 ml water add 10 ml 0.5M sodium carbonate and 10 ml 40% formaldehyde solution, and dilute the mixture to 100 ml. The reagent decomposes on standing.

Fuchsin (0.1%). Dissolve 0.1 g fuchsin (magenta, Colour Index 42500) in 100 ml water.

Fuchsin (0.015%). Dissolve 0.015 g fuchsin (magenta, Colour Index 42500) in 100 ml water.

Gallocyanine (1%).** Dissolve 1 g gallocyanine, $C_{15}H_{12}O_5N_2$, in 100 ml water.

Gold(III) chloride (0.33M for gold). Dissolve 13.3 g sodium tetrachloroaurate(III) dihydrate, $NaAuCl_4 \cdot 2H_2O$, in water and dilute to 100 ml. Alternatively, dissolve 6.57 g gold metal in 50 ml aqua regia, evaporate the solution to dryness, and dissolve the residue in 100 ml water.

Hexachloroplatinic(IV) acid (26% or 0.5M). Dissolve 2.6 g hexachloroplatinic(IV) acid, $H_2[PtCl_6] \cdot 6H_2O$, in 10 ml water.

Hydrazine sulphate (saturated).* To 2 g hydrazine sulphate, $N_2H_4H_2SO_4$, add 5 ml water and saturate the solution by vigorous shaking.

Hydrochloric acid (concentrated, 'fuming'). The commercial concentrated hydrochloric acid is a water-like solution with a characteristic smell, and is 'fuming' owing to the evaporation of hydrogen chloride gas. It has a density of 1.19 g cm^{-3}, contains 36.0% (w/w) HCl (or 0.426 g HCl per ml), and is approximately 11.7 molar. The reagent should be stored far away from concentrated ammonia to prevent the formation of ammonium chloride fumes. It should be handled with care, using eye protection.

Hydrochloric acid (6M, 1 + 1). To 500 ml water add 500 ml concentrated hydrochloric acid, and let the solution cool to room temperature.

Hydrochloric acid (3M). To 500 ml water add 265 ml concentrated hydrochloric acid, and dilute with water to 1 litre.

Hydrochloric acid (2M). Pour 170 ml concentrated hydrochloric acid into 800 ml water stirring constantly, and dilute with water to 1 litre.

Hydrochloric acid (0.5M). Dilute 4.5 ml concentrated hydrochloric acid with water to 100 ml.

Hydrochloric acid (0.1M).   Dilute 20 ml 0.5M hydrochloric acid with water to 100 ml.

Hydrogen peroxide (concentrated).*   Commercial concentrated hydrogen peroxide (so called '100 volume') solution contains 30% (w/w) of H_2O_2 and has a density of $1.10\,g\,cm^{-3}$. Sometimes it contains small amounts of sulphuric acid or organic material as stabilizer. It should be handled with care.

Hydrogen peroxide (10%).*   To 30 ml concentrated (30%) hydrogen peroxide add 60 ml water and mix. Dilute with water to 100 ml.

Hydrogen peroxide (3%, approx. M).*   Dilute 100 ml concentrated ('100 volume') hydrogen peroxide with water to 1 litre.

Hydrogen sulphide gas.   This gas can be obtained from a Kipp apparatus, using solid iron(II) sulphide, FeS, and 6M hydrochloric acid. The gas can be washed by bubbling it through water.

Hydrogen sulphide (saturated solution, about 0.1M).**   Saturate 250 ml water with hydrogen sulphide gas, obtained from a Kipp apparatus. The solution contains approximately $4\,g\,H_2S$ per litre. The solution will keep for about 1 week.

Hydroxylamine hydrochloride (10%).*   Dissolve 10 g hydroxylammonium chloride, $NH_2OH\cdot HCl$, in water and dilute to 100 ml.

8-Hydroxyquinoline (5%).*   Dissolve 5 g 8-hydroxyquinoline, C_9H_7ON, in a mixture of 90 ml water and 10 ml of M sulphuric acid. The reagent is stable for several months.

8-Hydroxyquinoline (2%, in acetic acid).*   Dissolve 2 g 8-hydroxyquinoline, C_9H_7ON, in 100 ml 2M acetic acid.

8-Hydroxyquinoline (1% in alcohol).*   Dissolve 1 g 8-hydroxyquinoline, C_9H_7ON, in 100 ml of 96% ethanol.

Hypophosphorous acid (50%).   The commercial reagent contains 49–53% pure HPO_2 and has a density of $1.2\,g\,cm^{-3}$. Dilute the solution with 4 volumes of water for the tests.

Indigo solution (1%).   Dissolve 0.1 g indigo, $C_{16}H_{10}O_2N_2$ (Colour Index 73000), in 10 ml concentrated sulphuric acid.

Indole (0.015%).   Dissolve 15 mg indole, $C_6H_4\cdot NH\cdot CH\cdot CH$, in 100 ml 96% ethanol.

Iodine (potassium tri-iodide) (0.05M).   Dissolve 12.7 g iodine, I_2, and 25 g potassium iodide, KI, in water, and dilute to 1 litre. Do not handle iodine with a metal or plastic spatula: use a spoon made of porcelain or glass.

Iodine (potassium tri-iodide) (0.005M).   Dilute 10 ml 0.05M iodine solution with water to 100 ml.

Iron(III) chloride (0.5M).   Dissolve 135.2 g iron(III) chloride hexahydrate, $FeCl_3\cdot 6H_2O$, in water, add a few millilitres concentrated hydrochloric acid if necessary, and dilute with water to 1 litre. If the solution turns dark, add more hydrochloric acid.

**Iron(III) periodate reagent.*   Dissolve 2 g potassium periodate, KIO_4, in 10 ml freshly prepared 2M potassium hydroxide solution. Dilute with water to 50 ml, add 2 ml of 0.5M iron(III) chloride solution and dilute with 2M potassium hydroxide to 100 ml.

Iron(II) sulphate (saturated).**   To 1 g iron(II) sulphate, $FeSO_4\cdot 7H_2O$, add 5 ml water. Shake well in the cold. Decant the supernatant liquid, which should be used for the tests. The reagent should be prepared freshly.

Iron(II) sulphate (0.5M).*   Dissolve 139 g iron(II) sulphate heptahydrate, $FeSO_4\cdot 7H_2O$, or 196 g iron(II) ammonium sulphate hexahydrate (Mohr's salt, $FeSO_4\cdot (NH_4)_2SO_4\cdot 6H_2O$) in a cold mixture of 500 ml water and 50 ml M sulphuric acid, and dilute the solution with water to 1 litre.

Iron(III) sulphate (25%).   Heat 5 g iron(III) sulphate, $Fe_2(SO_4)_3$, and 15 ml water. After complete dissolution and cooling, dilute with water to 20 ml.

Iron(III) thiocyanate (0.05M).   Dissolve 1.35 g iron(III) chloride hexahydrate, $FeCl_3\cdot 6H_2O$, and 2 g potassium thiocyanate, KSCN, in water and dilute the solution to 100 ml.

Lanthanum nitrate (0.1M).   Dissolve 4.33 g lanthanum nitrate hexahydrate, $La(NO_3)_3\cdot 6H_2O$, in 100 ml water.

Lead acetate (0.25M).   Dissolve 95 g lead acetate trihydrate, $Pb(CH_3COO)_2 \cdot 3H_2O$, in a mixture of 500 ml water and 10 ml glacial acetic acid, CH_3COOH, and dilute the solution with water to 1 litre.

Lead acetate (0.0025M).   Dilute 1 ml of 0.25M lead acetate solution with water to 100 ml.

Lead nitrate (0.25M).   Dissolve 82.8 g lead nitrate, $Pb(NO_3)_2$, in water and dilute to 1 litre.

Lime water (saturated).*   Shake 5 g calcium hydroxide, $Ca(OH)_2$, with 100 ml water. Allow to stand for 24 hours. Use the clear supernatant liquid for the tests.

Lithium chloride (M).   Dissolve 42.4 g anhydrous lithium chloride, LiCl, or 60.4 g lithium chloride monohydrate $LiCl \cdot H_2O$ in water and dilute to 1 litre.

Lucigenin (1%).   Dissolve 1 g lucigenin (bis-*N*-methylacridinium nitrate or *N,N'*-dimethyl-9,9'-biacridinium dinitrate) in water and dilute to 100 ml.

Magnesia mixture (0.5M for magnesium).   Dissolve 102 g magnesium chloride hexahydrate, $MgCl_2 \cdot 6H_2O$, and 107 g ammonium chloride, NH_4Cl, in water, dilute to 500 ml, add 50 ml concentrated ammonia and dilute the solution to 1 litre with water.

Magnesium chloride (0.5M).   Dissolve 101.7 g magnesium chloride hexahydrate, $MgCl_2 \cdot 6H_2O$, in water and dilute the solution to 1 litre.

Magnesium nitrate reagent (ammoniacal) (0.5M for magnesium).   Dissolve 128 g magnesium nitrate hexahydrate, $Mg(NO_3)_2 \cdot 6H_2O$, and 160 g ammonium nitrate, NH_4NO_3, in water, add 50 ml concentrated ammonia and dilute with water to 1 litre.

Magnesium sulphate (M).   Dissolve 246.5 g magnesium sulphate heptahydrate, $MgSO_4 \cdot 7H_2O$, in water and dilute the solution to 1 litre.

Magnesium sulphate (0.5M).   Dissolve 123.2 g magnesium sulphate heptahydrate, $MgSO_4 \cdot 7H_2O$, in water and dilute to 1 litre.

Manganese(II) chloride (saturated).   Heat 7 g manganese(II) chloride tetrahydrate, $MnCl_2 \cdot 4H_2O$, with 3 ml water on a water bath until complete dissolution. Allow to cool and use the clear supernatant liquid for the tests.

Manganese(II) chloride (saturated in HCl).   To 7 g manganese(II) chloride tetrahydrate, $MnCl_2 \cdot 4H_2O$, add 5 ml concentrated hydrochloric acid. Shake, allow to stand for 24 hours and use the clear, supernatant liquid for the tests.

Manganese(II) chloride (0.25M).   Dissolve 49.5 g manganese(II) chloride tetrahydrate, $MnCl_2 \cdot 4H_2O$, in water and dilute to 1 litre.

Manganese(II) nitrate–silver nitrate reagent.*   Dissolve 2.87 g manganese(II) nitrate hexahydrate, $Mn(NO_3)_2 \cdot 6H_2O$, in 40 ml water. Add a solution of 3.55 g silver nitrate, $AgNO_3$, in 40 ml water and dilute the mixture to 100 ml. Neutralize the solution with 2M sodium hydroxide until a black precipitate is starting to form. Filter and keep the solution in a dark bottle.

Manganese(II) nitrate–silver nitrate–potassium fluoride reagent.*   To 100 ml manganese(II) nitrate–silver nitrate reagent add a solution of 3.5 g potassium fluoride in 50 ml water. Boil, filter off the dark precipitate in cold, and use the clear solution for the tests.

Manganese(II) sulphate (saturated).   Shake 4 g manganese(II) sulphate tetrahydrate, $MnSO_4 \cdot 4H_2O$, with 6 ml water. Allow to stand for 24 hours. Use the clear supernatant liquid for the tests.

Manganese(II) sulphate (0.25M).   Dissolve 55.8 g manganese(II) sulphate tetrahydrate, $MnSO_4 \cdot 4H_2O$, in water and dilute to 1 litre.

Mannitol (10%).*   Dissolve 10 g mannitol, $C_6H_{14}O_6$, in water, and neutralize the solution with 0.01M sodium hydroxide against bromothymol blue indicator, until the colour of the solution just turns to green. Dilute with water to 100 ml.

Mercury(I) chloride (calomel) suspension.*   To 5 ml of 0.05M mercury(I) nitrate add 1 ml of 2M hydrochloric acid. Wash the precipitate 5 times with 10 ml water by decantation. Suspend the precipitate finally with 5 ml water.

Mercury(II) chloride.(5% in alcohol).   Dissolve 1 g mercury(II) chloride, $HgCl_2$, in 20 ml 96% ethanol.

Mercury(II) chloride (saturated). Shake 7 g mercury(II) chloride, $HgCl_2$, with 100 ml water. Allow to stand for 24 hours and use the clear supernatant liquid for the tests.

Mercury(II) chloride (0.05M). Dissolve 13.9 g mercury(II) chloride, $HgCl_2$, in water and dilute to 1 litre.

Mercury(I) nitrate (0.05M).* Dissolve 28.1 g mercury(I) nitrate dihydrate, $Hg_2(NO_3)_2 \cdot 2H_2O$, in a cold mixture of 500 ml water and 10 ml concentrated nitric acid. Dilute with water to 1 litre. Add 1 drop of pure mercury metal to prevent oxidation.

Mercury(II) nitrate (0.05M). Dissolve 17.1 g mercury(II) nitrate monohydrate, $Hg(NO_3)_2 \cdot H_2O$, in water and dilute the solution to 1 litre.

Mercury(II) sulphate reagent (0.2M) (*for Denigés's test*). To 40 ml water add cautiously 10 ml concentrated sulphuric acid with constant stirring. Dissolve in this mixture 2.2 g mercury(II) oxide.

Methanol (methyl alcohol). The commercial product is a clear, colourless liquid with a characteristic odour. It boils at 64.7°C and has a density of $0.79 \, g \, cm^{-3}$. It is **POISONOUS** (causing blindness and ultimately death) and should therefore never be tasted.

Methylene blue (0.1%). Dissolve 0.1 g methylene blue (Colour Index 52015) in 100 ml water.

1-Naphthylamine (0.3%) (*for the Griess–Ilosvay test*).* Boil 0.3 g 1-naphthylamine $C_{10}H_7NH_2$ with 70 ml water for a few minutes. Filter or decant the clear liquid. To the cold solution add 30 ml glacial acetic acid.

Nessler's reagent.* Dissolve 10 g potassium iodide in 10 ml water (solution *a*). Dissolve 6 g mercury(II) chloride in 100 ml water (solution *b*). Dissolve 45 g potassium hydroxide in water and dilute to 80 ml (solution *c*). Add solution *b* to solution *a* dropwise until a slight permanent precipitate is formed, then add solution *c*, mix, and dilute with water to 200 ml. Allow to stand overnight and decant the clear solution, which should be used for the test.

Nickel chloride (0.5M). Dissolve 119 g nickel chloride hexahydrate, $NiCl_2 \cdot 6H_2O$, in water and dilute to 1 litre.

Nickel ethylenediamine nitrate reagent.* This reagent should be prepared when required from 2 ml 0.5M nickel chloride solution by adding ethylenediamine dropwise, until a violet colour appears. The colour is due to the formation of the complex ion $[Ni(NH_2 \cdot CH_2 \cdot CH_2 \cdot NH_2)_3]^{2+}$.

Nickel sulphate (0.5M). Dissolve 140 g nickel sulphate reagent (approx. composition $NiSO_4 \cdot 7H_2O$) in water and dilute to 1 litre.

Nitrazine yellow (0.1%). Dissolve 0.1 g of nitrazine yellow (2,4-dinitrobenzeneazo)-1-naphthol-4,8-disulphonic acid, disodium salt (Colour Index 14890) in 100 ml water.

Nitric acid (concentrated). The commercial concentrated nitric acid is a water-like solution with a density of $1.42 \, g \, cm^{-3}$. It contains 69.5% (w/w) HNO_3 or 0.99 g $HNO_3 \, ml^{-1}$. It is approx. 15.6 molar. Nitrous fumes make the partially decomposed reagent reddish-brown. It should be handled with utmost care, wearing gloves and eye protection.

Nitric acid (83%). Mix 68.0 ml concentrated (70%) nitric acid with 66.2 ml fuming (95%) nitric acid.

Nitric acid (8M, 1 + 1). To 500 ml water add cautiously 500 ml concentrated nitric acid.

Nitric acid (2M). Pour 128 ml concentrated nitric acid into 500 ml water, and dilute to 1 litre.

Nitric acid (0.4M). Dilute 20 ml of 2M nitric acid with water to 100 ml.

2-Nitrobenzaldehyde (0.5%).* Dissolve 5 mg of 2-nitrobenzaldehyde, $NO_2 \cdot C_6H_4 \cdot CHO$, in 1 ml of 2M sodium hydroxide.

4-Nitrobenzene-azo-1-naphthol (0.5%).* Dissolve 0.5 g 4-(4-nitrophenylazo)-1-naphthol (magneson II), $NO_2 \cdot C_6H_4 \cdot N:N \cdot C_{10}H_6 \cdot OH$, in a mixture of 10 ml 2M sodium hydroxide and 10 ml water. Dilute the solution to 100 ml.

4-Nitrobenzene-azo-chromotropic acid (0.005%).* Dissolve 5 mg 4-(4-nitrophenyl-azo)-chromotrope, sodium salt (Chromotrope 2B, Colour Index 16575) in 100 ml of concentrated sulphuric acid.

4-Nitrobenzene–diazonium chloride reagent.** Dissolve 1 g 4-nitroaniline, $NO_2 \cdot C_6H_4 \cdot NH_2$, in 25 ml 2M hydrochloric acid and dilute with water to 160 ml. Cool, add 20 ml of 5 per cent sodium nitrite solution and shake until the precipitate dissolves. The reagent becomes turbid on keeping but can be employed again after filtering.

3-Nitrobenzoic acid (saturated).* Dissolve 1 g 3-nitrobenzoic acid, $NO_2 \cdot C_6H_4 \cdot COOH$, in 250 ml water by heating the mixture on a water bath. Allow to cool and filter.

Nitron reagent (5%).* Dissolve 5 g nitron ($C_{20}H_{16}N_4$) in 100 ml 2M acetic acid.

4-(4-Nitrophenylazo) resorcinol (0.025%).* Dilute 5 ml of the 0.5% reagent with water to 100 ml.

4-(4-Nitrophenylazo) resorcinol (0.5%).* Dissolve 0.5 g 4-(4-nitrophenylazo) resorcinol (magneson) $NO_2 \cdot C_6H_4 \cdot N{:}N \cdot C_6H_2(OH)_2CH_3$ in a mixture of 10 ml 2M sodium hydroxide and 10 ml water. After complete dissolution dilute with water to 100 ml.

4-Nitrophthalene-diazoamino-azobenzene, Cadion 2B (0.02%).** Dissolve 0.02 g Cadion 2B, $NO_2 \cdot C_{10}H_6 \cdot N{:}N \cdot NH \cdot C_6H_4 \cdot N{:}N \cdot C_6H_5$, in 100 ml 96% ethanol to which 1 ml 2M potassium hydroxide has been added.

1-Nitroso-2-naphthol (1%). Dissolve 1 g 1-nitroso-2-naphthol, $C_{10}H_6(OH)NO$, in 100 ml of 9M acetic acid. Instead of acetic acid, 100 ml 96% ethanol or acetone may be used.

Osmium tetroxide (1%). Dissolve 1 g osmium tetroxide (osmic acid), OsO_4, in 10 ml M H_2SO_4 and dilute with water to 100 ml. The solution is available commercially.

Oxalic acid (saturated). Shake 1 g oxalic acid dihydrate, $(COOH)_2 \cdot 2H_2O$, with 10 ml water, and use the clear supernatant liquid for the tests.

Oxalic acid (0.5M). Dissolve 63 g oxalic acid dihydrate, $(COOH)_2 \cdot 2H_2O$, in water and dilute to 1 litre.

Oxalic acid (2%). Dissolve 2 g oxalic acid dihydrate, $(COOH)_2 \cdot 2H_2O$, in 100 ml water.

Oxine. See 8-hydroxyquinoline.

Palladium chloride (1%). Dissolve 1 g palladium chloride, $PdCl_2$, in water and dilute to 100 ml.

Perchloric acid (concentrated). The concentrated perchloric acid solution is a clear, colourless liquid with a density of $1.54 \, g \, cm^{-3}$ and contains 60% (w/w) $HClO_4$ (or 0.92 g $HClO_4$ per ml). It is approximately 9.2 molar.

Perchloric acid (2M). Dilute 216 ml concentrated perchloric acid with water to 1 litre.

1,10-Phenanthroline (0.1%). Dissolve 0.1 g 1,10-phenanthroline hydrate, $C_{12}H_8N_2 \cdot H_2O$, in 100 ml water.

p-Phenetidine hydrochloride (2%).* Dissolve 2 g p-phenetidine, $C_2H_5 \cdot O \cdot C_6H_4 \cdot NH_2$, in 5 ml concentrated hydrochloric acid and dilute with water to 100 ml (complete dissolution may take place only after dilution).

p-Phenetidine (0.1%).* Dissolve 0.1 g p-phenetidine, $C_2H_5OC_6H_4NH_2$, in 100 ml 2M hydrochloric acid.

Phenolphthalein (0.5%). Dissolve 0.5 g phenolphthalein, $C_6H_4(COOH) \cdot CH(C_6H_4OH)_2$, in 100 ml 96% ethanol.

Phenylarsonic acid (10%).** Shake 1 g phenylarsonic acid, $C_6H_5AsO(OH)_2$, with 10 ml water.

Phenylhydrazine reagent.* Dissolve 1 g phenylhydrazine, $C_6H_5 \cdot NH \cdot NH_2$, in 2 ml glacial acetic acid.

Phosphomolybdic acid (5%).* Dissolve 0.5 g dodecamolybdophosphoric acid icositetrahydrate, $H_3PO_4 \cdot 12MoO_3 \cdot 24H_2O$, in 10 ml water. The solution does not keep well.

Phosphoric acid (concentrated).   Concentrated phosphoric acid is a viscous, clear, colourless liquid with a high density ($1.75 \, \text{g cm}^{-3}$). It contains 88 per cent (w/w) H_3PO_4 (or $1.54 \, \text{g} \, H_3PO_4$ per ml) and is approximately 16 molar.

Phosphoric acid (M).   Dilute 63.7 ml concentrated phosphoric acid with water to 1 litre.

Picrolonic acid (picrolic acid) (saturated).*   Shake 1 g picrolonic acid, $C_{10}H_8O_5N_4$, with 5 ml water, and use the clear supernatant liquid for the tests.

Potassium acetate (M).   Dissolve 98.1 g potassium acetate, CH_3COOK, in water and dilute to 1 litre.

Potassium antimonate (0.2M).   Dissolve 10.5 g potassium antimonate, $K[Sb(OH)_6]$, in water and dilute the solution to 200 ml. Alternatively, dissolve 6.47 g antimony(V) oxide, Sb_2O_5, in 100 ml 6M hydrochloric acid and dilute the solution to 200 ml. Neither solution keeps well.

Potassium benzoate (0.5M).   Dissolve 80.1 g potassium benzoate, C_6H_5COOK, in water and dilute to 1 litre.

Potassium bromate (saturated).   To 7 g potassium bromate, $KBrO_3$, add 100 ml water and shake. Allow to stand for 24 hours and use the clear, supernatant liquid for the tests.

Potassium bromate (0.1M).   Dissolve 16.7 g potassium bromate, $KBrO_3$, in water and dilute to 1 litre.

Potassium bromide (0.1M).   Dissolve 11.9 g potassium bromide, KBr, in water and dilute to 1 litre.

Potassium chlorate (saturated).   To 7 g potassium chlorate, $KClO_3$, add 100 ml water and shake. Allow to stand for 24 hours and use the clear supernatant liquid for the tests.

Potassium chlorate (0.1M).   Dissolve 12.6 g potassium chlorate, $KClO_3$, in water and dilute the solution to 1 litre.

Potassium chloride (saturated).   To 30 g potassium chloride add 70 ml water and heat on a water bath until all the solid dissolves. Pour the hot solution into a 100 ml reagent bottle and allow to cool. Some solid KCl will crystallize on cooling. Use the clear liquid for the tests.

Potassium chloride (M).   Dissolve 74.6 g potassium chloride, KCl, in water and dilute to 1 litre.

Potassium chromate (saturated).   To 60 g potassium chromate add 100 ml water and heat on a water bath. After complete dissolution allow to cool. Use the clear supernatant liquid for the tests.

Potassium chromate (0.1M).   Dissolve 19.4 g potassium chromate, K_2CrO_4, and dilute the solution to 1 litre.

Potassium cyanate (0.2M).   Dissolve 16.2 g potassium cyanate, $KOCN$, in water and dilute the solution to 1 litre.

Potassium cyanide (10%).**   **(POISON)** Dissolve 1 g potassium cyanide, KCN, in 10 ml water. Discard the solution immediately after use.

Potassium cyanide (M).*   **(POISON)** Dissolve 0.61 g potassium cyanide, KCN, in 10 ml water. Discard the reagent immediately after use.

Potassium cyanide (0.1M).*   **(POISON)** Dissolve 0.61 g potassium cyanide in water and dilute to 100 ml. The solution should not be stored for long because it takes up carbon dioxide from the air releasing hydrogen cyanide gas.

$$2CN^- + CO_2 + H_2O \rightarrow 2HCN\uparrow + CO_3^{2-}$$

so the solution smells of hydrogen cyanide.

Potassium dichromate (0.1M).   Dissolve 29.4 g potassium dichromate, $K_2Cr_2O_7$, in water and dilute to 1 litre.

Potassium fluoride (0.1M).   Dissolve 5.81 g anhydrous potassium fluoride, KF, in water and dilute to 1 litre.

Potassium hexacyanocobaltate(III) reagent (4%).   Dissolve 4 g potassium hexacyanocobaltate(III) (potassium cobalticyanide) $K_3[Co(CN)_6]$ and 1 g potassium chlorate, $KClO_3$, in water and dilute to 100 ml.

Potassium hexacyanoferrate(II) (potassium ferrocyanide) (0.025M).   Dissolve 10.5 g potassium hexacyanoferrate(II) trihydrate, $K_4[Fe(CN)_6] \cdot 3H_2O$, in water and dilute to 1 litre.

Potassium hexacyanoferrate(III) (potassium ferricyanide) (0.033M).* Dissolve 10.98 g potassium hexacyanoferrate(III), $K_3[Fe(CN)_6]$, in water and dilute to 1 litre.

Potassium hydroxide (2M).   To 112 g potassium hydroxide, KOH, add 50 ml water. Stir the mixture several times, until all the solid dissolves. The mixture warms up considerably. Allow to cool and dilute with water to 1 litre.

Potassium iodate (0.1M).   Dissolve 21.4 g potassium iodate, KIO_3, in 500 ml hot water. Allow to cool and dilute to 1 litre.

Potassium iodide (6M).**   Dissolve 5 g potassium iodide, KI, in 5 ml water.

Potassium iodide (10%).   Dissolve 1 g potassium iodide, KI, in water and dilute to 10 ml.

Potassium iodide (0.1M).   Dissolve 16.6 g potassium iodide, KI, in water and dilute to 1 litre.

Potassium nitrate (0.1M).   Dissolve 10.1 g potassium nitrate, KNO_3, in water and dilute the solution to 1 litre.

Potassium nitrite (saturated).**   To 2 g potassium nitrite, KNO_2, add 1 ml water and shake in the cold. Use the clear supernatant liquid for the test.

Potassium nitrite (0.1M).**   Dissolve 0.9 g potassium nitrite, KNO_2, in water and dilute to 100 ml.

Potassium periodate (saturated).*   To 0.1 g potassium periodate, KIO_4, add 20 ml water and heat on a water bath until dissolution. Allow to cool and use the clear supernatant liquid for the tests.

Potassium permanganate (0.02M).   Dissolve 3.16 g potassium permanganate, $KMnO_4$, in water and dilute to 1 litre.

Potassium permanganate (0.004M).   Dilute 4 ml 0.02M potassium permanganate solution with water to 20 ml.

Potassium peroxodisulphate (0.1M).*   Dissolve 27.0 g potassium peroxodisulphate, $K_2S_2O_8$, in water and dilute to 1 litre.

Potassium selenate (0.1M).   Dissolve 22.1 g potassium selenate, K_2SeO_4, in water and dilute to 1 litre.

Potassium sodium tartrate (10%).*   Dissolve 10 g potassium sodium tartrate tetrahydrate $KNaC_4H_4O_6 \cdot 4H_2O$ (Rochelle salt or Seignette salt) in water and dilute to 100 ml.

Potassium sodium tartrate (0.1M).*   Dissolve 28.2 g potassium sodium tartrate tetrahydrate $KNaC_4H_4O_6 \cdot 4H_2O$ (Rochelle salt or Seignette salt) in water and dilute to 1 litre.

Potassium sulphate (saturated).   Shake 10 g potassium sulphate, K_2SO_4, with 90 ml water. Allow to stand for 24 hours and use the clear, supernatant liquid for the tests.

Potassium tellurite (0.1M).   Dissolve 25.4 g potassium tellurite, K_2TeO_3, in water and dilute to 1 litre.

Potassium thiocyanate (10%).   Dissolve 1 g potassium thiocyanate in 10 ml water.

Potassium thiocyanate (0.1M).   Dissolve 9.72 g potassium thiocyanate, KSCN, in water and dilute to 1 litre.

Propylarsonic acid (5%).**   Dissolve 0.5 g propylarsonic acid, $CH_3CH_2CH_2AsO-(OH)_2$, in 10 ml water.

Pyridine.   The pure solvent is a clear, colourless liquid with a characteristic odour. It has a density of 0.98 g cm^{-3} and boils at 113°C.

Pyrogallol (10%).*   Dissolve 0.5 g pyrogallol, $C_6H_3(OH)_3$, in 5 ml water. The reagent decomposes slowly.

Pyrrole (1%).*   Dissolve 1 g pyrrole, C_4H_5N, in 100 ml of (aldehyde-free) 96% ethanol.

Quinaldic acid (1%).*   Dissolve 1 g quinaldic acid, $C_{10}H_7O_2N$, in 5 ml 2M sodium hydroxide and dilute with water to 100 ml.

Quinalizarin (0.05%, in NaOH).* Dissolve 0.05 g quinalizarin (1,2,5,8-tetrahydroxyanthraquinone) in a mixture of 50 ml water and 5 ml 2M sodium hydroxide. Dilute the solution with water to 100 ml.

Quinalizarin (0.05% in pyridine).* Dissolve 0.01 g quinalizarin (1,2,5,8-tetrahydroxyanthraquinone) in 2 ml pyridine, and dilute the solution with acetone to 20 ml.

Rhodamine-B (0.01%). Dissolve 0.01 g rhodamine-B, $C_{28}H_{31}N_2O_3$ (Colour Index 45170), in 100 ml water. A more concentrated reagent can be prepared by dissolving 0.05 g of Rhodamine-B in 100 ml solution containing 15 g of potassium chloride in 2M hydrochloric acid.

Rochelle salt. See potassium sodium tartrate.

Salicylaldehyde oxime (salicylaldoxime) (1%).* Dissolve 1 g salicylaldoxime, $C_6H_4 \cdot CH(NOH) \cdot OH$, in 5 ml cold 96% ethanol and pour the solution dropwise into 95 ml water at a temperature not exceeding 80°C. Shake the mixture until clear, and filter if necessary.

Sebacic acid (saturated). Mix 5 g sebacic acid $(C_4H_8COOH)_2$ with 95 ml water and shake. Allow to stand for 24 hours and use the clear supernatant liquid for the tests.

Selenious acid (0.1M). Dissolve 11.1 g selenium dioxide, SeO_2, in water and dilute to 1 litre.

Silver nitrate (20%). Dissolve 2 g silver nitrate, $AgNO_3$, in water and dilute to 10 ml.

Silver nitrate (0.1M). Dissolve 16.99 g silver nitrate, $AgNO_3$, in water, and dilute to 1 litre. Keep the solution in a dark bottle.

Silver sulphate (saturated, about 0.02M). To 1 g silver sulphate, Ag_2SO_4, add 100 ml water and shake. Allow to stand for 24 hours, and use the clear supernatant liquid for the tests. Alternatively, heat the mixture on a water bath until complete dissolution and allow to cool.

Silver sulphate (ammoniacal) (0.25M).** Dissolve 7.8 g silver sulphate, Ag_2SO_4, in 50 ml 2M ammonia solution, and dilute with water to 100 ml. The reagent should be prepared freshly and discarded after use. **Aged reagents might cause serious explosions.**

Sodium acetate (saturated). Dissolve 53 g sodium acetate trihydrate, $CH_3COONa \cdot 3H_2O$, in 70 ml hot water. Allow to cool and dilute to 100 ml.

Sodium acetate (2M). Dissolve 272 g sodium acetate trihydrate, $CH_3COONa \cdot 3H_2O$, in water and dilute to 1 litre.

Sodium arsenate (0.1M). Dissolve 31.2 g disodium hydrogen arsenate heptahydrate, $Na_2HAsO_4 \cdot 7H_2O$, in water to which 10 ml concentrated hydrochloric acid was added. Dilute the solution to 1 litre.

Sodium arsenite (0.1M). Dissolve 13.0 g sodium meta-arsenite, $NaAsO_2$, in water and dilute to 1 litre.

Sodium azide–iodine reagent.* Dissolve 3 g sodium azide, NaN_3, in 100 ml 0.05M iodine (sodium tri-iodide) solution.

Sodium carbonate (saturated, about 1.5M). Dissolve 4 g anhydrous sodium carbonate, Na_2CO_3, in 25 ml water.

Sodium carbonate (0.5M). Dissolve 53 g anhydrous sodium carbonate, Na_2CO_3, in water and dilute to 1 litre.

Sodium carbonate (0.05M). Dilute 10 ml 0.5M sodium carbonate with water to 100 ml.

Sodium carbonate–phenolphthalein reagent.* Mix 1 ml 0.005M sodium carbonate with 2 ml 0.5% phenolphthalein and dilute with water to 10 ml.

Sodium chloride (M). Dissolve 58.4 g sodium chloride, $NaCl$, in water and dilute to 1 litre.

Sodium chloride (0.1M). Dissolve 5.84 g sodium chloride, $NaCl$, in water and dilute to 1 litre.

Sodium citrate (0.5M).* Dissolve 147 g sodium citrate dihydrate, $Na_3C_6H_5O_7 \cdot 2H_2O$, in water and dilute to 1 litre.

Sodium disulphide (2M).* First prepare 100 ml 2M sodium sulphide. Add to this liquid 8 g finely powdered sulphur, and heat on a water bath for 2 hours. Stir the mixture from time to time. Let the yellow solution cool and filter if necessary. The composition of the product is uncertain; the formula Na_2S_x is often used and the substance called sodium polysulphide. With the amounts of reagents mentioned x is approximately equal to 2.

Sodium dithionite (0.5M).** Dissolve 10.5 g sodium dithionite dihydrate, $Na_2S_2O_4 \cdot 2H_2O$, in water and dilute to 100 ml. The solution should be freshly prepared.

Sodium ethylenediamine tetra-acetate (5%). Dissolve 5 g sodium ethylenediamine tetra-acetate dihydrate, $[CH_2N(CH_2COOH) \cdot CH_2 \cdot COONa]_2 \cdot 2H_2O$ (Na_2EDTA), in water and dilute to 100 ml.

Sodium fluoride (0.1M). Dissolve 4.2 g sodium fluoride, NaF, in water and dilute to 1 litre.

Sodium formate (M).* Dissolve 68 g sodium formate, $H \cdot COONa$, in water and dilute to 1 litre.

Sodium hexafluorosilicate (0.1M).* Dissolve 18.8 g sodium hexafluorosilicate, $Na_2[SiF_6]$, in water and dilute to 1 litre.

Sodium hexanitritocobaltate(III) (0.167M).* Dissolve 6.73 g sodium hexanitrito-cobaltate(III) (sodium cobaltinitrite) in 100 ml water.

Sodium hydrogen carbonate (10%). Dissolve 10 g sodium hydrogen carbonate in 80 ml cold water, by shaking. Dilute the solution to 100 ml.

Sodium hydrogen carbonate (0.5M). Dissolve 42.0 g sodium hydrogen carbonate, $NaHCO_3$, in water and dilute to 1 litre.

Sodium hydrogen sulphite (1.25%).** Dissolve 0.1 g sodium sulphite Na_2SO_3 in 8 ml 0.1M hydrochloric acid.

Sodium hydrogen tartrate (saturated).* To 10 g sodium hydrogen tartrate mono-hydrate, $C_4H_5O_6Na \cdot H_2O$, add 100 ml water and shake. Allow to stand for 24 hours and use the clear supernatant liquid for the tests.

Sodium hydroxide (concentrated). To 5 g solid sodium hydroxide, NaOH, add 5 ml water and mix. Dissolution is slow first but becomes rapid as the mixture gets hotter. Use the cold liquid for the tests.

Sodium hydroxide (5M). Dissolve 20 g sodium hydroxide in 30 ml water; after dissolution dilute the solution to 100 ml. Keep the solution in a plastic bottle.

Sodium hydroxide (2M). To 80 g solid sodium hydroxide, NaOH, add 80 ml water. Cover the mixture in the beaker with a watch-glass and mix its contents from time to time. The heat liberated during this process ensures a quick dissolution. Allow the solution to cool and dilute with water to 1 litre.

Sodium hydroxide (0.01M).* Dilute 0.5 ml 2M sodium hydroxide with water to 100 ml. The solution should be prepared freshly.

Sodium hypobromite (0.25M).** To 50 ml freshly prepared bromine water add 2M sodium hydroxide dropwise, until the solution becomes colourless (about 50 ml is needed).

Sodium hypochlorite (M).** This reagent is available commercially as a solution, or can be prepared in the laboratory by saturating 2M sodium hydroxide solution with chlorine gas. This operation must be carried out in a fume cupboard.

Sodium hypophosphite (0.1M).** Dissolve 0.88 g sodium hypophosphite, NaH_2PO_2, in water and dilute to 100 ml.

Sodium metaphosphate. Heat 1 g ammonium sodium hydrogen phosphate (microcosmic salt), $NH_4NaHPO_4 \cdot 4H_2O$, in a porcelain crucible until no more gases are liberated. Dissolve the cold residue in 50 ml water. The reagent must be prepared freshly.

Sodium metavanadate (0.1M). Dissolve 12.2 g sodium metavanadate, $NaVO_3$, in 100 ml M sulphuric acid and dilute with water to 1 litre.

Sodium nitrite (5%).** Dissolve 5 g sodium nitrite, $NaNO_2$, in water and dilute to 100 ml.

Sodium nitroprusside reagent. Rub 0.5 g sodium nitroprusside dihydrate, $Na_2[Fe(CN)_5NO] \cdot 2H_2O$, in 5 ml water. Use the freshly prepared solution.

Sodium 1-nitroso-2-hydroxy-naphthalene-3,6-disulphonate (1%).* Dissolve 1 g sodium 1-nitroso-2-hydroxy-naphthalene-3,6-disulphonate [Nitroso R-salt, $C_{10}H_4(OH)-(SO_3Na)_2NO$] in 100 ml of water.

Sodium oxalate (0.1M). Dissolve 13.4 g sodium oxalate in water and dilute to 1 litre.

Sodium peroxide (0.5M).** Dissolve 3.9 g sodium peroxide, Na_2O_2, in water and dilute to 100 ml. The solution must be prepared freshly.

Sodium phosphite (0.1M).** Dissolve 2.16 g sodium phosphite pentahydrate, $Na_2HPO_3 \cdot 5H_2O$, in water and dilute the solution to 100 ml. The reagent should be prepared freshly.

Sodium pyrophosphate. Heat 1 g disodium hydrogen phosphate, $Na_2HPO_4 \cdot 2H_2O$, in a porcelain crucible until no more water is liberated. Dissolve the cold residue in 50 ml water. The reagent must be prepared freshly.

Sodium rhodizonate (0.5%).* Dissolve 0.5 g sodium rhodizonate (rhodizonic acid, sodium salt, $C_6O_6Na_2$) in 100 ml water. The solution decomposes rapidly.

Sodium salicylate (0.5M). Dissolve 80 g sodium salicylate, $C_6H_4(OH)COONa$, in water and dilute to 1 litre.

Sodium selenate (0.1M). Dissolve 36.9 g sodium selenate decahydrate, $Na_2SeO_4 \cdot 10H_2O$, in water and dilute to 1 litre.

Sodium selenite (0.1M). Dissolve 17.3 g sodium selenite, Na_2SeO_3, in water and dilute to 1 litre.

Sodium silicate (M).* Dilute 200 ml commercial (30%) water glass solution with water to 1 litre.

Sodium succinate (0.5M). Dissolve 135 g sodium succinate hexahydrate, $(CH_2COONa)_2 \cdot 6H_2O$, in water and dilute to 1 litre.

Sodium sulphate (0.1M). Dissolve 32.2 g sodium sulphate decahydrate, $Na_2SO_4 \cdot 10H_2O$, in water and dilute to 1 litre.

Sodium sulphide (2M). To 16 g solid sodium hydroxide, NaOH, add 20 ml water. Cover the beaker with a watch glass and shake the mixture gently, until complete dissolution. Dilute the mixture with water to 100 ml. Take 50 ml of this solution and saturate with hydrogen sulphide gas. Then add the rest of the sodium hydroxide solution to the latter.

Sodium sulphite (0.5M).* Dissolve 6.3 g anhydrous sodium sulphite, Na_2SO_3, or 12.6 g sodium sulphite heptahydrate, $Na_2SO_3 \cdot 7H_2O$, in water and dilute to 100 ml.

Sodium tellurate (0.1M). Dissolve 27.4 g sodium tellurate dihydrate, $Na_2TeO_4 \cdot 2H_2O$, in water and dilute to 1 litre.

Sodium tellurite (0.1M). Dissolve 22.2 g sodium tellurite, Na_2TeO_3, in water and dilute to 1 litre.

Sodium tetraborate (0.1M). Dissolve 38.1 g sodium tetraborate decahydrate (borax), $Na_2B_4O_7 \cdot 10H_2O$, in water and dilute to 1 litre.

Sodium tetrahydroxo stannate(II) reagent (0.125M).** To 2 ml 0.25M tin(II) chloride add 2M sodium hydroxide solution with vigorous shaking, until the precipitate just dissolves (about 2 ml of the latter is needed). The reagent decomposes rapidly; it has to be prepared freshly each time.

Sodium tetraphenyl boron (0.1M).** Dissolve 3.42 g sodium tetraphenyl boron, $Na[B(C_6H_5)_4]$, in 100 ml water. The solution does not keep well.

Sodium thiosulphate (0.5M). Dissolve 124.1 g sodium thiosulphate pentahydrate, $Na_2S_2O_3 \cdot 5H_2O$, in water and dilute to 1 litre.

Sodium thiosulphate (0.1M). To 2 ml 0.5M sodium thiosulphate add 8 ml water and mix.

Sodium tungstate (0.2M). Dissolve 65.97 g sodium tungstate dihydrate, $Na_2WO_4 \cdot 2H_2O$, in water and dilute to 1 litre.

Starch solution.* Suspend 0.5 g soluble starch in 5 ml water, and pour this into 20 ml water, which has just ceased to boil. Mix. Allow to cool, when the solution becomes clear.

Strontium chloride (0.25M). Dissolve 66.7 g strontium chloride hexahydrate, $SrCl_2 \cdot 6H_2O$, in water and dilute to 1 litre.

Strontium nitrate (0.25M). Dissolve 52.9 g strontium nitrate, $Sr(NO_3)_2$, in water and dilute to 1 litre.

Strontium sulphate (saturated). Shake 0.1 g strontium sulphate, $SrSO_4$, with 100 ml water and allow to stand for at least 24 hours. Use the clear supernatant liquid for the tests.

Sulphanilic acid reagent (1%).   Dissolve 1 g sulphanilic acid (4-aminobenzenesulpho-nic acid, $H_2N \cdot C_6H_4 \cdot SO_3H$) in 100 ml 30% warm acetic acid and allow to cool.

Sulphur dioxide gas.   This gas is available commercially in a liquefied state in aluminium canisters, from which it can be taken. Alternatively it can be produced from sodium sulphite and 8M sulphuric acid. The solid reagent should be placed into a round-bottomed flask, which can be heated. The acid is kept in a funnel with a stopcock, inserted into one opening of the flask. By adding some sulphuric acid to the solid and by gentle heating, sulphur dioxide gas comes through the second opening of the flask and can be washed in concentrated sulphuric acid.

Sulphur dioxide (sulphurous acid), saturated aqueous solution.**   Bubble sulphur dioxide gas through a thin glass tube into 200 ml water, until the solution becomes saturated. The gas can be taken from a cylinder or can be generated from solid sodium sulphite and 8M sulphuric acid. The solution should be well stoppered.

Sulphuric acid (concentrated).   The commercial reagent is a colourless, oil-like liquid of a high density ($1.84 \, g \, cm^{-3}$). It contains 98% (w/w) H_2SO_4 (or 1.76 g H_2SO_4 per ml) and is approximately 18 molar. The reagent must be handled with utmost care, wearing rubber gloves and eye protection. When diluting, concentrated sulphuric acid must be poured slowly into water (*and never vice versa*), while stirring and, if necessary, cooling. Traces of the reagent have to be removed immediately from the skin or from clothing by washing with large amounts of water.

Sulphuric acid (12M, 3 + 2).   To 40 ml water add cautiously 60 ml concentrated sulphuric acid, stirring constantly. If the mixture becomes hot, set aside for a while to cool, and then continue mixing.

Sulphuric acid (8M).   To 500 ml water add slowly, stirring constantly, 445 ml concentrated sulphuric acid. (If the mixture becomes too hot, wait for 5 minutes before resuming the dilution.) Finally, when the mixture is cool again, dilute with water to 1 litre.

Sulphuric acid (3M).   To 50 ml water add, stirring constantly, 16.6 ml concentrated sulphuric acid, with greatest caution. When cool, dilute the solution to 100 ml.

Sulphuric acid (M).   Pour 55.4 ml concentrated sulphuric acid slowly into 800 ml cold water, stirring vigorously. Wear protective glasses and rubber gloves during this operation. Finally, dilute the cold solution with water to 1 litre.

Sulphurous acid.   See sulphur dioxide (sulphurous acid) saturated aqueous solution.

Tannic acid (5%).*   Dissolve 5 g tannic acid, $C_{76}H_{52}O_{46}$, in 100 ml water and filter the solution.

Tartaric acid (M).*   Dissolve 15 g tartaric acid, $C_4H_6O_6$, in water and dilute to 100 ml. The solution does not keep for very long (max. 1–2 months).

Tartaric acid (0.1M).*   Dissolve 15 g tartaric acid, $C_4H_6O_6$, in water and dilute to 1 litre.

Tartaric acid–ammonium molybdate reagent.   Dissolve 15 g tartaric acid, $C_4H_6O_6$, in 100 ml 0.25M ammonium molybdate reagent.

Thallium(III) chloride (0.2M).   Dissolve 45.7 g thallium(III) oxide, Tl_2O_3, in 50 ml of 6M hydrochloric acid. Evaporate the solution to dryness and dissolve the residue in water. Dilute the solution to 1 litre.

Thallium(I) nitrate (0.05M).   Dissolve 13.32 g thallium(I) nitrate, $TlNO_3$, in water and dilute to 1 litre.

Thallium(I) nitrate (saturated).   Shake 1 g thallium(I) nitrate $TlNO_3$ with 10 ml water. Use the clear supernatant liquid for tests.

Thallium(I) sulphate (0.025M).   Dissolve 12.62 g thallium(I) sulphate, Tl_2SO_4, in water and dilute to 1 litre.

Thiourea (5% in 5M hydrochloric acid).*   To 50 ml water add 43 ml concentrated hydrochloric acid. Allow to cool. Dissolve in this mixture 5 g thiourea, $CS(NH_2)_2$, and dilute the solution with water to 100 ml.

Thorium nitrate (0.1M).   Dissolve 58.8 g thorium nitrate hexahydrate, $Th(NO_3)_4 \cdot 6H_2O$, in a mixture of 500 ml water and 25 ml concentrated nitric acid, and dilute the solution to 1 litre.

Tin(II) chloride (saturated).* Shake 2.5 g tin(II) chloride dihydrate, $SnCl_2 \cdot 2H_2O$, in 5 ml concentrated hydrochloric acid. Allow the solid to settle and use the clear solution for the tests.

Tin(II) chloride (0.25M).* Dissolve 56.5 g tin(II) chloride dihydrate, $SnCl_2 \cdot 2H_2O$, in a cold mixture of 100 ml concentrated hydrochloric acid and 80 ml water, then dilute with water to 1 litre. To prevent oxidation keep a small piece of granulated tin metal at the bottom of the solution.

Tin(IV) chloride reagent.** Dissolve 5 g tin(IV) chloride, $SnCl_4$, in 5 ml water. Prepare the reagent freshly.

Titanium(IV) chloride (10%). To 90 ml of 6M hydrochloric acid add 10 ml liquid titanium tetrachloride, $TiCl_4$, and mix.

Titanium(III) sulphate (15%).** The commercial titanium(III) sulphate solution contains about 15 per cent $Ti_2(SO_4)_3$ and also free sulphuric acid. It has a violet colour. On standing it is slowly oxidized to colourless titanium(IV) sulphate.

Titanium(IV) sulphate (0.2M). Dilute 33 ml 15% (w/v) commercial titanium(IV) sulphate solution with M sulphuric acid to 100 ml. Alternatively, fuse titanium dioxide TiO_2 with a 12–15 fold excess of potassium pyrosulphate in a porcelain or platinum dish, and dissolve the residue in 3M sulphuric acid. If necessary, filter the solution, or use the clear, supernatant liquid for the tests.

Titan yellow (0.1%). Dissolve 0.1 g Titan yellow (Clayton yellow, Colour Index 19540) in 100 ml water.

Triethanolamine $N(CH_2 \cdot CH_2 \cdot OH)_3$. The commercial liquid is a viscous liquid with a density of $1.12 \, g \, cm^{-3}$.

Uranyl acetate (0.1M). Dissolve 42.4 g uranyl acetate dihydrate, $UO_2(CH_3COO)_2 \cdot 2H_2O$, in a mixture of 200 ml water and 30 ml concentrated acetic acid. After dissolution dilute the solution with water to 1 litre.

Uranyl magnesium acetate reagent.* Dissolve 10 g uranyl acetate dihydrate, $UO_2(CH_3COO)_2.2H_2O$, in a mixture of 6 ml glacial acetic acid and 100 ml water (solution *a*). Dissolve 33 g magnesium acetate tetrahydrate, $Mg(CH_3COO)_2 \cdot 4H_2O$, in a mixture of 100 ml glacial acetic acid and 100 ml water (solution *b*). Mix the two solutions, allow to stand for 24 hours, and filter.

Uranyl nitrate (0.1M). Dissolve 50.2 g uranyl nitrate hexahydrate, $UO_2(NO_3)_2 \cdot 6H_2O$, in water and dilute to 1 litre.

Uranyl zinc acetate reagent.* Dissolve 10 g uranyl acetate dihydrate, $UO_2(CH_3COO)_2 \cdot 2H_2O$, in a mixture of 5 ml glacial acetic acid and 20 ml water, and dilute the solution to 50 ml (solution *a*). Dissolve 30 g zinc acetate dihydrate, $Zn(CH_3COO)_2 \cdot 2H_2O$, in a mixture of 5 ml glacial acetic acid and 20 ml water, and dilute with water to 50 ml (solution *b*). Mix the solutions, add 0.5 g sodium chloride, NaCl, allow to stand for 24 hours and filter.

Zinc acetate (1%). Dissolve 1 g zinc acetate dihydrate, $Zn(CH_3COO)_2 \cdot 2H_2O$, in water and dilute to 100 ml.

Zinc hexammine hydroxide. To 25 ml 0.5M zinc nitrate add 12.5 ml of 2M potassium hydroxide. Filter, wash the precipitate with water, and dissolve the precipitate off the filter with 15 ml 7.5M ammonia. Pour the liquid on the filter several times, until complete dissolution.

Zinc nitrate (0.5M). Dissolve 149 g zinc nitrate hexahydrate, $Zn(NO_3)_3 \cdot 6H_2O$, in water and dilute to 1 litre.

Zinc nitrate (0.1M). Dilute 1 ml 0.5M zinc nitrate with water to 5 ml.

Zinc sulphate (0.25M). Dissolve 72 g zinc sulphate heptahydrate, $ZnSO_4 \cdot 7H_2O$, in water and dilute to 1 litre.

Zirconium nitrate reagent. Heat 10 g commercial zirconium nitrate, $ZrO \cdot (NO_3)_2 \cdot 2H_2O$, and 100 ml 2M nitric acid to boiling, stirring constantly. Allow to cool and set aside for 24 hours. Decant the clear liquid and use this for the tests.

Zirconyl chloride (0.1M). Dissolve 32.2 g zirconyl chloride octahydrate, $ZrOCl_2 \cdot 8H_2O$, in 100 ml 2M hydrochloric acid and dilute with water to 1 litre.

Zirconyl chloride (0.1%). Dissolve 0.1 g zirconyl chloride octahydrate, $ZrOCl_2 \cdot 8H_2O$, in 20 ml concentrated hydrochloric acid and dilute with water to 100 ml.

Zirconyl nitrate (0.1M). Dissolve 26.2 g zirconyl nitrate dihydrate, $ZrO(NO_3)_2 \cdot 2H_2O$, in 100 ml 2M nitric acid and dilute with water to 1 litre.

Solid reagents

Aluminium powder, Al
Aluminium sheet, Al
Ammonium acetate, CH_3COONH_4
Ammonium carbonate, $(NH_4)_2CO_3$
Ammonium chloride, NH_4Cl
Ammonium fluoride, NH_4F
Ammonium nitrate, NH_4NO_3
Ammonium peroxodisulphate, $(NH_4)_2S_2O_8$
Ammonium sodium hydrogen phosphate (microcosmic salt),
    $NH_4NaHPO_4 \cdot 4H_2O$
Ammonium thiocyanate, NH_4SCN
Arsenic(III) oxide, As_2O_3 **(POISON)**
Benzoic acid, C_6H_5COOH
Cadmium metal, Cd (granulated)
Calcium carbonate, $CaCO_3$
Calcium chloride anhydrous, $CaCl_2$
Calcium fluoride, CaF_2
Calcium hydroxide, $Ca(OH)_2$
Chromotropic acid (sodium salt), $C_{10}H_6O_8S_2Na_2$
Cobalt(II) acetate, $Co(CH_3COO)_2 \cdot 4H_2O$
Copper foil, Cu
Copper sheet or coin, Cu
Copper turnings, Cu
Devarda's alloy (50% Cu, 45% Al, 5% Zn)
Dimethylglyoxime, $CH_3C(NOH)C(NOH) \cdot CH_3$
NN-Dimethyl-*p*-phenylenediamine, $NH_2 \cdot C_6H_4 \cdot N(CH_3)_2$
Diphenylamine, $(C_6H_5)_2NH$
Disodium hydrogen phosphate, $Na_2HPO_4 \cdot 2H_2O$
Ethyl potassium xanthate, $C_2H_5O(CS)SK$
Hydrazine sulphate, $(N_2H_4) \cdot H_2SO_4$
Hydroquinone (quinol), $C_6H_4(OH)_2$
Hydroxylamine hydrochloride (hydroxylammonium chloride), $HO \cdot NH_3 \cdot Cl$
Iron (filings and nails), Fe
Iron wire, Fe
Iron(II) ammonium sulphate, $FeSO_4 \cdot (NH_4)_2SO_4 \cdot 6H_2O$
Iron(II) sulphate, $FeSO_4 \cdot 7H_2O$
Lead acetate, $(CH_3COO)_2Pb \cdot 3H_2O$
Lead acetate paper
Lead dioxide, PbO_2
Litmus (test) paper
Magnesium metal (powder), Mg
Magnesium metal (turnings), Mg
Manganese dioxide, MnO_2
Mannitol, $C_6H_{14}O_6$
Mercury metal, Hg
Mercury(I) chloride (calomel), Hg_2Cl_2
Mercury(I) nitrate paper

Mercury(II) oxide, HgO (red)

Microcosmic salt (see ammonium sodium hydrogen phosphate)

Oxalic acid, $(COOH)_2 \cdot 2H_2O$

Paraffin wax

Phenol, C_6H_5OH

Platinum sheet or foil, Pt

Potassium benzoate, C_6H_5COOK

Potassium bromate, $KBrO_3$

Potassium bromide, KBr

Potassium chlorate, $KClO_3$

Potassium chromate, K_2CrO_4

Potassium cyanide, KCN **(POISON)**

Potassium dichromate, $K_2Cr_2O_7$

Potassium dichromate paper

Potassium fluoride, KF

Potassium hydrogen sulphate, $KHSO_4$

Potassium hydrogen tartrate, $KHC_4H_4O_6$

Potassium iodide, KI

Potassium nitrate, KNO_3

Potassium nitrite, KNO_2

Potassium periodate, KIO_4

Potassium permanganate, $KMnO_4$

Potassium peroxodisulphate, $K_2S_2O_8$

Potassium sodium tartrate (Rochelle salt or Seignette salt),
  $COOK \cdot CHOH \cdot CHOH \cdot COONa$

Resorcinol, $1,3\text{-}C_6H_4(OH)_2$

Salicylic acid, $C_6H_4(OH)COOH$

Silica (precipitated), SiO_2

Silver sheet, Ag

Silver sulphate, Ag_2SO_4

Soda lime

Sodium acetate, $CH_3COONa \cdot 3H_2O$

Sodium bismuthate, $NaBiO_3$

Sodium carbonate (anhydrous), Na_2CO_3

Sodium chloride, NaCl

Sodium citrate, $Na_3C_6H_5O_7 \cdot 2H_2O$

Sodium dihydroxytartrate osazone $(NaO \cdot CO \cdot C{:}N \cdot NH \cdot C_6H_5)_2$

Sodium dithionite, $Na_2S_2O_4 \cdot 2H_2O$

Sodium fluoride, NaF

Sodium formate, HCOONa

Sodium hexafluorosilicate, $Na_2[SiF_6]$

Sodium hexanitritocobaltate(III), $Na_3[Co(NO_2)_6]$

Sodium hydrogen carbonate, $NaHCO_3$

Sodium hypophosphite, NaH_2PO_2

Sodium nitrite, $NaNO_2$

Sodium nitroprusside, $Na_2[Fe(CN)_5NO] \cdot 2H_2O$

Sodium oxalate (Sörensen's salt), $(COONa)_2$

Sodium perchlorate, $NaClO_4 \cdot H_2O$

Sodium peroxide, Na_2O_2

Sodium peroxoborate, $NaBO_3 \cdot 4H_2O$

Sodium phosphite, $Na_2HPO_3 \cdot 5H_2O$

Sodium salicylate, $C_6H_4(OH)COONa$

Sodium succinate, $(CH_2COONa)_2 \cdot 6H_2O$

Sodium sulphate (anhydrous), Na_2SO_4

Sodium sulphite (anhydrous), Na_2SO_3

Sodium tetraborate (Borax), $Na_2B_4O_7 \cdot 10H_2O$. The anhydrous (fused) sodium tetraborate, $Na_2B_4O_7$, is even more suitable for the borax bead tests

Sodium thiosulphate, $Na_2S_2O_3 \cdot 5H_2O$
Starch (potato-starch or soluble starch)
Sulphamic acid, $H_2N \cdot SO_3H$
Sulphur, S ('flowers of sulphur'), sublimed
Tartaric acid, $C_4H_6O_6$ [$HOOC \cdot CH(OH) \cdot CH(OH) \cdot COOH$]
Thiourea, $(NH_2)_2CS$
Tin, granulated, Sn
Turmeric (test) paper
Uranyl acetate, $UO_2 \cdot (CH_3COO)_2 \cdot 2H_2O$
Uranyl nitrate, $UO_2(NO_3)_2 \cdot 6H_2O$
Urea, $H_2N \cdot CO \cdot NH_2$
Zinc, granulated, Zn
Zinc powder, Zn

Index